T0392705

Sustainability and Health in Intelligent Buildings

Woodhead Publishing Series in Civil and Structural Engineering

Sustainability and Health in Intelligent Buildings

Riadh Habash

Woodhead Publishing is an imprint of Elsevier
50 Hampshire Street, 5th Floor, Cambridge, MA 02139, United States
The Boulevard, Langford Lane, Kidlington, OX5 1GB, United Kingdom

ISBN: 978-0-323-98826-1 (print)
ISBN: 978-0-323-98640-3 (online)

For information on all Woodhead publications
visit our website at https://www.elsevier.com/books-and-journals

Publisher: Matthew Deans
Acquisitions Editor: Gwen Jones
Editorial Project Manager: Mica Ella Ortega
Production Project Manager: Debasish Ghosh
Cover Designer: Miles Hitchen

Typeset by STRAIVE, India

Dedication

To nature and nurture: whatever these are
To my father, Wadie, and to the shadows of my mother's soul, Fadhila

Contents

Preface

People have been constructing buildings for thousands of years, so why the topics of "sustainability," "smart cities," and "intelligent buildings," as "humankind's highest invention" are of so much interest today? Initially, urbanization was relatively slow but accelerated quickly as economies have become more industrialized during the past hundred years, and as a result, cities began to boom. Nearly half of people worldwide now live in towns, cities, and megacities. They are attracted by opportunities for education, employment, and recreation. Consequently, the growth of these cities comes with challenges, but also with an enormous economic opportunity, if it is managed correctly. However, this rapid urbanization is making people spending the majority of their life indoors where humans are becoming indoor species within walled spaces. The difference, though, is only during recent decades, the building industry has undergone an unprecedented transformation by injecting digital technology to the forefront in planning, design, and development.

The modern built environment in general and smart growth communities in particular are the natural extensions of the above digital transformation toward a technology-based Industry 5.0 and human-centered Society 5.0. In this paradigm, the notions of "sustainable buildings" and "healthy buildings" represent a step forward toward industrial value creation and quickly show several coinciding purposes and objectives between the two. While both notions have earned interest in the academic, industry, and policy bodies, progress for the implementation of intelligence has been widespread and is rapidly advancing the built environment. Key innovation factors for such buildings include information and communication technology, embedded sensors and actuators, smart grid capabilities, intelligent control strategies, environmentally oriented architecture, and artificial materials including nanomaterials and metamaterials for facades and walls. By using these technologies, an enhanced level of indoor environmental quality, human health environment, and energy efficiency can be achieved.

Today, sustainability and health in buildings have been a combined priority that goes beyond illness prevention to include enjoyment and activities as well as physical and ecological spaces that may promote human potential. This priority can be realized while examining the technological side of the COVID-19 pandemic, where a complex interaction of factors will determine how this crisis will influence sustainability and health in the buildings sector. It may serve as a reset button for building green policies and as an opening to fast-track emphasis on the sustainable development goals, supported by the circular economy model as a key force of digital transformation. This has sparked a spotlight on the resilience of mankind and the built environment upon which buildings play a critical role in disease transmission. The demanding nature of a healthy indoor built environment is a critical part of any resilience thinking, where engineers, architects, and public health professionals are a key to pandemic solution evolution. Such strategy requires a lateral-thinking partnership of sustainability, health, intelligence, and urban planning. This partnership will convey a message of resilience, adaptation, and action by being able to change and renew in response to new circumstances as well as to create knowledge from this experience to be better prepared in the future. Accordingly, this book responds to further advancing knowledge creation with a balanced focus on technical details and the strength of systems thinking.

Features

- Introduces the totality of intelligence, sustainability, and health in the built environment that assists in understanding and designing buildings for a quality of life.
- Discusses the foundational aspects of healthy buildings with a special emphasis on assessing various aspects of indoor environmental quality and human health.
- Presents the road map for developing high-performance, energy-efficient buildings based on bioclimatic architecture design, clean energy resources, energy storage systems, electrification, and decarbonization.
- Describes the smart infrastructure in the built environment, particularly the electrical grid and advanced technology-enabled solutions including grid-interactive efficient buildings, load management strategies, digital ecosystem, in-building wireless, building Internet of Things, and smart net-zero energy buildings.
- Exhibits digital transformation, digital sustainability, and the applicability of existing intelligent technologies that incorporate distributed sensing, data storage, processing system to address the building as a human-cyber-physical system including digital twins, and building information modeling as well as safety, privacy, and security with the employment of blockchain technology.
- Explains various modeling and optimization techniques used in the design of controllers for major systems such as heating, ventilation, air conditioning, lighting, access, and safety where large buildings require sophisticated intelligent controls to reduce energy consumption and maintain a healthy indoor environment. This entails the integration of various proactive and predictive digital capabilities.
- Explores the notion of building bioelectromagnetics to ensure health and safety from human exposure to electromagnetic fields due to electrification and communication technologies that largely exist in the built environment.
- Provides a special emphasis on the evolution of engineering control strategies in preparedness, response, and recovery from pandemic crises including infectious diseases.
- Concludes with the fact that urbanization, together with a truly integrative triadic force of human, nature, and artificial intelligence, should be incorporated in planning to positively influence sustainable and healthy living and working conditions in times going forward.

Scope

If sustainability, health, and intelligence are to become positive agents of change in this world, they will have to find a reliable place in the educational curriculum and professional development. This book evolved around this notion based on information, discussion exchanges, and knowledge creation from various disciplines including architecture, engineering, computing, building science, and public health. The motivation is an intelligent built environment, with a focus on indoor environmental quality, human health and comfort, and energy efficiency.

The nine chapters of this book present the many emerging facets of sustainability and health in intelligent buildings with annotated references for additional topic information. Chapter 1 introduces the principles of sustainability, health, and intelligence in the built environment with a focus on the earth's circulatory and circularity systems. It facilitates a gradual movement of the building system hierarchy with a future vision on occupants. Building science knowledge is presented to inform design decisions to optimize building performance as well as human performance. The two major domains of intelligence in a cyber-physical building system, facade and technical services, are briefly described. The duality of sustainability and health in which each creates and conditions the other for an enhanced quality of life is extensively discussed. Further, various topics of the book and several international standards and rating systems that prioritize sustainability and health are outlined.

Chapter 2 discusses the foundation of healthy buildings with special emphasis on assessing indoor environmental quality including air and water quality, ventilation, thermal and lighting comfort, acoustic and noise reduction, security and privacy, and electromagnetic health hazard. It examines the basic scientific principles underlying this foundation including a range of approaches and analysis techniques for designing a comfortable indoor environment.

Chapter 3 presents the road map for developing high-performance, energy-efficient buildings based on bioclimatic architecture design and clean energy technologies. It introduces several sustainability design approaches for operational and embodied life cycle energy consumption efficiency, on-site distributed energy resources, energy storage systems, electrification and decarbonization, electricity supply in small and large buildings, power quality, and alternating current vs direct current systems.

Chapter 4 explains the smart infrastructure in the built environment particularly in terms of the electrical grid and energy efficiency which can be enhanced by integrating distributed energy generation technologies and demand flexibility into grid-interactive efficient buildings. Such integration is feasible with the deployment of smart electrical load management strategies and 5G/Internet of Things/artificial intelligence-based sensor and actuator wireless networks that offer a path toward Tactile Internet, Internet of Energy, smart urban mobility, net-zero energy buildings, and ultimately long-term sustainability and health.

Chapter 5 consolidates all aspects of digitization into one operable model. This is where a digital twin with a human-cyber-physical scenario comes in to enhance performance and improve living conditions in the built environment. The applicability of intelligent technologies that incorporate distributed sensing and big data addresses characteristics like energy efficiency, occupants' health, and comfort in buildings. It presents possible solutions for a safe and secure scenario, with the employment of blockchain technology as a measure to secure and control the building information modeling coupled with digital twinning.

Chapter 6 focuses on the role of modeling and design of highly integrated building automation and control systems in tackling challenges for sustainability and health. It raises questions and provokes a debate about data-driven and computational design innovation in buildings. Different approaches used in the design of heating, ventilation, air conditioning, and lighting systems are discussed. Research trends in advanced control systems that evolve around several approaches including model-based, model-predictive, data-based, agent-based, and exergy-based have been highlighted.

Chapters 7 explores an additional factor that should be considered in assessing indoor environmental quality, namely exposure to prolonged and excessive electromagnetic fields. This includes power frequency (50/60 Hz) fields due to electrification and radiofrequency fields from emerging wireless technologies where the existing scientific evidence indicates some noticeable biological and health effects. This suggests the need to address the safety concerns arising from the extra complexity of 5G technologies by managing and/or reducing field exposure in the built environment.

Chapter 8 addresses various hygiene control strategies to respond and recover from pandemic crises like infectious diseases. This involves engineering infection control strategies for indoor air quality such as ventilation, air circulation, filtration, and humidity as well as several environmental electromagnetic disinfection techniques. In addition, several intelligent solutions that can contribute to the management of the pandemic crises are described.

Chapter 9 discusses the role of urbanization in providing a context for individuals to live comfortably and for communities to thrive. The combination of urban intelligence and digital twins as a planning tool to optimize the performance of cities and communities coveys the message that urbanization requires artificial intelligence by using smart technologies and data-driven solutions to enhance human intelligence and nature intelligence, not to replace them. This approach promotes human-centric design into the built environment to facilitate sustainability, health, and various postpandemic experience.

Audience

Instead of saying this book is here to solve everything, it is appropriate to say how the book helps in generating and leveraging knowledge as a tool for critical reflection and eventually, leading by learning. Because of its comprehensive coverage and a large number of topics, the book will be useful as a primary reference for a course on intelligent buildings in a program like the relatively new architectural engineering, which requires engineering knowledge into architectural practice. Besides, the book will be necessary for all those who design, manage, and use buildings—in fact, anyone who is concerned about sustainable and healthy living. This will not make the reader an expert in designing future buildings, but rather make him/her capable of understanding better the basic ideas of the process and its application. Further, the book will inspire the reader to explore the exciting and rewarding process of integrating sustainability, health, and intelligence in the built environment.

Acknowledgments

I am grateful for the assistance given by Gandhi Habash, AIA, NCARB, LEED AP BD+C, Ryan Companies US, Inc., who largely contributed to the writing of this book. I thank my students and colleagues at the University of Ottawa, who have provided the reason and environment.

Abbreviations

AC alternating current
ADC analog-to-digital converter
AEC architecture, engineering, and construction
AI artificial intelligence
AIoT Artificial Internet of Things
ANN artificial neural network
ANSI American National Standards Institute
AQHI air quality health index
ASHRAE American Society of Heating Refrigerating and Air-Conditioning Engineers
B2B business-to-business
B2C business-to-consumer
BACS building automation and control system
BAS building automation system
BBCD building bioclimatic design
BCA Building and Construction Authority
BEAM Building Environmental Assessment Method
BEMS building energy management system
BICSI Building Industry Consulting Services International
BIM building information modeling
BIQ Building Intelligence Quotient
BMS building management system
BPI bipolar ionization
BPIE Buildings Performance Institute Europe
BREEAM Building Research Establishment Environmental Assessment Method
BRI building-related illnesses
BTO Building Technologies Office
Ca^{2+} intracellular calcium
CABA Continental Automated Buildings Association
CAD computer-aided design
CASBEE Comprehensive Assessment System for Built Environment Efficiency
CDC Centers for Disease Control and Prevention
CO2 carbon dioxide
CPS cyber-physical system
DC direct current
DER distributed energy resources
DGNB German Sustainable Building Council
DOE Department of Energy

EDGE excellence in design for greater efficiencies
EHS electromagnetic hypersensitivity
EIBG European IB Group
ELF extremely low frequency
EM electromagnetic
EPA Environmental Protection Agency
EROI energy return on investment
EV electric vehicle
EWICON electrostatic wind energy converter
FL fuzzy logic
GHG greenhouse gases
H2M human-to-machine
HCPS human-cyber-physical system
HCS human-cyber system
HEPA high-efficiency particulate air
HI human intelligence
HIL human in the loop
HVAC heating, ventilation, and air conditioning
IAQ indoor air quality
IARC International Agency for Research on Cancer
IB intelligent building
IBCP Intelligent Building Certification Program
IBN Institute of Building Biology
ICNIRP International Commission on Non-ionizing Radiation Protection
ICT information and communications technology
IEEE Institute of Electrical and Electronics Engineers
IEQ indoor environmental quality
IESNA Illuminating Engineering Society of North America
Industry 5.0 Industrial Revolution 5.0
IoE Internet of Energy
IoT Internet of Things
ISM industrial, scientific, and medical
ISO International Organization for Standardization
ITU International Telecommunication Union
IWBI International WELL Building Institute
LBC Living Building Challenge
LCA life cycle assessment
LED light-emitting diode
LEED leadership in energy and environmental design
LTE long-term evolution
M2M machine-to-machine
MBC model-based control
MERV minimum efficiency reporting value
ML machine learning

MMW millimeter wave
MPC model predictive control
NASA National Aeronautics and Space Administration
NI nature intelligence
NREL National Renewable Energy Laboratory
NZEB net-zero energy building
ORCID open researcher and contributor ID
OSHA Occupational Safety and Health Administration
PF power factor
PHIUS Passive House Institute US
PID proportional, integral, derivative
PLC programmable logic controller
PM particulate matter
QML quantum machine learning
QoE quality of experience
QoL quality of life
QoS quality of service
RBD responsive building design
REHVA Federation of European Heating Ventilation and Air Conditioning Associations
ROI return on investment
SARS severe acute respiratory syndrome
SBS sick building syndrome
SDG sustainable development goal
SROI social return on investment
THD total harmonic distortion
TI Tactile Internet
USGBC US Green Building Council
UV ultraviolet
UX user experience
VFD variable-frequency drive
VLF very low frequency
VPG vertical potential gradients
VPP virtual power plants
WHO World Health Organization
WPT wireless power transfer
WSAN wireless sensor and actuator networks

Building as a system

1

Riadh Habash
School of Electrical Engineering and Computer Science, University of Ottawa, Ottawa, ON, Canada

We shape our buildings; thereafter they shape us.

Winston Churchill

1.1 Sustainability by closing the loop

The practice of sustainability as a catchphrase advanced from the discipline of environmental science; however, various disciplines from business to technology and social sciences have always contributed by bringing critical knowledge and effective tools. Although a young concept, closing the loop of dynamic relationships is a growing cause-and-effect idea in the world of sustainability, the concept covers a little part of the sustainability challenge but is still robust and distinct enough to be comprehensible for the construction industry and the entire society. This is mostly related to the analysis of two types of closed-loop thinking, namely circulatory and circularity systems (Fig. 1.1).

1.1.1 Through a circulatory system

The earth's two most complex systems are the biosphere (natural environment) and the technosphere (built environment) as a complex subset of the biosphere created by humans. Biosphere as a term was first coined by the famous Austrian geologist Eduard Suess in 1875 and revived by a Russian scientist named Vladimir Ivanovich Vernadsky in 1926 (Smil, 2003). It represents the parts of the earth where life exists and is mostly powered by solar energy. The biosphere includes the circulation of resources like air, water, materials, energy, and the corresponding wastes and emissions.

Peter Haff (2014) defines technosphere as the sum of communication, transportation, bureaucratic, and other human-made systems that act to metabolize fossil fuels and other energy resources on earth. The technosphere is a system that humans in their social systems (sociosphere) have structured to meet their needs. It is complete with its dynamics and energy flows, self-contained, permanently balanced on the planet where life naturally exists. The biosphere unlike the technosphere is exceptionally good at recycling and decomposing its materials and contaminants from its natural resources, the fact that has persevered the earth for billions of years. However, the technosphere, by contrast, is deficient in recycling.

The technosphere with its anatomy continues to flourish as inputs and outputs are derived from the physical and digital ecosystems within the biosphere and formed

Sustainability and Health in Intelligent Buildings. https://doi.org/10.1016/B978-0-323-98826-1.00001-6

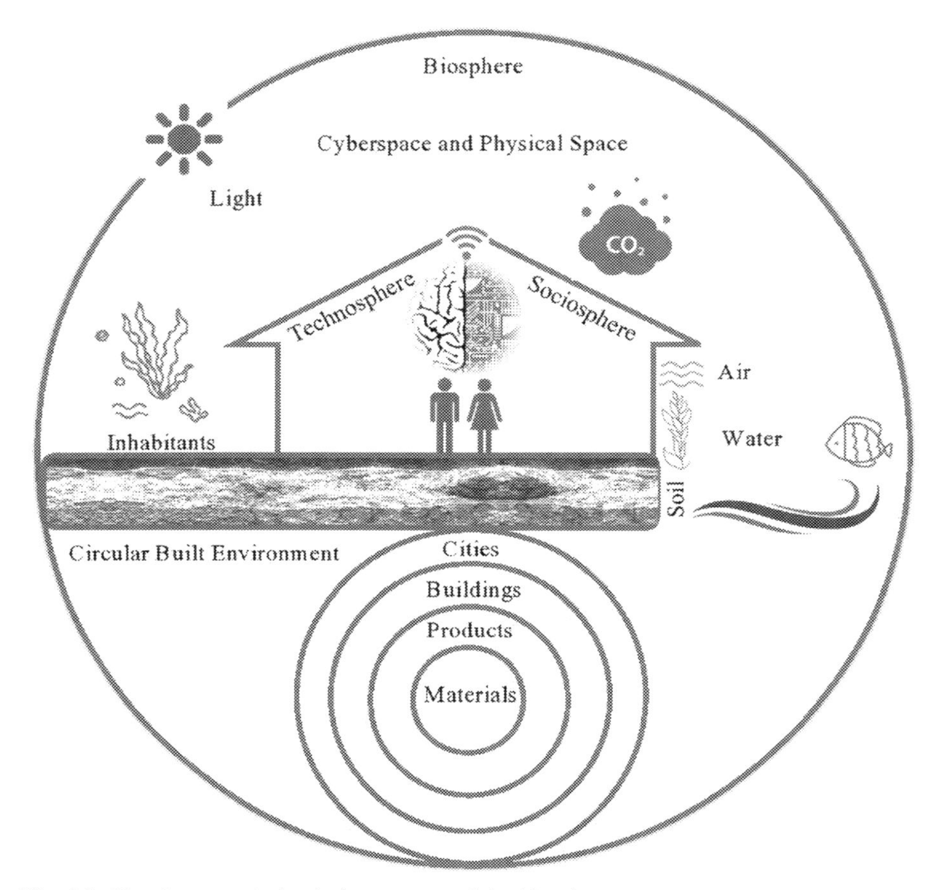

Fig. 1.1 Circulatory and circularity systems of the biosphere.

within the sociosphere. The widespread use of energy across various parts of the world has generated a paradigm of the technosphere which closely looks like the biosphere. However, the biosphere and technosphere are contradictory because technological growth happens on time scale orders of magnitude faster than biological growth. According to Zalasiewicz et al. (2016), the planet's technosphere now weighs some 30 trillion tons, a mass of more than 50 kg for every square meter of the earth's surface. Importantly, the human needs of cultural and social nature that reflect the relationship between humans and the technosphere are represented by the sociosphere.

A newcomer to the biosphere is the cyberspace that would be the world of the Internet involving computerized platforms, technological dependencies, and the corresponding electromagnetic (EM) fields. This implies the worldwide digital connectivity and corresponding Internet of Things (IoT). The above reflects the vision that technology is being integrated into one sphere, increasingly controlled via machine agents that stem from the advancement of artificial intelligence (AI).

Elements of intelligence include the ability to reason, plan, solve problems, understand complex ideas, learn from experiences, make decisions, and adapt to changing environments. By compacting activities and data into this ecosystem, it is not only impacting certain aspects of life but entire life in general. However, this digital transformation may make the built environment including buildings, communities, and cities smarter and more sustainable.

The phenomenon of an intelligent building (IB) is promoted within the notion of "smart city" that was materialized during the 1980s. Both built on technology and focused on outcomes, may be described by a concept that mimics a living system. This is a breathing circulatory machine that is very much like a human body, with a heart and soul, living and performing activities and desires where the citizens are the brain. Buildings inside these cities look like muscles, open and public spaces resemble the lungs in cleaning the air, and roads look like veins and arteries. In a healthy city, urban systems such as the built environment may be conceived as living subsystems with information and smart infrastructure as their circulatory systems within the digitally transformed infrastructure. The generated body of data and live analysis run all over them, reach to intelligent components supporting the built environment which turn data into nutrient information and knowledge to continually work effectively in conjunction with one another to improve liveability and resilience.

Cities today are enormously vibrant and multifaceted multidimensional systems that are gradually consistent as a consequence of globalization and advances in information and communication technology (ICT). These cities have been designed by humans and merged to manifest as the technosphere. The interaction between city systems in the technosphere and ecosystems in the biosphere is causing an exceptional interruption and disintegration of both. They exist as a dependent subset of nature's closed natural circulatory cycles, and thus always exert an influence on each other (Ratcliffe and Krawczyk, 2011). On the other hand, buildings are no longer simply bricks and mortar, they are living systems, breathing organisms that require unique expertise and technical knowledge (Josal, 2017). From the perspective of environmental impact, the most significant technosphere system is the building. Its primary resources include the building façade (envelope), inhabitants (humans, animals, and/or plants), technical systems (electrical/mechanical/electronic systems), a site with its landscape and services infrastructure, and external environment (landscape, weather, and micro-climate) (Kesik, 2016). To achieve a well-performing technosphere, it is helpful to think of buildings and cities as living systems that rely on a healthy circulation of harmonized resources within a circular economy that contributes to a reduction in energy consumption and increase in the efficiency of resource use.

1.1.2 Toward a circularity system

Circularity is emerging as a novel way of thinking. It entails a shift in the mindset of the building industry facilitated by technology. It is a transdisciplinary and systemic approach concerning various levels from material, components to buildings, and cities. Various technological, sustainability, and stakeholder aspects are of major importance for the success of circularity.

The practice of circularity is focused on the technosphere, on resource cycles, on a human construct designed to support the conversion of raw materials for human consumption. It influences a sustainable world, but not all sustainability programs contribute to circularity; however, both share common visions. Sustainability and circularity in design and construction mean thinking carefully about future use; however, the intended design of a system is what divides circularity from sustainability. Circularity is centered on three approaches: minimizing resource use, reducing energy and materials loops, and avoiding loss of economic and ecological value.

The building sector generates approximately one-third of all waste worldwide, much of which ends up in a landfill. It does not have to be this way, but due to the design of production facilities, it is possible to use materials that might then be landfilled or downcycled. This is not necessarily just about new technologies, but about rethinking how to design to meet the need for growth while at the same time reducing negative environmental and social impacts. It requires understanding the building sustainability assessment tools that play significant roles to promote and guide sustainability at the building level (Habash, 2017). These tools are a roadmap for designers to consider utilizing reusing and recycling or renewable energy where such practices were employed many centuries ago.

The cost of circularity is relatively high as the concept is in its evolution cycle. Moving from a traditional linear system of typical design to a closed-loop circular system where the disposal stage of building products could be fed back into the material extraction and processing phase, is a growing idea in the world of sustainability (Fig. 1.2). The long-lasting design is key in the transition toward circularity where

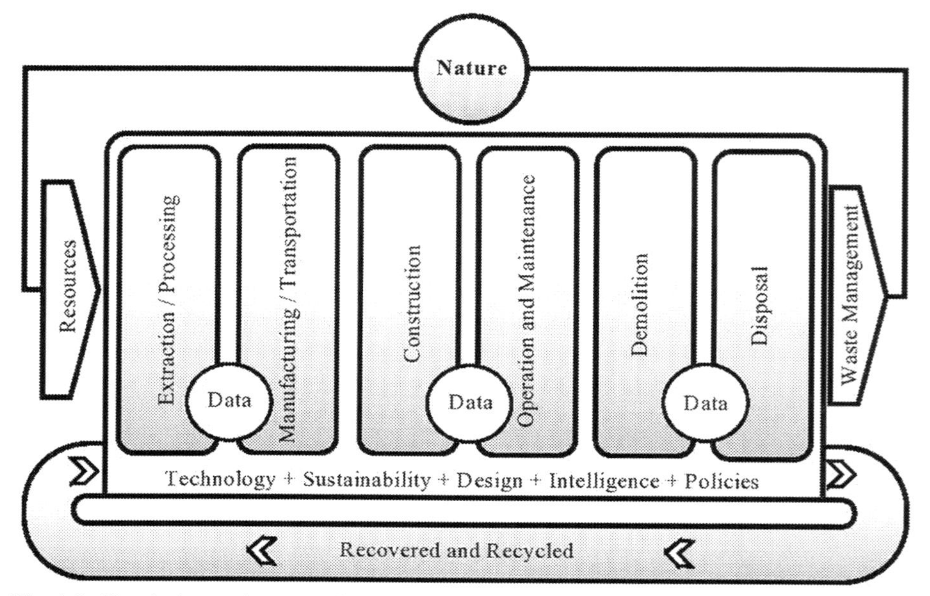

Fig. 1.2 Circularity performance in the life-cycle design of a building.

buildings should be constructed for a whole lifecycle. This is best understood by looking into living systems that function optimally because each of their components fits into the whole.

Circularity can have positive impacts on materials, energy, and water which are associated with physical resource flows. Cycling these resources will avoid unnecessary resource depletion and environmental pollution (DGBC, 2008). With the current linear economy (take, make, dump) model and the rising population in cities, there will be rising pressure on limited natural resources and climate change. Long-lasting design is key in the transition toward circularity where buildings should be constructed for a whole lifecycle. A proof of the impact linear tradition has on the environment and humankind is growing, and life-as-usual is no longer a feasible choice. Technology innovations, sustainability, design, intelligence, and government policies will accelerate the progress toward maximizing process and materials efficiency. Therefore, cities need to be transformed into spaces where humans and technologies can all harmonize together rather than being in a struggle with one another. The most significant change in recent times has been the awareness of the ecological impacts of buildings. Today, society recognizes that the "built environment as a technosphere system" does not arbitrarily end at the property domain, and may have far-reaching environmental and social implications.

1.2 Intelligence for sustainability and health

Just four walls and a roof, the minimum requirement of a building, have become the eventual articulation of civilization's creativity. Over the past 40 years, several developments have emerged that challenge the wisdom of just "doing it the conventional way." Today, the transdisciplinary fields of sustainability and health, as well as the disciplines of architecture, engineering, and building science, are all grown up and the prudence of their alliance will illuminate the path for an improved built environment. This means more than just places and spaces but a healthy and energy-efficient environment.

The above transition raises the following question: how to maintain sustainable and healthy living in growing cities? Being able to answer this question is important for at least the population growth of cities is economically essential in itself. By 2050, the number of people living in cities will reach more than 6 billion (Macomber, 2013). Yet the urban areas are already overcrowded and, particularly in developing countries, suffer from shortages of essential services. This challenge requires urban transformation through planning that incorporates regulations and decisions about implementing sustainable projects.

Fortunately, part of the answer for the above roadmap is the combined force of three emerging trends that can empower a change toward a sustainable and healthy built environment, namely, sustainability, health, and intelligence (human, nature, and artificial) as reflected in Fig. 1.3. This triadic relationship is integrated with resilience and systems thinking as well as their corresponding interdisciplinary practices. None of these forces should be underestimated in designing and constructing

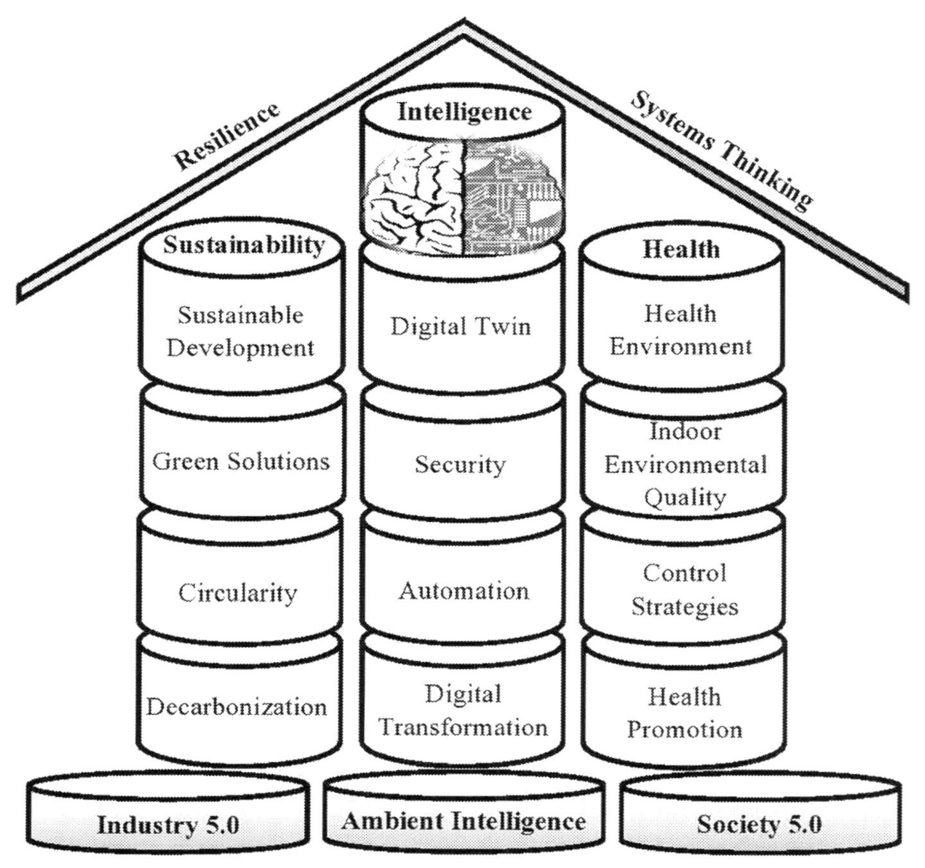

Fig. 1.3 Triadic relationship between sustainability, health, and intelligence to empower the built environment.

buildings. Systems thinking is a lens through which the whole can be seen. It is a framework to understand interrelationships and solve complex problems.

Intelligence, in general, encompasses learning from experience, recognizing, planning, and solving problems. This applies to the building sector, where intelligence is much more sophisticated throughout ambient intelligence which refers to a futuristic vision that presumes technology will be a vital part of human interactions to the extent that technology will recede and become invisibly integrated into daily activities. Ambient intelligence may be materialized through the integration of Industrial Revolution 5.0 (Industry 5.0) chronologically succeeding steam engines, band conveyer, programmable logic controller (PLC), cyber-physical systems (CPS) and Society 5.0 (chronologically succeeding hunting, farming, industry, information; first introduced by the Japanese Government in 2019). Together, emerging universal trends and forward-looking exercises will reinforce the role and the contribution of digitization and automation in the industry to society. These two concurrent concepts offer mass customization, functionality, and resource efficiency in a circular economy platform that is restorative and generative by design. This platform contributes to refining the collaborative interactions between humans and machines with a high level of convergence between cyberspace and physical space.

The sustainability contribution of ambient intelligence in buildings may be partially classified as a design for circularity including all life-cycle stages of material extraction, manufacturing, construction, building operation, demolition, and disposal as well as the corresponding transportation emission. The incorporation of more intelligence highly relies on harnessing the power of digital technology over the building, user, and safety systems by deploying advanced artificial IoT (AIoT), big data, cloud computing, control, and automation.

1.3 Hierarchy of buildings development

The building sector currently covers a residential and commercial floor area of 150 billion square meters (m^2), but this is projected to increase to 270 billion m^2 by 2050 (IRENA, 2018). Globally, the construction industry generated $1.39 trillion in revenue in 2018 within a total built real estate valuation of $200 trillion (Deloitte, 2018). This industry is witnessing a strong shift of emphasis from "energy efficient design" to "context and user aware design" and from "creating static buildings" to "designing interactive systems" with long-term building performance in mind. This fact raises the question: what does it imply to be a conventional, high-performance, smart, intelligent, or circular building? They are all colloquial terms describing ever-growing opportunities to improve upon buildings with lots of overlap (Ghaffarianhoseini et al., 2016). Fig. 1.4 reflects on the above question and shows the gradual progress of the building hierarchy with a vision where the human being is the focal point.

1.3.1 Conventional buildings

In its widest sense, the term "conventional" refers simply to "traditional" methods of construction based on knowledge passed from generation to generation. However, it may also be used to refer to traditional materials like stone, mud, or wood buildings that are not modern. Conventional buildings are usually those with the sum of three independent systems and are not designed for energy efficiency. These include building systems (heating, cooling, ventilation, lighting, and lifts), user systems (voice, data, and video communication), and life safety systems (access, video surveillance, and fire alarm).

The legal systems in various countries set up regulations (codes and standards) that require most buildings should meet a certain level of performance that reflects on the health, safety, and well-being of the occupants. The construction regulations ensure compliance throughout the life of the building. Technically, they depend largely on the thermal properties of the static materials composing the solid area of the enclosing

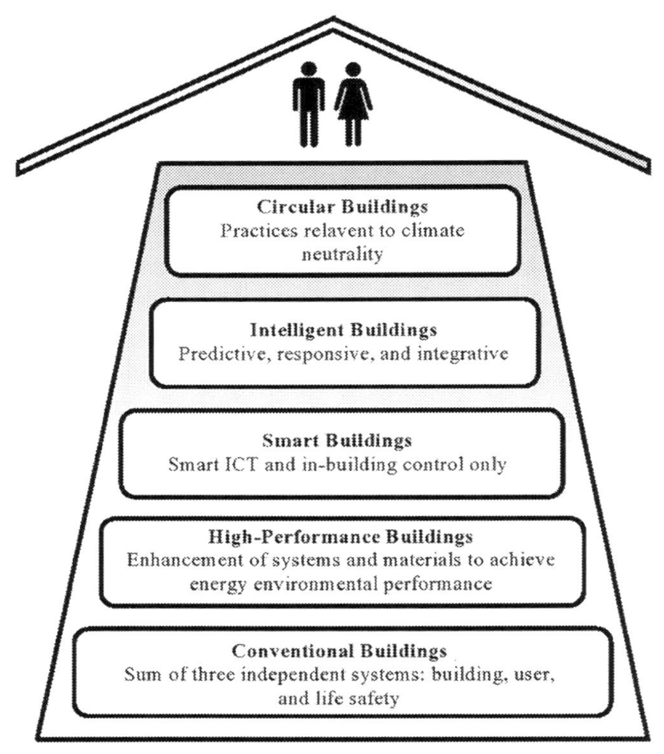

Fig. 1.4 Gradual progress of buildings with a vision on occupants.

walls. Currently, the conventional building is an open project for improvement not only to save energy and produce fewer gas emissions but also to provide healthier living and comfort to building occupants.

1.3.2 High-performance buildings

Starting in the 1960s, the building industry slowly became concerned with the performance of buildings. However, the progress in this regard is slow although has been around for decades. This is attributed to the enabling technologies that have slowly improved over the last decade. Practically, contemporary interest in high-performance buildings as a design paradigm has been around since the late 1990s. It emerged from an increasing interest in sustainability as a defining socio-economic issue. This paradigm was first used in the whole building design guide to shape the concept of a highly efficient and sustainable type of building. During that time, there was little mention of sustainability and health standards and practices to lead the construction industry. Today, there exist several organizations and programs that help firms to commercialize and deploy established technologies for high-performance buildings in nearly every type of structure, from schools and to hospitals, industrial plants, and other kinds of facilities. For example, The 2005 US Energy Policy Act directed the National Institute of Building Sciences to "explore the potential for accelerating the development of consensus-based voluntary standards to set requirements for less resource-intensive, more energy-efficient, high-performance buildings."

Moving from conventional to high-performance buildings requires a shift from notions on building design and construction to the inclusion of functionality, serviceability, durability, life-cycle performance, and occupant comfort. The logic for high-performance means buildings that maintain optimized configurations relative to changing environmental conditions to function more effectively (Hoberman and Schwitter, 2008) with fewer resources, and provide a better place for people to work and live. It incorporates building materials, construction techniques, structure strength, daylight, natural ventilation, fire resistance, durability, and the design of buildings to meet firmer environmental standards.

The practical key to high performance is to focus on the building envelope including walls, roof, and foundation. This requires a complete approach, with a high strategy and components integration with a network of building blocks including lifecycle analysis, indoor environmental quality (IEQ), high-efficiency equipment, control and automation, renewable energy, safety and security, energy analysis tools, and commissioning. Today high-performance buildings are encouraged by various green building assessments and ratings in a way that contributes to organizational and occupant value.

1.3.3 Smart buildings

Smart means common sense and awareness of the environment. Within the design disciplines, it has most frequently been cited with materials and surfaces (Addington and Schodek, 2009). The term "smart building" was created in the 1980s after ICT changed what could be done in the built environment. At the most basic level, smart buildings provide useful building services that make occupants productive at the lowermost economical and environmental impact over the building lifecycle.

A smart building is very energy efficient and covers its energy demand to a large extent by on-site renewable energy sources. It stabilizes and drives faster decarbonization of the energy system through energy storage and demand-side flexibility (De Groote et al., 2017). Enabled by technology, a smart building uses sensors, actuators, sophisticated control, and automation to gather, analyze, and share information about what happens in the building to optimize the building's performance while enhancing occupants' comfort and using less energy than a conventional building. Whereas conventional buildings have systems operating independently, smart buildings use ICT to connect building systems to optimize operations and whole-building performance. Smart buildings also allow operators and occupants to interface with the building, providing visibility into its operations and actionable information. Besides, smart buildings can communicate with the smart power grid, a feature that is becoming increasingly important for utility demand response deployment.

The most fundamental features of smart technologies are advanced sensors, actuators, and connected devices that respond to various environmental stimulus. Constant monitoring helps amendment adjustments that may control conditions across a whole building.

1.3.4 Intelligent buildings (IB)

A literature investigation into the IB reveals that there is not a commonly recognized definition for it. During the 1990s, the IB definition expanded to include many aspects related to a cohesive linkage between users, technical systems, and environment as well as the quality of life (QoL). This may be shown by reference to the Center for IB 1995 Working Group W98 stating that "an intelligent building is a dynamic and responsive architecture that provides every occupant with productive, cost-effective and environmentally approved conditions through continuous interaction among its basic elements including places, process, people, management, and the interrelation between them" (Clements-Croome, 2004).

The IB Institute in 1989 defines IB as "one which provides a productive and cost-effective environment through optimization of its four basic elements including structures, systems, services and management and the interrelationships between them." Later, the European IB Group in 1998 defined IB as "one that creates an environment which maximizes the effectiveness of the building's occupants, while at the same time enabling efficient management of resources with minimum lifetime costs of hardware and facilities." The above two definitions are performance- and operation-based with a focus on comfort, adaptability, reduced lifecycle costs, and enhanced control over available resources. The Japanese IB Institute identifies an IB by its service functions of communication, automation, and its convenience for intelligent activities. The Chinese IB Design Standard (GB/T50314–2000) states that the IB provides automation, communication network systems, and an optimal composition that integrates the structure, system, service, and management to provide efficiency, comfort, convenience, and safety to users.

Kroner (1997) argues that these buildings are technically enhanced where instead of architecture, the building services became intelligent with a little effect on user comfort. According to Atkin (1988), an IB is a system that knows what is happening inside it and immediately outside, a building that decides the most efficient way to provide an appropriate environment for its occupants and responds quickly to their requests. Wigginton and Harris (2002) indicate that there are over 30 definitions for an IB. They state that the term should be used with caution and its meaning should be clarified for the respective context. However, two different interpretations of the notion should be distinguished from each other. First, the notion refers to intelligent design and solutions. Second, it refers to structures that provide additional intelligent features in the building's operation phase. Furthermore, Clements-Croome (2011) developed the following definition "An IB is responsive to the requirements of occupants, organizations, and society. It is sustainable in terms of energy and water consumption besides being lowly polluting in terms of emissions and waste: healthy in terms of well-being for the people living and working within it; and functional according to the user needs."

With so many definitions for the IB, it is hard to propose a unique description. Professionals from various building sectors also have different views on the concept of IB. However, this is also not very essential; rather, there is one important question to ask: how to make a building intelligent in real life? Answering this question requires realizing IB as an improving transdisciplinary phenomenon that involves multiindustrial system engineering domains. The idea of continuous improvement comes in many managerial methodologies such as Kaizen, Six Sigma, and Lean which have been around for a while. The IB intends to put these methodologies to work to improve the occupant living experience. To meet this end, the IB requires the right combination of architecture, engineering, environment, ICT, automation, management, economics as well as systems and techniques that gather data and synthesize it in a way that may be effective.

1.3.5 Circular buildings

A circular building can be defined as a climate-neutral building that is constructed, used, and reused without unnecessary resource depletion, environmental pollution, and ecosystem degradation. It is constructed in an economically responsible way and contributes to the health and well-being of people and the biosphere. A circular building should ensure the preservation and growth of natural capital (biodiversity) and social capital (health and well-being, culture, and society) (DGBC, 2008). The circular building ideally contributes to sustainable production and consumption in all lifecycle phases including conception, design, planning, construction, use, operation, and deconstruction. These phases should be planned and implemented with the least environmental or social impact.

Several impact areas determine the design of circular buildings including materials, energy, water, biodiversity, and human culture. Optimizing building material takes place through smarter design and production, avoiding toxic materials, reusing and recycling materials instead of sourcing new ones, and assuring quality and safety. Each building is responsible for applying systematic consideration to the production of the net required energy, designing more efficient systems, and taking into account reducing GHG emissions. Water can be made wholly circular by matching water use in buildings to the local water cycle, by appropriate conservation and rainwater harvesting, besides retrieving all nutrients and materials from wastewater (DGBC, 2008). Human cultures and social structures are important to maintain and stimulate diversity and complexity.

Circular buildings are typically healthier and more comfortable than their conventional equivalents and might often be cost-competitive. The architecture, engineering, and construction (AEC) community should be able to benefit from the possibilities of interaction between sustainability and circulatory in developing a design practice that reflects the health and well-being needs of people and the spirit of digital transformation. Embracing industrial intelligence and digitization, innovation and education, as well as acquiring information about material flow in the building ecosystem can significantly bring down costs and increase the adoption of circularity. Appropriate building performance objectives involve several interfaces between the building, its occupants, and the natural and built environment.

To incentivize owners and designers of buildings to take action in sustainability, circularity, and health, it is essential to link them with intelligence. However, the cost of intelligence is a major challenge of building development. For example, the average cost to deploy a building automation system (BAS) is at least $2.50 per square foot and can be as high as $7.00 per square foot, equivalent to at least $250,000 for a 100,000 square foot building (Rawal, 2016). The very high cost of traditional BAS means the return of investment (RoI) which is a challenge for all but the largest buildings. Further, instead of evaluating based on short-term cost-effectiveness, it is necessary to use a longer-term metric, namely, social return on investment (SROI), which is strongly related to nonfinancial aspects but to the well-being of occupants directly or indirectly.

1.4 Duality of sustainability and health

The sustainability concept is both very ancient and relatively modern. It has been a design element throughout history where the increasing demand for sustainable design and green architecture has focused the attention toward the paradigm of smart cities and IBs. However, this paradigm is not about sustainability only, but about an alliance between sustainable development and health promotion which makes a duality that may be adopted as a starting point for the realization and enablement of higher QoL. Fig. 1.5 reflects the duality of sustainability and health to function optimally in the built environment; they must be working in synchronous with ambient intelligence. This relationship can be seen in the powerful symbol of yin yang where a strong balance is needed, the most essential activity to sustain life.

1.4.1 Sustainable development (SD)

The concept of sustainability takes its roots in German forestry and lumber businesses at the beginning of the eighteenth century. At that time, the German lumber business community agreed on a limitation of cutting wood to an amount that will be

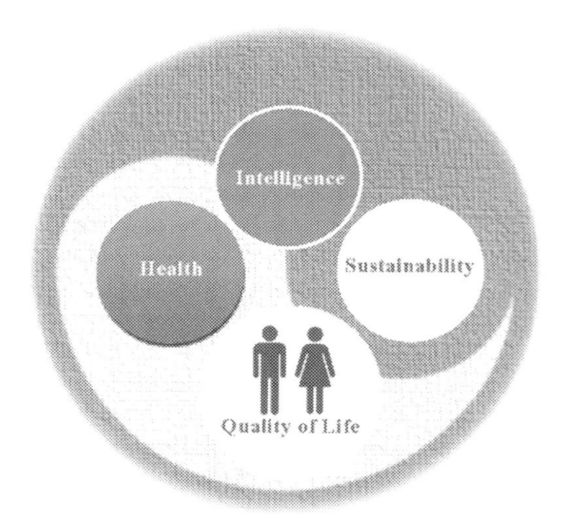

Fig. 1.5 Quality of life as a focal point for sustainability, health, and intelligence.

compensated by foresting every year. Hans Carl von Carlowitz (1645–1714) called this principle Nuchhaltige Entwicklung, which translates in English to SD as referenced in the "United Nations' Our Common Future." It is also known as the Brundtland Report, a broad concept reconsidered and reinterpreted since it first informed the building industry in 1987.

Sustainability has multiple dimensions and transcends multiple disciplines; therefore, design in this scenario requires cross-disciplinary expertise covering the interaction between various aspects. Within the scientific community, the concept of sustainability and what should be sustained is by no means agreed on and is subject to value judgments, up to be interpreted as a shared ethical belief. It may be identified as environmental, economic, social (Habash, 2017) and can be categorized as internal and external. Internal sustainability incorporates functionalities to achieve the goals for proper system operation. While, external sustainability includes three major conformity elements, namely economical, social, and environmental.

The SD concept is more than just sustainability. Its movement has been evolving worldwide for almost 30 years, causing significant changes in building delivery systems in a relatively short period. Although engineering is not one of the main elements of sustainability, it is indirectly linked to each. That is, engineering uses resources to drive much if not most of the world's economic activity, in virtually all economic sectors. Sustainable construction, a subset of SD, addresses the role of the built environment in contributing to the overarching vision of sustainability.

The SD goals (SDGs) came into force as a nonbinding international agreement in 2016. They are focused on six elements: dignity, basic needs of people, prosperity, planet, partnerships, and justice. Around these six elements, there are currently 17 global goals with 169 specific targets to be achieved by 2030, with an overarching aim of promoting SD by harmonizing the interconnected elements of economic growth, social inclusion, and environmental protection (Habash, 2017). In June 2020, Barcelona Institute for Global Health started an internal participatory process aimed at establishing priorities concerning the 2030 Agenda. The result is an institutional strategy on SDGs 2020–2030, which is expected to be adopted in 2021.

The SDGs are an indivisible and interdependent set of goals and the discussion on SD at Rio + 20 introduced the notion of health co-benefits (Ramirez-Rubio et al., 2019). Among the 17 SDGs, several are built environment-related: SDG3, good health, and well-being; SDG7, affordable and clean energy; SDG9, industry, innovation, and infrastructure; SDG11, sustainable cities, and communities; and SDG13, climate action (Fig. 1.6). However, more are gaining traction within the broader strategic framework of the AEC and manufacturing industries, namely SDG5, gender equality, and SDG15, life on land. The SDGs denote an exceptional prospect to promote public health through an integrated approach to public policies across different sectors. Further, the EU adopted a new circular economy action plan in 2020. This plan establishes the groundwork for the EU green deal. Construction and buildings, power, electronics and ICT, batteries, and green vehicles are some of the key areas of focus (Royan, 2021).

The practice of SD provides engineers with indications of techniques and processes that when used in real-life lead to better outcomes. Technology is the foundation of SD

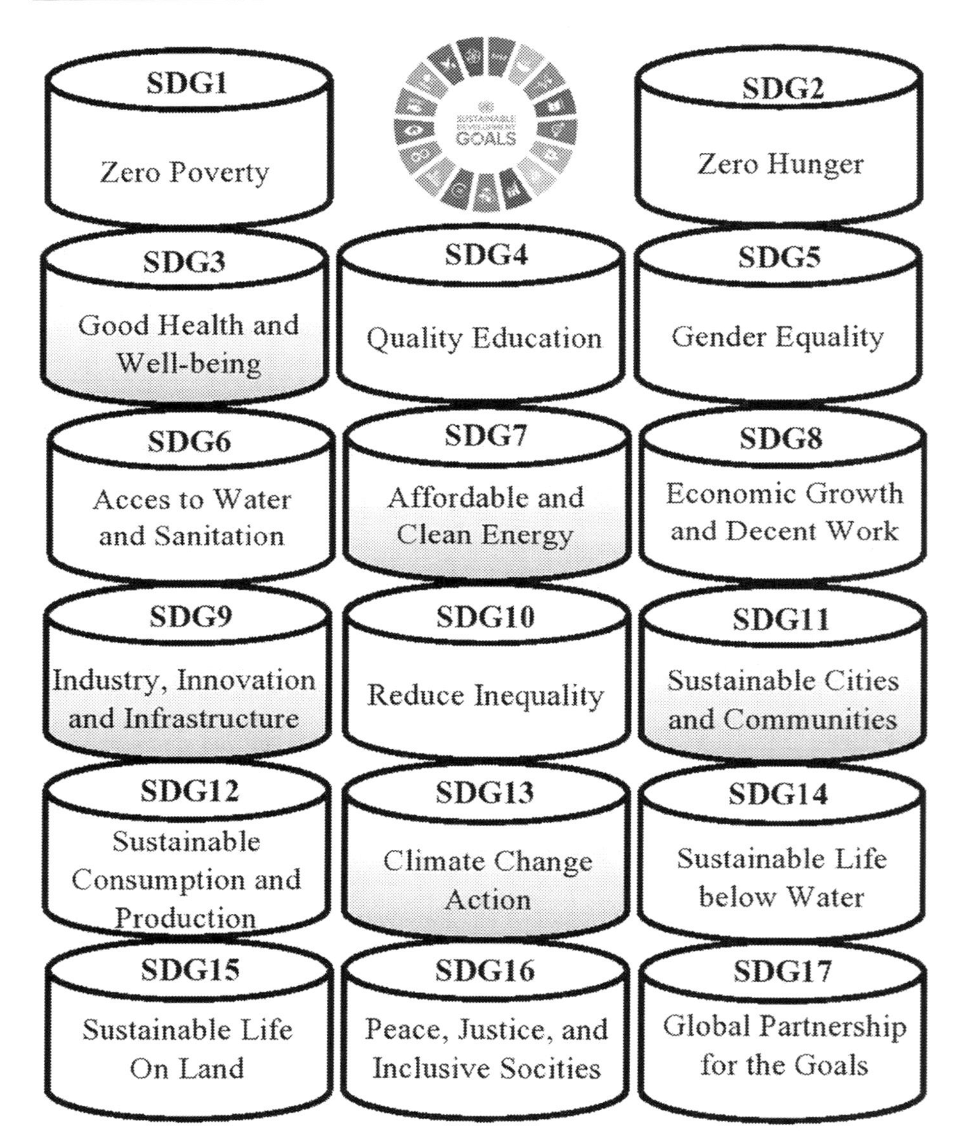

Fig. 1.6 The UN sustainable development goals.

and engineers provide the interface between its determinants including economy, society, and environment. To become sustainable, architects and engineers must act as leaders who recognize the world's needs and act accordingly (Habash, 2017). They should work together to help communities access better services such as cleaner energy, water, healthcare, transportation, and ICT. They can accomplish all that with the minimal use of natural resources in the design process while paying attention to the environment and social values.

Design in the context of sustainability may be described as a new possibility, which is expected to allow the achievement of a preferred situation. It is increasingly being viewed as a significant facilitating factor for sustainability because the design function is a joint point for decisions about a significant set of human and material resource flows. Throughout history, a Persian-designed windcatcher (of Pharaonic origin) was introduced as an architectural device, which achieves thermal comfort inside buildings. Sustainable design was also prevalent during Roman Antiquity, where in addition to building aqueducts, the Roman used geothermal energy to heat homes and baths.

Currently, there is increasing recognition that urban sustainability is tied directly to the QoL, which is improved by convenient, efficient, and accessible buildings, public transportation, recreation areas, and suitable educational and health services (Habash, 2017). Beyond this, buildings should be healthy places to live and work in, be equipped with reliable technology, meet the requirements and abide by regulations, respond to the needs of the occupants, be adaptable and robust, and give value for money. Such buildings will contain a variety of systems that can work together satisfactorily with an effective integration between the supply (designers, builders, and manufacturers) and demand (developers, owners, and occupants) as well as between the occupants, the systems, and the building. In such a scenario, systems thinking is essential in planning, design, and management, together with the ability to create and innovate while remaining practical.

1.4.2 Health environment

The concept of health is understood in a wider context as the absence of disease. The World Health Organization (WHO) now defines health not as the absence of ill-health but as a state of complete physical, mental and social wellness. The notion of well-being consists of two main elements, feeling good and functioning well. The definition of health has been changing and now includes an awareness of the interrelationships between social and psychological, as well as medical factors.

The relationship between architecture, engineering, and health has generally received little attention, beyond the design needs of healthy buildings. One important challenge is the quantification of health and well-being, and then the assessment of the overall health performance of building design (Nilon et al., 2003). It is apparent from the available research literature that there are no particular or universal design solutions to ensure that every health parameter is optimized and that building occupants and wider population will grow healthy. As a minimum requirement, designers should ensure that direct physical health parameters achieve a level that is considered good enough to avoid ill health.

The energy crisis in the United States between 1973 and 1974 had a major effect on the way people lived, drove, and built their homes. The high cost of both heating and cooling homes required action, and some of the action taken was ill-advised or failed to consider healthy housing concerns. Sealing homes and using untried insulation materials and other energy conservation actions often resulted in major and sometimes dangerous buildups of indoor air pollutants. Sealing buildings for energy efficiency and using building materials containing urea-formaldehyde, vinyl, and other new plastic surfaces, new glues, and even wallpapers created toxic environments. The results of these actions are still being dealt with today (Kaufman, 2006).

1.4.3 Quality of life (QoL)

The dual notion of sustainability and health is multifaceted since the indoor climate is closely related to life management. Given the deep link between buildings, life, and health, there are many pathways through which enhancements in IEQ will lead to advances in the QoL. IEQ and QoL are different but both allow the comfort and productivity of individuals to be increased much more. This relationship may have a positive impact on the built environment. The key to serving all the above in addition to energy efficiency in one whole design thinking approach is the initiation and development of new ideas that describe how intelligent control acts as a moderator for linking. Although energy efficiency has become secondary, it is still a significant, co-benefit of health.

With QoL exactly at the center of consideration in the built environment, there has never been a more critical time for the AEC and public health professionals to be responsive to occupant health and well-being. In addition to being responsive to occupants and their surrounding environment, buildings should be goal aware, which includes a holistic set of objectives including air and water quality, energy efficiency, thermal and light comfort, productivity, etc. By harnessing forecasts of future weather-related or infectious disease conditions, buildings will not only anticipate how to make better use of resources and create better conditions for occupants, but it will also guarantee compliance with advancing public health recommendations.

1.5 Building as a responsive and adaptive system

Architecture is a changing topic and is often referred to as a "third skin," delivering shelter and space for a human activity where the biological concept has become significant in the modern architectural debate (Gruber and Gosztonyi, 2010). The integration of responsive components in architecture strengthens strong indoor-outdoor connections and advances the potential to improve the experience of the building by giving expression to changeable aspects of the environment, therefore, seeing the building as an organism responding to human and environmental needs.

1.5.1 Responsive buildings

Responsive architecture, a term dating back more than 40 years, has its roots in the writings of Greek American architect Nicholas Negroponte and founder of MIT's Media Lab. Beginning with "The Architecture Machine (1970)," Negroponte proposed that responsive architecture is the natural product of the integration of computing power into the built environment (Sterk, 2003). He also extends this belief to include the concepts of recognition, intention, contextual variation, and meaning into computed responses and their successful and ubiquitous integration into architecture (d'Estrée, 2005). Negroponte, also proposed that advances in artificial intelligence (AI) and the miniaturization of components would soon give rise to buildings capable of intelligently recognizing the activities of their users and responding to their needs, as well as changes in the external and internal environment (Meagher, 2015).

Responsive architecture today is a developing field in practice. It may be conceptualized as a system that causes change to its environment (Meyboom et al. 2010). It involves measuring actual environmental conditions via sensors to enable buildings to adapt their form, shape, color, or character responsively via actuators. Responsive architecture is distinguished from other forms of interactive design by incorporating intelligent and responsive technologies into the core elements of a building's structure.

Responsive building design (RBD) (also known as smart, climate-adaptive, or intelligent) is one of those flexible alternative approaches and has been a popular topic in literature (Taveres-Cachat et al. 2019). Despite the minor semantic differences introduced, most responsive building technologies can be described as an extension of the definition for climate adaptive building shells given in Loonen et al. (2013). The core concept is the result of architects and engineers being inspired to design buildings that could express similar responses to the ones found in plants, or that could imitate human physiological responses like sweating or shivering (Taveres-Cachat et al. 2019). To replicate such functionalities, responsive building rely on integrated technologies that are designed to enable the building to respond to a range of triggers (stimuli), using a combination of passive, active, and/or cognitive control strategies. This design approach is particularly interesting given that building envelopes have a significant impact on the performance of buildings (IEA 2013).

RBD may present interesting new architectural features in architecture due to its dynamic aspects. A building envelope can then be thought of as having multiple configurations, depending on the time of the day, the season, and the use, which may result in a certain architectural quality (Meagher, 2015). In RBD, climate analysis is performed to determine the weather patterns, climate, soil types, wind speed and direction, solar orientation, water flows, habitat, and geology of the site. Envelopes in different climate zones require different assemblies to minimize unwanted energy loss. When designing such envelopes, factors such as insulation, vapor barriers, and air barriers will vary radically depending on weather conditions. RBD utilizes data on the region's weather patterns and accounts for the intensity of the sun, wind, rainfall, and humidity.

1.5.2 Adaptive buildings

The term "responsive" is often used interchangeably with "adaptive," but most simply it is used to describe, "how natural and artificial systems can interact and adapt" (Beesley et al. 2006). Still, most building professionals have a difficult time making the conceptual and practical leap to the notion of performance-based or adaptive buildings (Hoberman and Schwitter 2008).

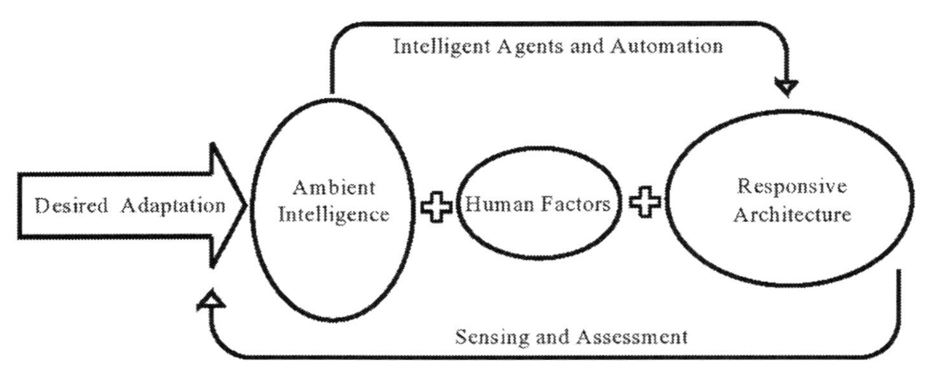

Fig. 1.7 Responsiveness and adaptation in the feedback loop.

The capacity of adaptation is an important feature in current definitions of the IB. An intelligent mechanism of response and feedback needs to be implanted in architecture, with real-time response and invention. The intelligence may be realized as a built-in capacity to respond to environments and demands (Böke et al., 2019). It is a complex phenomenon with a multilayered nonlinear process controlled by mechanisms as well as ambient intelligence and human factors including occupant behavior as shown by Fig. 1.7.

One of the most influential writers in the theory of responsive buildings is Cedric Price, whose "Fun Palace" project defined the model building as capable of adapting to the needs of its users. This building, with its temporary circulation and enclosures suspended from a space frame superstructure, was intended to radically transform the experience of architecture and was a strong influence on "High Tech" architecture such as the Centre Georges Pompidou (1977) of Renzo Piano and Richard Rogers in Paris which follows the spirit of the Fun Palace with its visible expression of independence between structure, services, and skin to increase flexibility and adaptation to change over time (Meagher, 2015). The Fun Palace project was never realized but is still considered a conceptual example of contemporary dynamic architecture.

The process of the adaptive feature and data flow through the IB façade should be treated as possessing smart features and some cognitive abilities to perceive and analyze data then respond with the appropriate action. Wigginton and Harris (2002) confirm that an IB can adapt to conditions and requirements to create interior comfort with low energy consumption. They complement the ability to learn. Buildings that continuously attune their configurations by changing environmental conditions use less energy, offer more occupant comfort, and feature better overall space efficiency than static buildings (Hoberman and Schwitter 2008).

1.6 Building as a human-cyber-physical system (HCPS)

The notion of "Industry 5.0" initiated in Germany and the notion of "Society 5.0" initiated in Japan represent the merge of cyberspace (computation, communication, control) with the physical space (natural and human-made systems governed by the laws of physics) beside the human dimension involved as well as the skills included and required. The human dimension as stakeholders, in the role of either occupants, designers, managers, or operators, will always play a significant role in future buildings. These global initiatives are closely associated with ICT, IoT, Internet of Energy (IoE), AI, and big data. The key idea is to stabilize human and technology engagement, taking advantage of industrial CPS progresses and human competencies recognized and employed through HCPS principles.

1.6.1 Digital transformation

Electrification empowered labor-saving machines to be spread across the world, while digitization, a recent revolution, facilitated computing and automation in efficient ways. Now, with the universal interconnections of Industry 5.0, industry, society, and governance are interacting even without human involvement. With AI, systems are subject to the continuing process of digital transformation, which is divided into several categories including optimization and customization of production; automation and adaptation; human–machine interaction; value-added services and businesses; and automatic data exchange (Shahzad et al. 2020). In total, digital transformation and smart embedded systems are at the heart of the HCPS.

At the core of the digital ecosystem is mechatronics, a discipline that reflects interdisciplinary real-life cases effectively. It is the synergic design of computer-controlled systems that integrates mechanical, electrical, and computing tools for the manufacture of products and processes (Habash and Suurtamm 2010). With the developments of mechatronics equipment inside buildings, the living quality and productivity will enhance while diminishing environmental effects. Today, IoT is a universal mechatronics technology that intends to enable virtually everything in the environment to be interconnected electronically using ever-cheaper electronics and an expanded Internet. It is the notion that every sensor and actuator in the building can be allocated an IP address and be directly connected to the Internet. The digital ecosystem, on the other hand, represents the movable networks of relating mechatronics systems and integrating data of all the incorporated systems into a single platform.

Digital transformation optimizes systems and processes to make workflows easier and more efficient. A culture of transformation is one where professionals are enabled to explore new ways of doing things efficiently. By incorporating a digital mindset, processes can be automated to pull better insights from the data and enable teams to concentrate on priorities. A central idea behind the efficiency is that savings can be accounted for at the system level. While embracing digital technologies is demanding, the advantages are well recognized by most practices. The building sector is, therefore, transforming in the shape of electrification, decarbonization, decentralization, and digitization, especially for renewable clean energy, smart grids, AI, and big data. Fig. 1.8 reflects the impact of the above transformation in the building sector. Managing the above transformation involves several domains including customer, competition, data, innovation, and value.

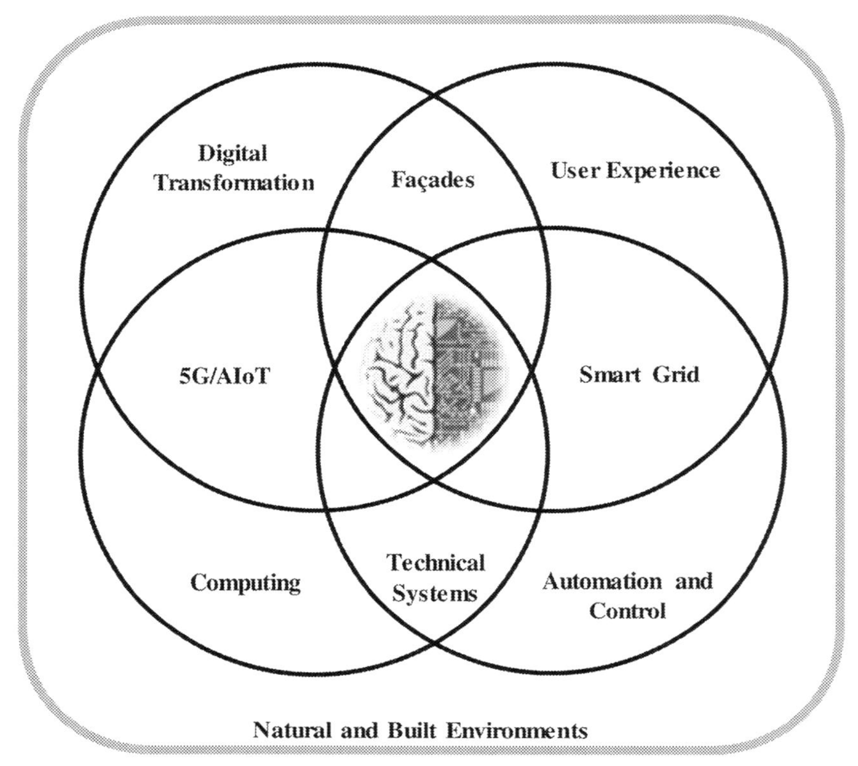

Fig. 1.8 Digital transformation in the built environment.

1.6.2 Façades and smart surfaces

The façade defines the shape and appearance but also determines the energy efficiency, the user comfort, and the functionality provided by the building (Böke et al., 2019). It has been a significant element of human settlements since the rise of civilization. It plays a dominant role in the exchange of heat and fresh air, provides views and daylight, and protects the indoor environment and occupants against extremes of temperature, solar radiation, water, and wind (Xing et al. 2018). Importantly, the façade outlines the architectural identity.

Regarding façades, the notion of "skin," whether single skin or double skin, is significant. The origin of this designation is the correlation to the human epidermis. The human skin is realized as a whole without distinction into components such as a wall or roof, and it has self-regulating properties. It recognizes changing conditions or requirements of the body and reacts to them independently. A similar understanding of the building façade is associated with the self-regulation between exterior and interior (Wigginton and Harris 2002). The guiding notion of a "thinking skin" reflects the idea of responsive and adaptive façades based on collecting environmental information and controlling the physical components using digital controllers on the cyber level (Böke et al., 2019). Such façade may be controlled to respond actively to changes in the external climate and be adapted to its environment through perception, reasoning, and action to its capability of dynamic adjustments.

An intelligent façade entails an integrated set of systems to perform a function of building enveloping with a protective layer that may increase the users' comfort while reducing power consumption. Innovations in façade technology are usually based on increased matter and energy transfer. Better energy management seems to be possible with the implementation of phase change materials that can store energy for delayed release (Gruber and Gosztonyi 2010).

The characteristics of smart materials can be described as immediacy (real-time response), transiency (responsive to more than one environmental state), self-actuation (internal intelligence), selectivity (a response is discrete and predictable), and directness (a response is local to the activating events) (Addington and Schodek 2009). Smart materials and surfaces can play a considerable role in intelligent, adaptive, and responsive envelopes because of the above intrinsic properties. Such materials are responsive to an external stimulus to control temperature changes or EM fields including solar radiation such as electrochromic glass with glazing material in curtain walls, windows, shading devices, and more applications. They can transform their physical properties and/or shape, or exchange energy without requiring an external source of power. Hence, they are attractive to building designers who aim to enhance performance while at the same time reduce energy consumption (Malik 2017). Occasionally, smart materials are used as mass materials, other times their applications can be found in covering the outer layer of the material that may strongly influence the building performance (Klooster 2009). Examples of smart materials used in building skins include aerogels, phase-changing materials microencapsulated wax, thermochromic polymer films, metamaterials, and building integrated photovoltaic.

1.6.3 Technical systems

Technical systems may include notions such as a plant, machine, and/or device, while intelligent technical systems refer to the application of CPS to networked production facilities in Industry 5.0. The total of all transformations inside the system is the process. Today, technical systems are a further development of mechatronics regarding their information processing and learning ability within the context of buildings. This is ensured using ICT that allow the exchange of information between systems and their subsystems (Böke et al. 2019). Practically, technical systems coordinate all integrated building operations, typically through a single control interface. These include building systems such as heating, ventilation, and air condition (HVAC), energy management, and lighting; user systems including video surveillance, voice and data communications; Life and safety systems including fire, security, access, and intrusion detection (O'Neil King 2016). Major technical systems in a building are shown in Fig. 1.9.

Mechatronics caters to the needs of a growing number of industrial products and processes that involve the integrated use of mechanical and electronic components, as

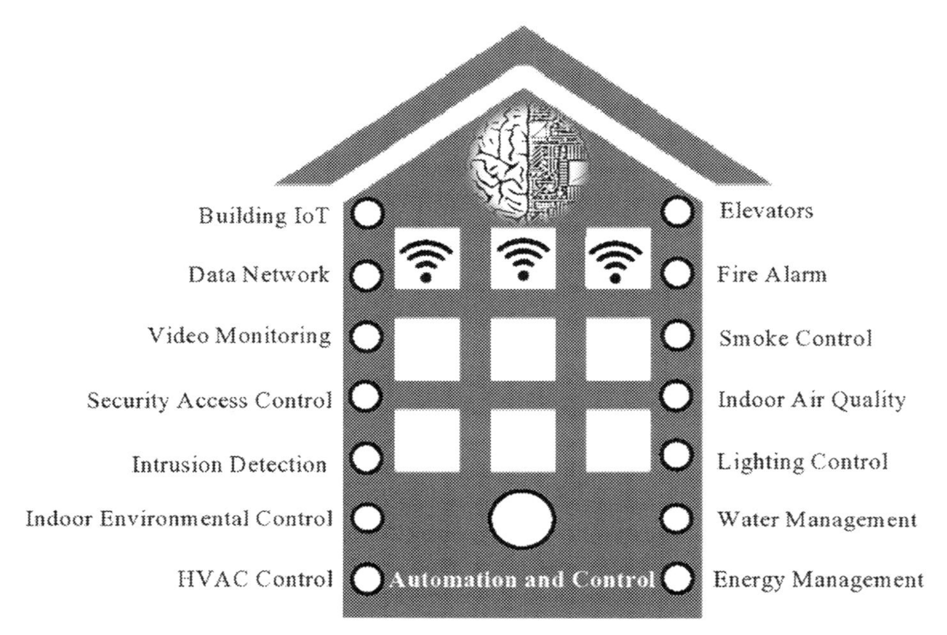

Fig. 1.9 Major technical systems in a building.

well as control software in their design and development (Habash et al., 2011). Such systems with integrated microcomputers are defined as embedded systems, which are differentiated by their computing power (16-bit, 32-bit, or more) and are used for the interaction with the sensors and actuators. The opportunity to embed intelligence into any component of a building system and merging the physical world of objects and humans , transforms the urban environment into artificial ecosystems of interconnected, interdependent, and intelligent digital organisms.

Automation is an important factor for the building industry. Such a requirement has led to several building automation networks and communication protocols being established. Generally, building automation begins with the control of mechanical, electrical, and plumbing systems including HVAC systems. Other systems that are often controlled include lighting systems, power monitoring, security close circuit video, fire alarm system, and elevators.

1.7 Resilience and systems thinking

2020 denoted the start of a new era, carrying with it a crisis that severely affects human health and well-being. An understanding of the COVID-19 pandemic that is challenging the world is under investigation. However, it should not be forgotten that everything in this world is interrelated. Today, more than ever, the AEC and health professionals have a profound opportunity to convey to the whole world, the necessity for built environment resiliency to keep people healthy and safe, businesses to grow and communities to thrive. As the COVID-19 pandemic advanced, it is necessary to realize how to design and develop a built environment that supports the hygiene system. It is time to launch a research investigation to evaluate the risk of airborne transmission of viruses within buildings and how to mitigate those risks. Emergency efforts that are rightly focused on dealing with the immediate impacts to slow the spread of the disease and ensure strong enough engineering control strategies to respond and recover from the disease should be investigated. These include series of engineering procedures including ventilation, air recirculation, filtration, humidity, and disinfection. On the risk management side, common strategies include naturally ventilated buildings.

Although urbanization and growth have provided several socioeconomic benefits, they have also brought several unwanted side effects such as air pollution, noise, heat, limited biodiversity, and social isolation. Therefore, the design of future built environment should be consistent with a high standard of sustainability and must not only be the tropicalization of buildings but also a system to control the impact of infectious diseases, which addresses factors related to the transmission of infections within buildings, including prevention, monitoring, and mitigation measures as well as to boost the mental and emotional well-being of occupants.

At this moment, the AEC community should jointly contribute its collective power of knowledge to understand, prevent, and mitigate future pandemics and increase the built environment resilience. This pandemic must be considered with a direct connection to the crises and with a universal approach that invites all professionals to conceive a new healthy living and working order. The goal is to create a better balance between nature and society with a renewed perspective on sustainability and health. The human element should be brought into the design process for greater environmental and societal resilience, as they are more robust and less vulnerable.

The focus should shift from surviving to thriving by identifying both the negative and positive aspects of space design as well as by responding and improving the many conditions that stakeholders face via a wide connection network. Building occupants should experience increased health, productivity, and well-being from the enhanced environmental quality and other value-added building services. This kind of coordinated action will require both resources and communications infrastructures as well as the need for advanced controls that enable the management of these resources and functions. This coveys the message that digital transformation may positively influence the built environment to be anticipatory and participatory by learning from data which makes the environment smarter, resilient, and engaging. Such tools will be essential in creating sustainable building spaces that enable communities to be more healthy.

Finally, technology is at the heart of the built environment where more people than ever before are embracing smart technologies and harnessing the power of data while digital literacy is becoming an effective economic and social platform for practical-life applications but more than that is the smart holistic-systems thinking.

1.8 Programs, standards, and rating systems

Standards, codes, and guidelines related to building practices are created through consensus processes by national and international organizations. Starting from the first known housing code of Hammurabi (King of Babylonia, 1792–1750 BCE) until today, standards and codes can restrain creativity but are necessary to set a minimum level of expectation and obey health and safety requirements. However, for quality, the design should exceed the above requirements.

At present, IBs in several developed countries have entered a relatively mature stage of development. They all have relatively sound standards, evaluation systems, and management institutions, which may grant important assurance and foundation. In this section, several programs, standards, and rating systems that prioritize sustainability and health are outlined.

1.8.1 American Society of Heating, Refrigerating and Air-Conditioning Engineers (ASHRAE)

ASHRAE is a nonprofit organization that develops and publishes standards for the heating, ventilating and air conditioning industry. ASHRAE has more than 57,000 members in more than 132 countries worldwide. Its members are composed of engineers, architects, mechanical contractors, building owners, equipment manufacturers' employees, and others concerned with the design and construction of HVAC systems in buildings.

1.8.2 Federation of European Heating, Ventilation and Air Conditioning Associations (REHVA)

REHVA's mission is to develop and disseminate economical, energy-efficient, safe and healthy technology for mechanical services of the building. This federation represents a network of more than 100,000 engineers from over 25 countries. It serves its members in the fields of HVAC services by facilitating knowledge exchange, supporting the development of related EU policies and their national-level implementation. Recently, ASHRAE and REHVA signed a memorandum of understanding to amplify their abilities to serve their members and the general public while simultaneously eliminating duplication and conflicts.

1.8.3 Architecture 2030

According to "architecture2030.com," cities are responsible for over 70% of global energy consumption and CO_2 emissions. Architecture 2030's mission is to transform the built environment from a major emitter of greenhouse gases (GHG) to a central solution to the climate emergency by empowering educators to shape the next generation of architects by providing global initiatives, educational content, tools, resources, and programs. This initiative realizes the value of engaging the world community in solving the climate crisis through the built environment and building sector.

1.8.4 Institute of Building Biology + sustainability (IBN)

IBN is an independent institute that provides among many educational programs and public outreach research and training in the area of building and sustainability. In addition, IBN gives consulting services for new construction, renovation, remediation; assessment of building materials, equipment, homes, and properties; air pollutant and mold analysis; audits of building specifications, service specifications, building plans; and inspections of homes, offices, and properties according to the standard of IBN testing methods.

1.8.5 For health: Healthy buildings

Healthy Buildings is an educational and research program at the Harvard T.H. Chan School of Public Health with a mission to improve the lives of all people, in all buildings. Its team created the nine foundations of a healthy building as a standardized, holistic approach to understanding how buildings impact the people inside them. In any indoor space like offices, homes, schools, airplanes, these foundations can be assessed via health performance indicators. This is called "Buildingomics," the totality of factors in the built environment that influence human health, well-being, and productivity of people who work in those buildings.

1.8.6 Healthy building network (HBN)

HBN defines the leading edge of healthy building practices that increase transparency in the building industry, reduce human exposures to hazardous chemicals, and create market incentives for healthier innovations in manufacturing. Its mission focuses on three fronts: research and policy; data tools and education; and capacity building. HBN works to reduce toxic chemical use, minimize hazards, and eliminate exposure, especially to those chemicals of concern deemed unnecessary or fail to improve product performance. It promotes the development of affordable green chemistry solutions that support a healthy, successful, and circular economy.

1.8.7 International rating systems

In addition to the major standards discussed above, Table 1.1 outlines several international building design systems that set up their criteria through a nationalistic focus, keeping local standards and codes in mind.

Table 1.1 International rating systems.

Rating system	Objectives
ANSI/BICSI-007	Design and implementation practices for IBs is the American National Standards Institute (ANSI)/Building Industry Consulting Service International (BICSI) standard intended for worldwide adoption. It specifically covers telecommunication cabling design for IBs. The standard includes requirements and recommendations for the design of cable pathways, room layouts with examples, and grid layouts to accommodate building services devices. Also, it includes requirements for documentation, commissioning, and management of an IB.
BCA GreenMark, Singapore	Launched by the Building and Construction Authority (BCA) in 2005 to promote environmental awareness in the construction and real estate sectors. It rates buildings according to five key criteria including energy efficiency, water efficiency, environmental protection, IEQ, and other innovative features that contribute to better building performance. The program outlines a six-step scheme that also offers cash incentives to developers, especially focused on addressing improvements to existing construction in areas such as energy use reduction and materials conservation.
BEAM, Hong Kong	The Building Environmental Assessment Method (BEAM) is a comprehensive standard and supporting process covering all building types, including existing and newly constructed mixed-use complexes. It is an initiative that assesses, improves, certifies, and labels the environmental performance of buildings. It is a voluntary program developed in partnership with and adopted by the industry. BEAM is intended to: stimulate demand for more sustainable buildings in Hong Kong and other regions, giving recognition for improved performance and minimizing false claims.
BPIE, Europe	The Buildings Performance Institute Europe (BPIE) is a leading independent center of expertise on energy performance of buildings with a mission to advance the transition toward climate-neturality in Europe and Globally.
BREEAM, UK	The Building Research Establishment Environmental Assessment Method (BREEAM) was launched in 1989 in the UK, it measures sustainable value in six different schemes, ranging from energy to ecology. Each of these categories addresses the most influential factors, including low impact design and carbon emissions reduction; design durability and resilience; adaption to climate change; and ecological value and biodiversity protection. Assessment and certification can take place at several stages in the built environment life-cycle, from design and construction through to operation and refurbishment.

Table 1.1 Continued

Rating system	Objectives
BIQ, Canada	Developed for the Continental Automated Buildings Association (CABA), the Building Intelligence Quotient (BIQ) rating system aims to encourage the design of more intelligent buildings by offering a standardized and fact-based way to answer the question, "How intelligent is the building?" The BIQ is an online tool that allows facility managers to assess the intelligence of their buildings by answering a set of 315 questions covering all aspects of facility operations. The managers can assess how well their systems are operating and sharing information. The BiQ and other CABA programs contribute to the notion of feedback and intelligence.
CASBEE, Japan	The Japanese Comprehensive Assessment System for Building Environmental Efficiency (CASBEE) is composed of four assessment tools corresponding to the building life-cycle. The assessment tools are for predesign, new construction, an existing building, and renovation, to serve at each stage of the design process. Each tool is intended for a separate purpose and target user and is designed to accommodate a wide range of uses (offices, schools, apartments, etc.) in the evaluated buildings. It covers the assessment fields of energy efficiency, resource efficiency, local environment, and indoor environment.
Chinese IB Design Standard	This system-based standard (GB/T50314–2000) describes IBs as those buildings which provide building automation, office automation, and communication network systems, and an optimal composition integrates the structure, system, service, and management, providing the building with high efficiency, comfort, convenience, and safety to users.
DGNB/BNB, Germany	Established in 2007, the German Sustainable Building Council (DGNB), which is directed at nonresidential, commercial buildings. BNB is another building assessment system, which is used only to assess government buildings. DGNB/BNB is the newest major building assessment system, and it differs significantly from systems employed by other countries. Its assessment system uses a top-down approach in its design with three sustainability points of evaluation including ecology, economy, and socio-culture.
Energy Star, US	Since 1992, the Energy Star program has certified energy-efficient products, homes, and commercial buildings. The program, administered by the US Environmental Protection Agency (EPA), has developed important benchmarks for everything from lighting to data centers, driving innovation and helping save energy and water.

Continued

Table 1.1 Continued

Rating system	Objectives
EDGE	Excellence in Design for Greater Efficiencies (EDGE) is a green building certification system for new residential and commercial buildings in more than 100 emerging markets. The program, which engages financiers, developers, regulators, and homeowners, shows property developers how fast and affordable it is to construct resource-efficient buildings, enabling them to pass the value directly to building owners and tenants. EDGE enables design teams and project owners to assess the most cost-effective ways to incorporate energy and water-saving options into their buildings.
Estidama, UAE	Pearl Rating System for Estidama, which means sustainability in Arabic, is intended to be the initiative that will transform Abu Dhabi into a model of sustainable urbanization. It aims to create more sustainable communities, cities, and global enterprises and to balance the four pillars of Estidama: environmental, economic, cultural, and social. It aims to address the sustainability of a given development throughout its life-cycle from design through construction to operation. Accordingly, three rating stages have been established: design, construction, and operational. Within each stage, there are both mandatory and optional credits and credit points are awarded for each optional credit achieved. To achieve a Pearl rating, all the mandatory credit requirements must be met.
Green Globes, US and Canada	An online assessment protocol, rating system, and guidance for green building design, operation, and management. It is interactive, flexible, affordable, and provides market recognition of a building's environmental attributes through third-party assessment. In general, it can be used to rate the sustainability of buildings for three aspects of energy, water, and resources.
Green Star, Australia	Green Star is the major Australian green building assessment scheme and is similar in many respects to BREEAM and LEED in its approach and structure. This building assessment system awards from one to six green stars, but only those buildings with ratings of four to six green stars have significance for being considered high-performance buildings.
IWBI	The international WELL health-safety rating standard of the International WELL Building Institute (IWBI) was established in 2013. It rates buildings and awards a score based solely on the health and comfort of occupants. It combines design and construction best practices with proven, evidence-based strategies for improving health and wellness. It follows comprehensive and interdisciplinary approaches that are necessary to meaningfully address the complex issues of human health and well-being. The WELL standard examines 22 features across five core areas

Table 1.1 Continued

Rating system	Objectives
	including cleaning and sanitization procedures, emergency preparedness programs, health service resources, air, water quality management, stakeholder engagement, and communication.
IBCP, Korea	Korea's IB Certification Program (IBCP) integrates architecture, electricity and electronics, information and communication, mechanical equipment, energy, and environmental systems to provide a comfortable, safe, and environmentally sustainable built environment.
LEED, USA	In 2000, the US Green Building Council (USGBC) developed the Leadership in Energy and Environmental Design (LEED), certification program with the intent of promoting the principles of SD. Since that time, LEED has become the industry standard for high-performance green buildings. It focuses on the sustainability of the building and infrastructure, including energy efficiency, materials, and the environment. LEED provides a point system to score green building design and construction. The rating system is categorized into five basic areas including sustainable sites, water efficiency, energy and atmosphere, materials and resources, and IEQ.
LBC, US and Canada	The Living Building Challenge (LBC) originated in 2005 as an extension of programs by the Cascadia Green Building Council in the northwestern US and western Canada. It is a green building certification program and sustainable design framework that visualizes the ideal for the built environment. It prioritizes sustainability, looking to create buildings that produce more energy than they use, and which have a positive impact on the people that use them. The LBC is a holistic standard, pulling together the most progressive thinking from the worlds of architecture, engineering, planning, interiors, landscape design, and policy. It consists of seven performance categories, or "petals" including place, water, energy, health + happiness, materials, equity, and beauty (LBC 2019).
PHIUS, US	The Passive House Institute US (PHIUS) administers a climate-specific passive building standard and certification system that was developed under a DOE/Building America grant specifically to address complex US climates. Buildings designed and built to the PHIUS+ 2015 Standard consume about 86% less energy for heating and 46% less energy for cooling; enables the supply of 100% fresh air, and does not allow to recirculate air (depending on climate zone and building type) when compared to a code-compliant building. This standard is applied internationally with certified projects in South Korea, Japan, China, and Israel. In North America, PHIUS is the leading educational institute with the most certified passive building professionals trained in North America (Vierra, 2019).

References

Addington, M., Schodek, D., 2009. Smart Materials. Birkhauser T. Press, Berlin.

Atkin, B., 1988. IBs–Applications of IT and Building Automation to High Technology Construction Projects. Halsted Press, New York.

Beesley, P., Hirosue, S., Ruxton, J., 2006. Responsive Architectures, Subtle Technologies. Riverside Architectural Press, Cambridge.

Böke, J., Knaack, U., Hemmerling, M., 2019. State-of-the-art of intelligent building envelopes in the context of intelligent technical systems. Intell. Build.Int. 11 (1), 27–45.

Clements-Croome, D.J., 2004. Intelligent Buildings: Design, Management and Operation. Thomas Telford, London.

Clements-Croome, D., 2011. Sustainable intelligent buildings for people: a review. Intell. Build. Int. 3 (2), 67–86.

d'Estrée, T., 2005. Building upon negroponte: a hybridized model of control suitable for responsive architecture. Autom. Constr. 14 (2), 225–232.

De Groote, M., Volt, J., Bean, F., 2017. Smart Buildings Decoded. https://urban-leds.org/wp-content/uploads/2019/resources/guidance_and_tools/Buildings_and_constructions/Smart%20buildings%20decoded%20-%20BPIE%20-%202017.pdf.

Deloitte, 2018. Global Construction Industry Overview: Insight Into Strategies, Trends, And Market Size. https://www2.deloitte.com/us/en/pages/energy-and-resources/articles/global-construction-industry-overview.html.

DGBC, 2008. A Framework for Circular Buildings: Indicators for Possible Inclusion in BREEAM. https://circulairebouweconomie.nl/wp-content/uploads/2019/09/A-Framework-For-Circular-Buildings-BREEAM-report-20181007-1.pdf.

Ghaffarianhoseini, A., Berardi, U., AlWaer, H., Chang, S., Halawa, E., Ghaffarianhoseini, A., Clements-Croome, D., 2016. What is an intelligent building? analysis of recent interpretations from an international perspective. Archit. Sci. Rev. 59 (5), 338–357. https://doi.org/10.1080/00038628.2015.1079164.

Gruber, P., Gosztonyi, S., 2010. Skin in architecture: towards bioinspired facades. WIT Trans. Ecol. Environ. 138, 503–513. https://doi.org/10.2495/DN100451.

Habash, R., 2017. Green Engineering: Innovation, Entrepreneurship, and Design. CRC Taylor and Francis, Boca Raton.

Habash, R.W.Y., Suurtamm, C., 2010. Engaging high school and engineering students: a multifaceted outreach program based on a mechatronics platform. IEEE Trans. Educ. 53 (1), 136–143.

Habash, R.W.Y., Suurtamm, C., Necsulescu, D., 2011. Mechatronics learning studio: from "play and learn" to industry-inspired green energy applications. IEEE Trans. Educ. 45 (4), 667–674.

Haff, P.K., 2014. Technology as a geological phenomenon: implications for human well-being. Geol. Soc. Lond., Spec. Publ. 395 (1), 301–309.

Hoberman, C., Schwitter, C., 2008. Adaptive Structures: Buildings for Performance and Sustainability. https://www.di.net/articles/adaptive-structures-building-for-performance-and-sustainability.

IEA, 2013. Transition to Sustainable Buildings—Strategies and Opportunities to 2050. http://www.oecd.org/greengrowth/transition-to-sustainable-buildings-9789264202955-en.htm.

IRENA, 2018. Global Energy Transformation: A Roadmap to 2050. International Renewable Energy Agency, Abu Dhabi. https://www.irena.org/-/media/Files/IRENA/Agency/Publication/2018/Apr/IRENA_Report_GET_2018.pdf.

Josal, L., 2017. The Future of Architecture Practice. https://www.di.net/articles/future-architecture-practice/.
Kaufman, T., 2006. House history and purpose. In: Healthy Housing Reference Manual. Centers for Disease Control and Prevention and U.S. Department of Housing and Urban Development. Healthy Housing Reference Manual. US Department of Health and Human Services, Atlanta. https://www.cdc.gov/nceh/publications/books/housing/cha01.htm.
Kesik, T.J., 2016. Building Science Concepts. http://www.wbdg.org/resources/building-science-concepts.
Klooster, T., 2009. Smart Surfaces—and their Application in Architecture and Design. Birkhäuser, Basel.
Kroner, W.M., 1997. An intelligent and responsive architecture. Autom. Constr. 6 (5–6), 381–393.
Loonen, R.C.G.M., Trcka, M., Costola, D., Hensem, J.L.M., 2013. Climate adaptive building shells: state-of-the-art future challenges. Renew. Sust. Energ. Rev. 25, 483–493.
Macomber, J., 2013. Building Sustainable Cities. https://hbr.org/2013/07/building-sustainable-cities.
Malik, R., 2017. Intelligent building facades. Int. J. Civ. Eng. Technol. 8 (5), 1340–1346.
Meagher, M., 2015. Designing for change: the poetic potential of responsive architecture. Front. Archit. Res. 4, 159–165.
Meyboom, A., Wojtowicz, J., Johnson, G., 2010. ROBO Studio: Towards Architectronics, Proceedings of the 15th International Conference on Computer Aided Architectural Design Research in Asia CAADRIA 2010. Association for Research in Computer-Aided Architectural Research in Asia (CAADRIA), Hong Kong.
Nilon, C.H., Berkowitz, A.R., Hollweg, K.S., 2003. Ecosystem Understanding is the Key to Understanding Cities: Introduction. Springer International Publishing AG, New York.
O'Neil King, R., 2016. Cyber security for intelligent buildings. Eng. Technol. Ref. 1–6. https://doi.org/10.1049/etr.2015.0115.
Ramirez-Rubio, O., Daher, C., Fanjul, G., Gascon, M., Mueller, N., Pajin, L., Plansencia, A., Rojaz-Rueda, D., Thondo, M., Nieuwenhuijsen, M.J., 2019. Urban health: an example of a "health in all policies" approach in the context of SDGs implementation. Glob. Health 15, 87. https://doi.org/10.1186/s12992-019-0529-z.
Ratcliffe, J., Krawczyk, E., 2011. Imagineering city futures: the use of prospective through scenarios in urban planning. Futures 43 (7), 642–653.
Rawal, G., Costs, Saving, and ROI for Smart Building Innovation. https://blogs.intel.com/iot/2016/06/20/costs-savings-roi-smart-building-implementation/#gs.xtxoze.
Royan, F., 2021. Digital Sustainability: The Path to Net Zero for Design and Manufacturing and Architecture, Engineering, and Construction (AEC) Industries. https://damassets.autodesk.net/content/dam/autodesk/www/campaigns/emea/docs/FS_WP_Autodesk_DigitalSustainability.pdf.
Shahzad, Y., Javed, H., Farman, H., Ahmad, J., Jan, B., Zubair, M., 2020. Internet of energy: opportunities, applications, architectures and challenges in smart industries. Comput. Electr. Eng. 86. https://doi.org/10.1016/j.compeleceng.2020.106739.
Smil, V., 2003. The earth's Biosphere. MIT Press, Cambridge.
Sterk, T., 2003. Building upon negroponte: a hybridized model of control suitable for responsive architecture. In: Digital Design. 21th eCAADe Conference Proceedings., ISBN: 0-9541183-1-6, pp. 407–414.
Taveres-Cachat, E., Grynning, S., Thomsen, J., Selkowitz, S., 2019. Responsive building envelope concepts in zero emission neighborhoods and smart cities-a roadmap to implementation. Build. Environ. 149, 446–457.

Vierra, S., 2019. Green Buildings Standards and Certification Systems. https://www.wbdg.org/resources/green-building-standards-and-certification-systems.

Wigginton, M., Harris, J., 2002. Intelligent Skins. Routledge, London.

Xing, Y., Jones, P., Bosch, M., Donnison, I., Spear, M., Ormondroy, G., 2018. Exploring design principles of biological and living building envelopes: what can we learn from plant cell walls? Intell. Build. Int. 10 (2), 78–102.

Zalasiewicz, J., Williams, M., Waters, C.N., Barnosky, A.D., Palmesino, J., Ro Nnskog, A.-S., Edgeworth, M., Neal, C., Cearreta, A., Ellis, E.C., Grinevald, J., Haff, P., Ivar Do Sul, J.A., Jeandel, C., Leinfelder, R., JR, M.N., Odada, E., Oreskes, N., Price, S.J., Revkin, A., Steffen, W., Summerhayes, C., Vidas, D., Wing, S., Wolfe, A.P., 2016. Scale and diversity of the physical technosphere: a geological perspective. Anthr. Rev. https://doi.org/10.1177/2053019616677743. November:1–4.

Building as a living system

2

Riadh Habash
School of Electrical Engineering and Computer Science, University of Ottawa, Ottawa, ON, Canada

For the resilient, sustainable cities we all want and need, urban plans need to be designed, evaluated and approved using a health lens.

Layla McCay

2.1 Building as a living organism

Building biology considers the home as a living organism and sets nature as the standard for a healthy human environment. For example, baubiologie or the study of natural building considers exterior walls a third skin, much like clothing. Using the same breathability principle that makes organic cotton and wool more comfortable than most synthetic fabrics, the baubiologie looks for wall materials that let vapor pass through without synthetic vapor barriers (Baker-Laporte, 2009). Baubiologie also uses building strategies that work with nature such as passive solar gain, high-mass interior walls, cross ventilation, and shading to reduce dependence on mechanical solutions.

2.1.1 Eliminate, isolate, and ventilate

Buildings like organisms have fostered a wide range of approaches to manage a changing external and internal environment including skins and body shells which act as separating and linking structures required to shield, restrain and control activities of life from external influences and other organisms.

For many centuries doctors and public health workers have understood the role of buildings in causing or exacerbating human diseases. As early as several centuries, the famous Greek "father of medicine" Hippocrates was aware of the adverse effects of polluted air in crowded cities and mines, and Biblical Israelites understood the dangers of living in damp housing (Sundell, 2004). The famed 19th-century nurse Florence Nightingale was conscious of the relationship between health and the dwellings of the population.

Given the fact that the majority of people are now spending as much as 90% of their lives inside buildings, so every occupant deserves healthy, affordable, comfortable, and durable environment. So, what does a healthy building means for occupants? Initially, health is a state of total physical, mental, and social well-being, and not just the absence of disease or sickness (Panagopoulos et al., 2015). Therefore, it is more important than ever to design buildings and indoor spaces that improve occupant health and well-being.

Sustainability and Health in Intelligent Buildings. https://doi.org/10.1016/B978-0-323-98826-1.00002-8

In the 1980s, John Bower, an American healthy home pioneer, formulated a method to make energy-efficient, light-frame homes healthier through his three-part formula: eliminate, isolate, and ventilate. He calls for eliminating as many pollutants as possible from within the building, isolating occupants from synthetic materials, toxic glues, and wraps found in the walls of conventional homes, and adding mechanical ventilation to ensure fresh air replaces the pollution and moisture occupants create (Baker-Laporte, 2009). This phrase is commonly used to summarize the steps for improving air quality. It expresses a reliable framework for ensuring a healthy indoor environment. This is a challenge for architects and engineers who try to design for the future since they face a great deal of uncertainty.

2.1.2 Sick building syndrome (SBS)

The concept of healthy buildings has been growing precedence since the recognition of SBS, which emerged in the 1970s, where people experience health issues or discomfort with no specific cause but can be attributed to time spent in a particular building. As defined by the WHO in 1984, SBS refers to the nonspecific set of health effects associated with time spent in a particular building (For Health, 2017). For SBS to exist, at least 20% of occupants of a building must experience symptoms that persist for at least 2 weeks. SBS symptoms can be categorized into respiratory symptoms (e.g., cough, shortness of breath, or chest tightness), mucosal irritation (e.g., dry throat, eye irritation), neurotoxic symptoms (e.g., headache, dizziness) dermal symptoms (e.g., skin irritation including skin rashes and dry itchy skin) (Thach et al., 2019), and thermal discomfort. Poor ventilation promotes SBS symptoms stemming from extended exposure to poorly ventilated spaces (For Health, 2017) and chemical contaminants from indoor and outdoor sources.

Solutions to SBS usually include pollutant source removal or modification, increasing ventilation rates, air cleaning, and occupant education. Research investigation in SBS has focused on physical or environmental conditions including air quality, thermal comfort, noise, and lighting levels. However, some individual and psychosocial factors may be involved such as occupant's perception of and satisfaction with the building's indoor environment.

2.1.3 Building-related illness (BRI)

BRIs are defined as a heterogeneous group of disorders whose etiology is linked to the environment of essentially modern air-tight buildings. These are typically characterized by features such as sealed windows, and dependence on HVAC systems for air circulation. BRIs apply to those adverse health effects for which there is a well-defined link between environmental agents in a specific building and the resultant health disorders.

BRIs usually appear in nonindustrial office buildings, but it is also found in apartment buildings, single-family homes, schools, and libraries (Kubba, 2017). Typical examples of BRIs include legionella infection, occupational asthma, inhalational fever, muscle aches, hypersensitivity pneumonitis, and other flu-like symptoms. Unlike SBS, the causes of BRIs can be determined and are typically related to allergic reactions and infections. Some BRIs are attributed to higher building temperature, higher humidity, and poor ventilation, typically with a failure to incorporate enough fresh air from the outdoors. To prevent BRIs, it is necessary to identify and remove sources of air pollution in the building.

2.2 Foundations of healthy buildings

A healthy building refers to an emerging area of interest that sustains the physical, psychological, and social health and well-being of people in the built environment. The benefits of healthy buildings can go beyond current concerns of reducing the potential for infection and cross-contamination to improve occupant comfort, satisfaction, engagement, and productivity while still meeting energy efficiency and sustainability goals.

2.2.1 *Indoor environmental quality (IEQ)*

IEQ may be defined as the various living conditions inside the building. It is a combined analysis of an occupant's response to the built environment. That means the capability to recognize and identify the quality of air, water, temperature, sound, light, and other components. Throughout the efforts of education and standard institutions as well as various rating programs, IEQ is being placed at the center of a healthy built environment due to related impacts on the comfort and well-being of people.

Although it seems obvious, comfort does not have a specific definition. Unfortunately, it is somehow a tenuous concept and may change from one person to another (Lotfabadi and Hançer, 2019). Comfort has traditionally been studied from the perspective of the physics of the environment and the physiology of the occupant, in terms of thermal comfort, acoustical quality, air quality, and visual quality. Standards and codes for each of the factors have been established, and technologies are being engineered to satisfy such standards in a presumably energy-efficient manner. In the IEQ literature, comfort is viewed from a physiological-technological perspective and described through the following parameters: visual (view, illuminance, and reflection), thermal (air velocity, humidity, and temperature), acoustical (control of unwanted noise, vibrations, and reverberations), and air quality (smells, irritants, outdoor air, and ventilation) (Ortiz et al., 2017). Similarly, exposure to electric and magnetic fields and currents with power frequency (50/60 Hz) applications including "dirty power" and ground current connection as well as to RF fields from ICTs are also associated with occupant health and comfort. The above qualities are the hallmark of sustainable design because they take advantage of freely available, unlimited resources.

The effects of poor IEQ are well documented in scientific investigations. Similarly, IEQ guidelines help occupants feel and perform at their best, with subsequent health, well-being, and productivity benefits. To achieve the quality or functional goal, a building must meet the basic demands of comfort which rely on architecture, engineering, design, and construction to provide spaces with the convenience of motion (ergonomics, circulation), air (temperature, humidity, air circulation, and pollution

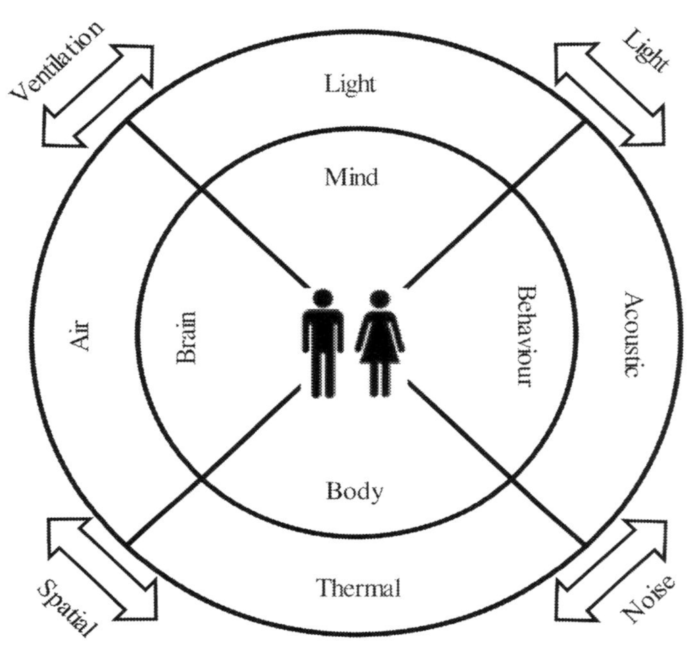

Fig. 2.1 Attributes of the indoor environmental quality and their relation to the human health environment.

levels), the eyes and ears (lighting, acoustics). The above attributes of IEQ and their relation to the human health environment are shown in Fig. 2.1.

Opportunities for improvements of IEQs fall into the following basic categories: good building design, including passive systems and landscaping; improved building envelope, including roofs, walls, and windows; improved equipment for heating and cooling air and removing humidity; thermal energy storage that can be a part of the building structure or separate equipment; improved sensors, control systems, and control algorithms for optimizing system performance (DoE, 2015). Both building designs and the selection of equipment depend on the climate where the building operates.

2.2.2 Human health environment

Human health is vital for human performance. Architects should strive to create spaces that properly drive performance through strategic and structured utilization of the built environment that stresses rigorous analysis of social, physiological, and psychological impacts. Humans' perception of the built environment is based on the ability to realize neighboring environmental forces affecting bodily senses (Youssef, 2021). Knowledge stored in the memory affects human behavior by way of predictions.

The provision of IEQ at levels that reinforce productive human habitation both complements and supports the environmental and economic goals for sustainable buildings. Poor IEQ is an important public health risk worldwide (Leech et al., 2002). Hence, IEQ in buildings has a significant impact on the human health environment.

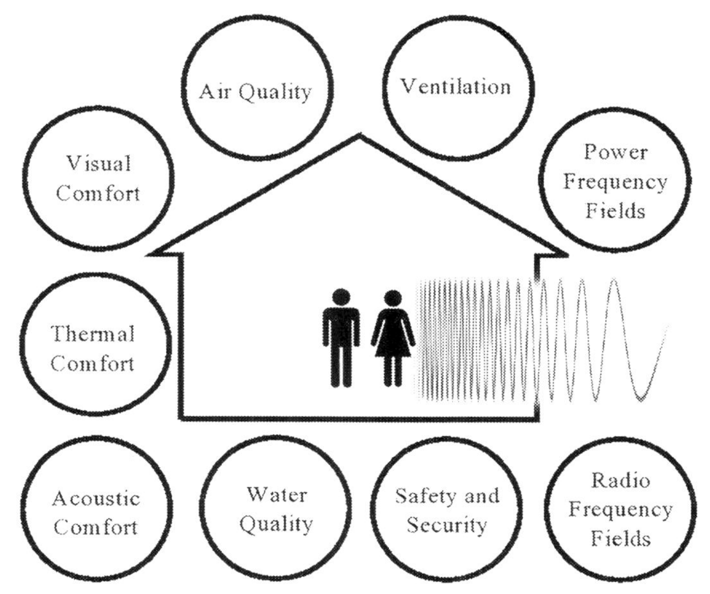

Fig. 2.2 Factors impacting human health environment in a building.

Additionally, IEQ is considered to be an independent factor in parallel to the energy efficiency of sustainable building design. For modern building controls, IEQ is often employed as a key component to optimizing building energy management.

Healthy buildings should improve the human health environment by impacting positively on IEQ and enhancing several factors such as air and water quality, thermal, lighting, and acoustic comfort in addition to making safe and secure places as shown in Fig. 2.2. What distinguishes this figure is the inclusion of EM (power frequency and RF fields) as an additional IEQ factor that impacts occupants' health. The ability to control these factors can lead to higher occupant satisfaction. Since IEQ is rooted in all aspects of building design and operations, any alteration to the design or operational practice may have considerable impacts on living conditions. Each of the given factors will be discussed in detail in the following sections.

2.3 Indoor air quality (IAQ)

Most infections of diseases occurred indoors, while outdoor infections have been much fewer. Therefore, a paradigm shift toward a better air environment in buildings should be a future priority. In this regard, IAQ will be a top subset of IEQ and a major public health issue. It is a main human health concern, where clean air, both outdoor and indoor, is a critical factor of a healthy life and a person's comfort. It depends on the presence and abundance of chemical and biological pollutants in the environment in gas, liquid, or solid states (Allen et al., 2015). According to the WHO, "more than three million people die prematurely each year because of ambient air pollution caused by high concentrations of small and fine particulate matter (PM) (Pringle, 2016)." The EPA reported that indoor air pollution is 25–100 times worse than outdoor air. It defines and monitors six "criteria pollutants" in ambient air: PM, ground-level ozone, sulfur dioxide, nitrogen dioxide, carbon monoxide, and lead. PM includes a variety of tiny airborne solids and droplets emitted by cooking and home heating, vehicles, factories, wildfires, and other anthropogenic and natural sources. Additional pollutants from outdoor sources include methane, volatile organic compounds, pollen, metals, and other sulfur and nitrogen oxides (Schmidt, 2019).

2.3.1 Air quality index (AQI)

The traditional AQI is based on air quality standards and takes into consideration both environmental and human health concerns. IAQ metric should identify when the air quality is unacceptable and should be based on its effects on human health and comfort. Poor IAQ is usually the result of sources that emit pollutants (gases or particles) into the air. The lower its value, the better the air quality is. An AQI below 50 indicates quite good air quality while an AQI over 300 means that the air contains hazardous concentrations of harmful substances. The most popular AQI is the one developed by the EPA. It calculates the AQI for several major air pollutants regulated by the Clean Air Act including ground-level ozone, carbon monoxide, sulfur dioxide, nitrogen dioxide, as well as particle matter.

As the IAQ of a building is one of the most important factors in maintaining health and productivity, HVAC and IAQ monitoring systems will help minimize occurrence cases of SBS, BRI, and other illnesses. Controlling IAQ involves integrating three main strategies. First, manage the sources of pollutants either by removing them from the building or isolating them from people through physical barriers. Second, dilute pollutants and remove them from the building through ventilation. Third, use filtration to clean the air of pollutants.

The air quality health index (AQHI) reflects the current knowledge of the health effects associated with air pollution. While both the traditional AQI and the AQHI have their respective strengths, AQHI has been designed to capture more information about health risks from air pollution on any given day. To obtain a comprehensive picture of the IAQ in a building it would be necessary to measure a range of contaminants as shown in Table 2.1.

2.3.2 Indoor pollutants

Outdoor air pollutants can enter buildings through open doors and windows, ventilation systems, and cracks in building structures. Some pollutants come indoors through building materials and occupant activities such as cooking, cleaning, and deodorizing products. Air pollutants can be found in all indoor spaces and are absorbed into occupant bodies through breathing, swallowing, and even some enter through the skin (Kubba, 2017). Long-term exposure to unhealthy air can cause respiratory problems, which can trigger other health symptoms. Common indoor pollutants that pose risks to human health are shown in Table 2.2.

Table 2.1 Common indoor types of contaminant particles.

Indoor particles	Approximate size in microns (μm)
Viruses including larger droplet nuclei	0.01–0.03
Bacteria including larger droplet nuclei	0.01–1
Combustion particles	0.05–10
Ozone and terpene formed aerosols	0.1
Tobacco smoke	0.01–0.8
Outdoor fine particles (metals, sulfates)	0.1–2.5
Asbestos	0.25–1
Fungi	0.5–30
Mold	2–20
Plant spores	10–70
Dust	0.05–40
Cement dust	0.05–50
Paint pigments	0.05–5
Asphalt paving	0.5–50
Pollen	10–100
Pet dander and allergen	0.3–100

Several factors influence how indoor air pollutants impact occupants. In offices, schools, and residences alike, these pollutants may come from printer emissions, pests, and rodenticides, cleaning supplies, personal care products, paint, pollen, and fungal spores. Outdoor pollutants, like PM that have a diameter of less than 2.5 μm (PM2.5), which is about 3% the diameter of a human hair, can penetrate indoors through several routes, one of which is through the mechanical system if the air stream is not properly filtered (Logue et al., 2011). Such particles may come from power plants, motor vehicles, airplanes, wood-burning, volcanic eruptions, and dust storms. Due to their various adverse effects, PM2.5 can inflict on a large number of people, which is one of the pollutants closely monitored by health organizations worldwide.

Particles of dust, dirt, or other substances may be drawn into the building from outside and can also be produced by activities that occur in buildings, like sanding wood or drywall, printing, copying, operating equipment, and smoking. Many contaminants reside in the dust and lead to exposure in three different ways including inhalation of resuspended dust, direct dermal absorption, or ingestion from hand-to-mouth behaviors (Ferro et al., 2004). People have a personal cloud of resuspended dust around them as they go about daily activities (Rodes et al., 1991). It is estimated that adults ingest up to 100 mg of house dust per day and children up to 200 mg per day (EPA, 1997). Higher ingestion rates in children are due to the greater amount of time they spend in contact with the floor and other surfaces, and the higher frequency of hand-to-mouth behavior. Indoor dust is the primary route of exposure for the lead from lead-based paint, which can accumulate in dust from flaked paint or dirt tracked in from outdoors (For Health, 2017). More than 21.2 million Americans live in counties with unhealthy levels of year-round particulate pollution (ALA, 2020).

Table 2.2 Common indoor pollutants.

Inorganic substances	**Asbestos and lead are among the leading indoor contaminants whose exposure can create significant health risks.**
Combustion by-products	Gases and tiny particles are created by the incomplete burning of fuels. There are many combustion by-products including, fine PM (which comes from dust, fly ash, soot, smoke, pollen, spores, fibers, pet dander, aerosols, fumes, and mists), carbon monoxide, nitrogen oxides, and tobacco smoke.
Organic pollutants	The main organic compounds include volatile organic compounds and particulate organic materials. These are coming from the emission of a mix of over 250 different chemicals such as formaldehyde, pesticides, limonene, ethanol, acetone, benzene; and from electrical and electronic equipment in the home and office such as computers, printers, and photocopiers.
Biological contaminants	Mold and mildew, viruses, bacteria, cat saliva, dust mites, pollen, and animal allergens arising from various sources such as microbiological contamination, and the remains and dropping of pests such as cockroaches. Some biologic pollutants, such as measles, chickenpox, influenza, and viruses including coronavirus are transmitted through the air. Although they are now preventable with vaccines, they remain of concern in crowded indoor conditions and may be affected by ventilation levels in buildings.
Radiological hazards	Radon, a radioactive noble gas formed by the decay of uranium, thorium, and radium that naturally emanates from soil and rocks. Outdoors, radon exposure is not a matter of concern because it is rapidly diluted to low levels by circulating air. It can enter buildings through cracks and fissures in the foundation, rising indoor concentrations to toxic levels in buildings that are not sufficiently ventilated, and increased risk of serious illnesses, such as lung cancer. Radon is considered the second leading cause of lung cancer, behind smoking (For Health, 2017). Radon with a concentration between 60 and 200 Bq/m^3 is considered a severe anomaly, while considered an extremely anomaly when the concentration is above 200 Bq/m^3 (IBN, 2015).

2.4 Healthy ventilation

Ventilation is the movement of outdoor air into a building or a space while the stale air is removed. It is a significant method of removing contaminants from buildings and ensuring good IAQ as well as thermal comfort. The main purpose of ventilation is to provide healthy air for breathing by diluting the pollutants originating in the building like carbon dioxide (CO2) and product-generated pollutants like VOCs. Therefore, improving ventilation systems plays a significant role not only in keeping a better indoor climate for the occupants and reducing the possibility of health problems.

2.4.1 Natural and sustainable ventilation

In ancient times, buildings whether a Babylonian palace, an Egyptian temple or a Roman castle, were ventilated naturally using either "badgeer/malqafs" (wind shafts/towers) or some other natural innovative methods since mechanical systems did not exist at the time (Kubba, 2017). By the late 1800s, ventilation technologies had progressed where buildings required different minimum ventilation rates for different space-use types and occupant densities (DOE, 2015).

Natural ventilation does not need any type of energy but works due to natural forces such as winds, breezes, thermal buoyancy, or combined as driving forces. They can drive outdoor air through purpose-built, building envelope openings. As a rule of thumb, the wind-driven natural ventilation rate through space may be calculated as follows (WHO, 2021):

$$\text{Ventilation rate } (l/\text{s}) = k \times \text{wind speed } (\text{m}/\text{s}) \times \text{smallest opening area } \left(\text{m}^2\right) \times 1000 \left(l/\text{m}^3\right) \tag{2.1}$$

where $k = 0.05$ in the case of single-sided ventilation and $k = 0.65$ in the case of cross ventilation. Such natural ventilation is characterized by using the natural driving forces to supply oxygen-rich air and remove the amount of polluted air, which is necessary to maintain a good thermal indoor climate and a good IAQ (IED, 2009). In general, the natural ventilation of buildings depends on the weather, building configurations, and human behavior.

It is acceptable for low-rise buildings where fresh air may infiltrate the building unintentionally through leaks or controlled ventilation. In spaces with lower ventilation rates, it is needed to consider enabling cross-ventilation rather than single-sided ventilation and/or installation of wall or window air extractors or whirlybirds (WHO, 2021). Not only does this make the indoor environment comfortable to live and work in, but the decreased pollutants can lower an array of harms. However, natural ventilation is not without its problems, particularly in places with cold winters and hot summers. Intensified ventilation increases energy consumption when unconditioned, outside air must be heated or cooled as it replaces conditioned indoor air that is being exhausted.

A natural ventilation cooling strategy that serves as the alternative to the air-conditioning system has been effectively employed in high-rise office and school buildings (Cuce et al., 2019). However, the utilization of natural ventilation in buildings has architectural effects on facades and the organization of the interior spaces. Historically, the office building topology emerged during the 19th century where the indoor environment in these buildings was controlled by passive means using operable windows for natural ventilation and keeping cool. This building type became particularly common in Chicago during the office building boom that followed the Great Fire of 1871 (Wood and Salib, 2012). During the oil crises in 1973, western countries aimed to reduce energy consumption mainly by increasing the insulation level of the building envelop and reducing the air infiltration level by sealing the buildings, the fact that impacted negatively on the health and comfort of the occupants as well as the productivity of the office workers.

2.4.2 Mechanical ventilation

Most large buildings have mechanical ventilation systems combined with heating or cooling systems that are typically designed and operated to heat and cool the air, as well as to draw in and circulate outdoor air (Kubba, 2017). Historically, the widespread of mechanical ventilation goes back to the 1950s due to cheap energy and the availability of technical systems. According to the consideration of passive design measures to provide a proper indoor environment were no longer a major concern for architects and engineers (Wood and Salib, 2012).

In mechanical ventilation, the airflow path can be controlled as the air is moved through ducts, filters, and fixtures using fans run by electricity. In such a case, a building's mechanical system is designed to bring in outdoor air, filter it, and deliver it to occupants (Shendell et al., 2004). Ventilation rate and airflow direction are significant elements for assessment and evaluation before starting any action on the ventilation system. The rate of ventilation is controlled by the need to remove indoor pollutants, not to provide oxygen to replace that consumed.

Balanced ventilation systems are the most typical layout of mechanical ventilation in buildings. The exhaust fan is supplemented with an inlet-air fan, creating a balance between the incoming air and the exhaust air. Due to the need to control both the incoming and outgoing air, two duct-work systems are needed. Balanced ventilation has the advantage of good possibilities for heat recovery from the exhaust air, thereby saving energy for preheating the incoming air. It can be used in almost all types of buildings and due to the need for heat recovery in low-energy buildings (IED, 2009). The key is to integrate ventilation with HVAC and other systems to work together to effectively ventilate spaces and minimize sources of indoor pollution. In addition to specifying higher ventilation rates, improved maintenance of the system is required because substandard ventilation often occurs in buildings where HVAC systems are either neglected or inadequately maintained (For Health, 2017).

Finally, the efficiency of the mechanical ventilation systems depends mainly on the recovery of the heat from the stale air that is being exhausted from the building. These systems require careful design and the problems involved in poor maintenance or incorrect management may cause a high concentration of infectious droplet nuclei and eventually end up in an increased risk of disease transmission.

2.4.3 Hybrid and intelligent ventilation

Technically, natural ventilation systems may be classified into simple natural ventilation and technology-based natural ventilation. The latter may be aided by mechanical systems (i.e., hybrid) and require control systems. Such systems may have the same drawbacks as mechanical ventilation systems; however, they may have the benefits of both natural and mechanical systems.

In addition to mechanical and control components, modern ventilation systems use smart monitoring, AI-based prediction, and automatic control to detect concentrations of contaminants and establish dynamic ventilation to make appropriate adjustments to ventilation frequency and rates. Such systems are known as demand-controlled ventilation, which allows the system to accommodate times when space is partially occupied, with occupancy sensors, CO2 sensors, or a time-of-day schedule. For example, smart IAQ sensors can be used to analyze and monitor the levels of airborne particulate transmission. Innovative IEQ management systems and IAQ monitoring stations in the built environment using IoT, AI analytics can provide tools to identify, monitor, and improve IAQ with an up-to-date overview of the situation.

A roadmap for buildings to become sustainable and healthy requires consideration of an open barrier between the indoor and outdoor climate by making use of natural ventilation as well as integration of human and technology intervention. Today, air heat-recovery systems are most often used to preheat and/or precool ventilation air to reduce the system load. Some heat-recovery ventilators are often referred to as energy-recovery ventilators. To realize these benefits integration must occur between the HVAC system that employs intelligent control techniques and automated windows via the building automation system (BAS), which will have an enhanced ability to monitor and control the windows for natural and mechanical ventilation, allowing for a complementary hybrid system to the HVAC, rather than two systems that operate independently. In general, the engagement of natural ventilation for periods when the external conditions permit, such as night flushing, reduces energy consumption, equipment maintenance, space, and working costs of mechanical equipment. Such engagement, however, makes use of dedicated ventilation elements to exploit the natural driving forces and to support the airflow (supply or exhaust) through the building. This includes wind scoops and towers, double façades, atriums, chambers, etc. Today, several examples of large buildings that engage natural ventilation strategies exist worldwide. Examples include the Post Tower in Bonn, Germany, and Manitoba Hydro Place in Winnipeg, Canada.

In sustainable design practice, building rating systems, such as WELL, LEED, BREAM, and many others, consider IAQ as one of the critical indexes and specify exposure thresholds for pollutants to ensure the healthy indoor environment requirements to be experienced.

2.5 Thermal comfort

Human thermal comfort is a condition of mind for expressing satisfaction. It is a combination of a subjective sensation and evaluation (feeling) and objective interaction with the environment (heat and transfer rates) regulated by the brain. It has been one of the most difficult aspects of designing buildings. It depends on many quantifiable factors like air temperature, air movement, and humidity that in turn depend on variables like personal activities, clothing, and psychological well-being. The long-term thermal comfort of a space is based on several approaches, the thermo-physiological sensation of the thermal environment and the adaptive approach which might be psychological.

2.5.1 Environmental and individual variables

Thermal comfort depends on several physical variables, environmental and individual. The environmental variables result from a combination of four heat balance factors and four thermal variables as shown in Fig. 2.3. Heat balance consists of heat loss (transmission, ventilation, and filtration), and heat gains (sun and internal loads). It involves four major factors including conduction, convection, evaporation, and radiation. Thermal comfort variables are influenced by physical factors like air temperature, mean radiant temperature, air velocity, and humidity. Air temperature is also affected by the people inside the building and their activity. Radiation is the largest component of thermal comfort which comes into the building from windows and walls.

In general, buildings absorb thermal energy through convection and radiation from the outdoor environment (Raja et al., 2001). Its intensity depends on several parameters, such as the number of surfaces and their area, materials' heat capacities and massiveness, thermal conductivity, density, and thickness, and so on (Kenisarin and Mahkamov, 2016). In this regard, one of the most effective parameters is the building's surface dimensions. In other words, based on heat transfer principles, the thermal inertia of a building is directly dependent on its volume. For this reason, building standards and regulations allow designers to reduce ceiling heights and decrease surface sizes while influencing indoor air temperature and reducing the indoor environment's thermal inertia (Guimarães et al., 2013).

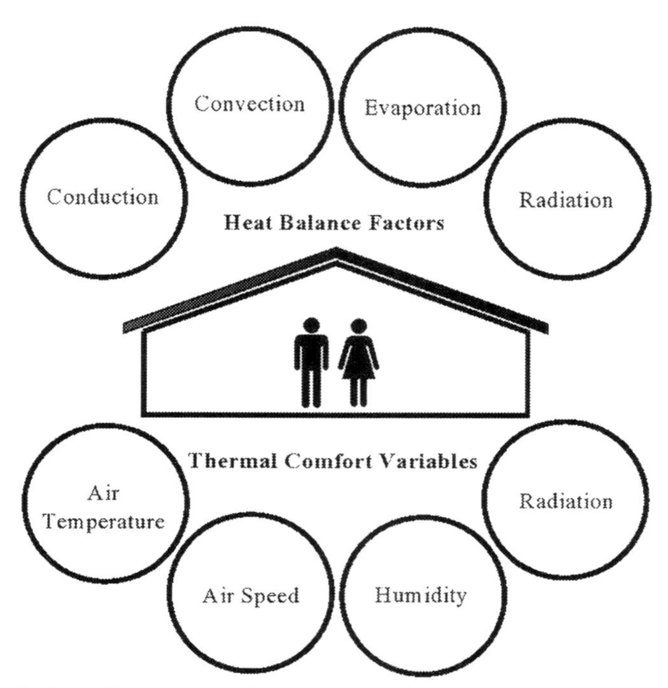

Fig. 2.3 Heat balance factors versus thermal comfort variables.

On the individual level, thermal comfort varies significantly from person to person due to factors like metabolic activity level and thermal clothing insulation. Deep body temperature is between 36.5°C and 37.5°C and the average skin temperature is between 33°C and 35°C when the skin surface is free of sweat and the muscles are not contracting the fact that allows heat outflow. Besides, age and risk groups, changing from indoors to outdoors, clothing differences among seasons, personal preferences, and the state of mind may have an influence.

2.5.2 Thermal health

Sometimes, thermal comfort may be replaced with the more commonly used "thermal health" (For Health, 2017), which encompasses all of the impacts of thermal conditions on health, including mortality that goes beyond just comfort. It is well documented that thermal conditions are integral to the occupant experience. In general, thermal comfort is experienced via some conscious interactions between three personal and environmental factors: physiological by the way people work and interact with the environment, physical by the main parameters of the surrounding environment, and socio-psychological by the social environment people live in. Thermal discomfort, on the other hand, is not just a lack of satisfaction with the ambient temperature but reflects a situation where there is a potential threat to health that is when the surrounding temperature falls below 18°C or rises above 24°C for some time. This range is based on the WHO's guidance on thermal comfort for the home environment.

In a study conducted in office buildings across Europe, the foremost complaint reported by employees was the thermal comfort of their workspaces. Many reported that ventilation, temperature control, and humidity were all factors that contributed significantly to work experience and task capabilities (Sakellaris et al., 2016). When indoor environments are too warm, there is evidence of increases in sick building syndrome symptoms, negative moods, heart rate, respiratory symptoms, and feelings of fatigue (Lan et al., 2011). Temperature and humidity may also influence disease transmission (Lowen et al., 2007). Cold and dry environments have been found to facilitate the spread of the influenza virus because low humidity levels permit virus particles to stay in the air longer and low temperatures prolong the virus shedding period. As temperatures fall below 18°C, the potential impact on health increases in severity (Howieson and Hogan, 2005). The most serious heat-related condition is heatstroke. It occurs when the human body becomes unable to control its temperature where the body's temperature increases quickly above 40°C, the sweating mechanism declines, and the body is unable to cool down.

2.5.3 Moisture and humidity

Moisture and mold are prevalent problems in buildings worldwide. In general, there are two main sources of moisture: internal and external. Internal includes evaporation, desorption, and people (breath, clothes); external includes ventilation, infiltration, and permeation (Raish, 2009). Humid indoor spaces cultivate the growth and spread of viruses and bacteria. Controlling moisture indoors can limit the spread of these infectious diseases and also limit mold, dust mite growth.

Temperature and RH contribute to virus transmission indoors affecting virus viability, droplet nuclei forming, and susceptibility of occupants' mucous membranes. The transmission of some viruses in buildings can be limited by changing air temperatures and humidity levels (REHVA, 2020). Relative humidity is a major factor that affects air temperature. It is the percentage of water vapor saturation that is in the air. This can be affected by the convection in space; the warm air enters the space and rises to the ceiling, pushing the cold air downward. RH levels that are too high can contribute to the growth and spread of unhealthy biological pollutants (EPA, 2020). Managing water entry points through the building envelope, by regular inspection and maintenance of the roof, and penetrations in the wall such as caulking around windows and doors is critical to minimize water intrusion. In buildings, molds reproduce through the accumulation of spores, tiny cells that float continuously through indoor and outdoor air. Areas typically exposed to mold in buildings are on carpets, ceiling tiles, insulation materials, wood, areas behind wallpaper, or in HVAC systems (For Health, 2017).

Well-designed building envelopes and foundations may greatly reduce moisture infiltration, but residual moisture transfer coupled with moisture generated by people and building operations will continue to make moisture removal a priority in building energy systems. Membrane technologies allow water vapor to pass but block the passage of dry air or can be used to separate moisture from the air using only the difference in vapor pressure, passing thermal energy from outgoing to incoming air (DoE, 2015). These systems can be integrated with evaporative cooling stages to provide both dehumidification and chilling.

2.6 Acoustic comfort

Providing acoustic comfort means minimizing interfering noise and promoting occupants' comfort, well-being, productivity, and ability to communicate.

2.6.1 *Noise pollution*

Noise is defined as an "unwanted or disturbing sound" that interferes with normal activities such as work, sleeping, and conversation. Sound waves combine and reach a listener via numerous direct and indirect pathways. Noise may be classified into two types: airborne and impact noise. Airborne noise is transmitted through the air as the medium while impact noise is produced from an impact on a building structure (For Health, 2017). Noise enters building interiors from outside sources while indoor noise can be generated from a building's mechanical systems.

Acoustic comfort is the outcome of the right balance of acoustic conditions in an environment, mainly the complete absence of noise. It is the well-being and feeling of a building's occupants regarding the acoustic environment. Acoustic comfort is affected by the levels and the nature of the sound experienced in space. Providing acoustic comfort consists of minimizing intruding noise and maintaining satisfaction.

Noise pollution is a major environmental problem affecting people around the world. Acoustic comfort is affected by the levels and the nature of the sound experienced in space. A healthy human ear is sensitive to a very wide range of frequencies. The effect of noise pollution on human health is described in various forms including irritation reaction, sleep disturbance, cardiovascular disease, high blood pressure, headaches, hormonal changes, psychosomatic illnesses, sleep disorders, loss of efficiency, decreased physical and mental performance (Münzel et al., 2014). Multiple studies on the nonauditory effects of noise exposure have observed that increased noise levels are associated with higher systolic and diastolic blood pressure, changes in heart rate, and hypertension (Van Kempen and Babisch, 2012; For Health, 2017).

2.6.2 Acoustic performance

Building noise can be controlled in different ways; however, approaching this problem during the design phase is the simplest, but has not become very popular. The building acoustical design is the science of controlling noise inside a building. This includes the minimization of noise transmission from one space to another, which can be achieved by following design and construction techniques especially for walls, windows, doors, floors, and building materials. Noise control is an important consideration in the design, operation, and construction of most buildings, and may have a significant impact on occupants' health and well-being, communication, and productivity. It can be influenced by the geometry and volume of space as well as the sound absorption, transmission, and reflection characteristics of surfaces enfolding the space.

Technically, the above is conceptualized in the term of "acoustic performance" which refers to the effectiveness of a building's acoustic design and the capacity of space to provide an acoustic environment appropriate to its intended use (Wilson, 2020a). However, another approach has gained popularity, according to which performance simulations are used to drive the design process. This approach is known as acoustic performance-based design which can be subdivided into two subclasses according to the way the design optimization process is conducted. In "formation models," the modifications are applied manually by the operator, while in "generative models," the design proposals are directly optimized by the computer. Currently, several commercial acoustic simulation tools are available (e.g., Odeon, CATTAcoustic, Pachyderm Acoustics, etc.), allowing professionals to estimate the performance of design proposals using the geometrical acoustic method (Badino et al., 2020). The acoustic performance of a given environment is calculated based on its geometrical characteristics and the acoustic properties of the materials used in the structure surfaces.

Traditionally, the method for mitigating noise intrusion is by introducing the rating system called sound transmission class, which essentially describes the ability of a particular material to resist airborne sound. For example, each wall of a room may be given a certain class rating, which indicates an average quantum of external noise that could potentially be isolated through it. Other actions that can help reduce noise include installing new, quieter systems and appliances and isolating existing machines to reduce noise and water passing through pipes. In specifying building designs, acousticians, architects, and engineers adopt an approach to use the ambient noise level to specify their desired design outcome. Ambient noise level is described as a background sound level for a desired indoor functional space. The sound spectrum must stay clear without portions distorted or omitted where different materials absorb sound frequencies differently. In reality, the desired acoustic comfort of a particular space depends on a complex combination of sound isolation of partitioning elements, indoor acoustics design, and finishes. It requires a more comprehensive method of measurement and design that provides a better reflection of the total noise levels of that space.

2.7 Visual comfort

Healthy lighting for humans and the environment involves not only which type of lighting to choose but also occupant habits, and largely includes building design.

2.7.1 Natural light

In ancient times, sunlight and its lunar reflections provided the bulk of the visible spectrum for human beings (Kostoffa et al., 2020). Light plays a critical role in human life and daily affairs. It permits one to function at a basic level and it also plays a key factor in our psychological and physiological health. Natural light is the source with the highest energy efficiency, and today no light sources match it.

Natural light provides positive psychological, mental, and physiological effects on building occupants (Hosseini et al., 2019). It is a combination of all the indirect and direct sunlight available during the day. It connects people with the outdoors, provides an essential nutrient, and makes interior spaces glow with natural beauty. The human body uses light through three systems: visual for eyesight, forming for circadian rhythm regulation, and skin for the synthesis of vitamin D and other composites (Kubba, 2017). Along with the potential of decreased demand for both artificial lighting and air cooling, the field of architectural daylighting was revived in the 1990s after years of neglect. The functionality of a building is dependent on the quality of its lighting. To safely and comfortably perform their duties, occupants need lighting that provides adequate visibility without causing discomfort or distraction. High-quality lighting, beyond providing basic comfort, can keep and promote health. Light has acute effects on cognitive function and sleep. Keeping the circadian sleep–wake cycle in proper alignment to obtain adequate levels of sleep is essential to maintaining good cognitive function. Circadian rhythms influence basic cognitive processes like attention, working memory, hormone regulation, sleep–wake cycles, alertness, mood and performance patterns, immune function, and reproductive function, executive functions, and learning and memory can be impaired when the sleep–wake cycle is disrupted (Miller et al., 2018). The first principle is to ensure exposure to a distinct and stable 24-h light-dark cycle each day, with bright days and nights, to ensure proper synchronization of circadian rhythms (Cho et al., 2015). Certain types of lighting can disrupt melatonin production in the body. Therefore, it is preferred to take advantage of natural daylight whenever they can, to further support healthier lifestyles.

There are two main ways to bring light into space: from the top through a roof or ceiling (top lighting) or from the side through windows which are the most common method of admitting light into spaces. Bringing light through the top normally delivers a more even distribution of light than bringing light through windows, where light levels drop with distance from the window wall. In spaces with high ceilings, skylights can light a large portion of the floor area. Natural light provided from the side (windows) can make a major contribution not only to the ambiance of indoor environments but to reducing a building's demand for artificial light.

2.7.2 Artificial light

Electric lighting became available about 140 years ago. At around 1880 Thomas Edison in the US and Joseph Swan in the United Kingdom introduced electrical light bulbs to the market. The artificial lighting industry was born, and it slowly revolutionized the building industry. The electric lamp provided people with full control over lighting inside their homes and working places (Savic, 2020). Electric lighting is considered an integral part of life for most people today and the industry revolutionized buildings where interior spaces became less dependent on daylight, which allowed for more experimental setups and extended living and working times.

Artificial light as a supplement to daylight replicates daylight conditions as much as possible through electrification which consists of networked light-emitting diodes (LED) and linear fluorescent luminaires with advanced sensing and control capability (Ece, 2018). Artificial lighting renders the colors poorer than daylight, as they do not hold the full-color register. Incandescent bulbs are the type of artificial light with the highest color rendering. To achieve such an environment several factors should be considered including the strength of lights, distribution of lights, glare, color rendering, and adjustments.

2.7.3 Lighting design

Lighting design can have either positive or negative effects on people, especially in the spaces where many people spend a lot of time, like homes, hospitals, schools, offices, and care environments (Wilson, 2020b). Therefore, the functionality of a building is largely dependent on the quality of its lighting which affects not only the performance and productivity of occupants but also, more critically, their well-being.

Effective sustainable lighting design reduces energy consumed by a building. The main strategies for improving the efficiency and quality of artificial lighting are good building and lighting design, window, and window covering technologies (such as blinds and diffusers), lighting sensors and controls, and LED lighting devices (DoE, 2015). Lighting designers should avoid overlighting, block flicker and light interlopes, and consider likely impacts of lighting on plants and animals of the surrounding ecosystem, as well as on adjacent buildings and places. In this regard, intelligent lighting systems are healthier, affordable, and more flexible, the facts that make them the future of lighting. Kinetic façade systems can contribute to better illumination levels in space through daylight harvesting, in turn reducing the reliance upon artificial lighting systems and saving energy (Kensek, 2014). These daylight-redirecting technologies promise to boost interior natural light while reducing the need for artificial lighting. This includes tinted, reflective or clear glass based on the building orientation and automated window roller shades.

Designing for natural light has for many years been a challenge that designers often face, due to the fluctuations in light levels, colors, and direction of the light source. Designers should adopt practical design strategies for a sustainable daylighting design that will increase visual comfort by applying three primary techniques (Kubba, 2017):

- **Architectonic:** Architectural practices have a diverse team of consultants and design tools that enable the architect to undertake the complex daylighting analysis.
- **Environmental:** By using the natural forces that impact design, resource, and energy conservation.
- **Human Factors:** Designers need to achieve the best lighting levels possible while avoiding glare and high-contrast ratios. These can usually be avoided by not allowing direct sunlight to enter a workspace, e.g., through the use of shading devices.

Building and window design that utilizes natural light will lead to conserving electrical lighting energy, shaving peak electric loads, and reducing cooling energy consumptions. The proper sizing and design of skylights and light tubes may let light in a way that keeps indoor space warm in winters and cool in summers (Fig. 2.4). At the same time, lighting designers should work with manufacturers to decrease the embodied energy and the corresponding carbon harm of fixtures. The ASHRAE Standard 90.1 provides a basis for improving the efficiency of buildings by encouraging occupants to use lighting more carefully and to automate processes to dim the lighting when there is enough daylight or turn off lighting when not in use.

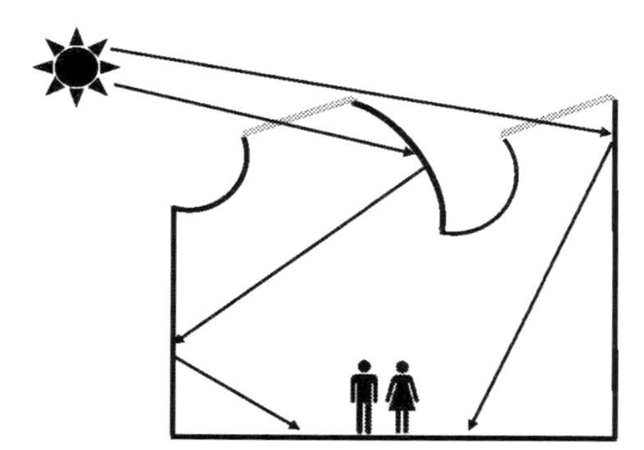

Fig. 2.4 Redirecting sunlight for providing adequate daylight for occupants.

2.8 Water quality

Treated water is usually low in biological contaminants because of chlorination, but it is not well screened for industrial and hazardous waste, although chlorine itself is a microbial poison. However, the built environment's plumbing infrastructure may affect water quality as 2015 events in Flint, Michigan, taught the world. Lead pipes or solder either within buildings or connecting buildings to water, mains can contaminate water coming into homes with enough lead to cause permanent harm to children's brains and also affect adult health (CHE, 2020). Lessons learned from the Flint case inform cities and communities that cheap sensors can reveal hazardous chemicals or even leaks in the water supply. That is an easy way to inform people that something dangerous has been discharged into the water or air and authorities should react to possible health hazards.

Microbial contamination is by far the largest contributor to the global burden of waterborne disease and outbreaks of waterborne illnesses continue to occur worldwide. In general, there are several key ways water quality may be compromised. First, wear and tear on service pipes can exacerbate corrosion, dissolving metals due to chemical reactions between water and plumbing fixtures, influencing the extent to which lead, copper, and other metals can contaminate drinking water. For example, lead may be present in service pipes due to lead piping and solder, particularly in water with high acidity, low mineral content, and hot water systems (Hanna-Attisha et al., 2015). Second, drinking water may be contaminated by improper treatment, poor maintenance of distribution systems, malfunctioning wastewater treatment systems, accidental sewage releases, and heavy metals from manufacturing processes (For Health, 2017). Third, the duration of time that water is stored within a system before being used may impact its quality.

Drinking-water distribution systems in large buildings are usually long and complex with a large number of branch mains. Control over such distribution systems is also hard to sustain. Letting water sit in pipes for long periods can create water quality problems in these buildings. Storage for long periods can damage plumbing materials and compromise the safety of drinking water by diminishing the efficacy of disinfecting agents (such as chlorine) and contributing to the growth of microorganisms that pose a risk to human health (Miranda et al., 2007). The water quality index is considered the most effective method of measuring water quality. WQI is defined as the degree of contamination existing in the water. Thus, highly polluted water will have a WQI close to or equal to 0%, and 100% denotes water in excellent condition.

Successful management of drinking water systems in buildings requires a complete consideration of the system including the range of potential hazards and the risks that may happen during delivery and use of water by occupants and visitors to buildings. One way to assure the quality of drinking water is by having purification systems to remove chemical and biological contaminants, which should be thoroughly and periodically maintained. Typically, purification systems for large municipal water systems can be standard off-the-shelf systems. Smart water management denotes the integration of systems and the implementation of multiple measures to monitor, control, and regulate the usage and quality of water resources and maintain the connected equipment.

Water treatment is used in some buildings to either enhance untreated independent water supplies or to improve treatment applied by the drinking water provider. Common modes of treatment may include filtration, disinfection, and softeners. According to the international Well rating system, potable (drinking) water used in the facility must be filtered, once again meeting specific standards. It cannot be hard and must not contain suspended soils, dissolved materials, chlorine, or fluoride.

2.9 Electromagnetic environment

Adequate indication now exists to imply that another indoor factor may be added to the list of foundations of healthy buildings. This factor is the long-term and excessive exposure to EM fields in buildings, which is a direct product of the ICT and one of the greatest technological enablers of our time (Habash, 2020). The vision of expanded ICT including IoT in the built environment may involve risks to public health by extensive exposure to pervasive manmade EM fields. Often little thought is given to their potential positive or negative health effects, a shortfall that can be actively addressed. It is therefore very important to investigate how healthier next-generation EM-built environments and technologies can be created.

Life on earth developed in an environment of natural EM fields. However, over the past century, this natural environment has changed with a fast-growing spectrum of artificial EM fields. Today, the earth's environment consists of natural and artificial or man-made EM fields. One of the main sources of natural EM fields is sun radiation where incident power density upon a human body ranges between 8 and 24 mW/cm^2 depending on the season, atmospheric conditions, geographical location, etc. Other natural EM sources incorporate the earth's electric and magnetic fields including the magnetic field of the earth, electric fields caused by changes in the clouds, the static electricity of two objects rubbing each other, or the electric and magnetic fields caused by thunderstorms and lightning.

The frequency spectrum of EM fields extends from extremely low frequency (ELF) fields at power frequency (50 or 60 Hz) and very low frequencies (VLF) to radiofrequency (RF) and microwaves to visible light and beyond to the substantially higher values of ultraviolet (UV) light, X-ray, and gamma-ray. In this regard, a distinction should be made between nonionizing versus ionizing radiation that has enough energy to displace electrons and ionize atoms and molecules (Habash, 2020).

The WHO/International Agency for Research on Cancer (IARC) already classifies EM fields "as possibly carcinogenic to humans (Group 2B)." Some experts suggest RF radiation should be upgraded to Group 2A, "probably carcinogenic," or even Group 1 "carcinogenic" (Habash, 2020), and there is some evidence that excessive exposure to EM fields might have a negative health impact. Therefore, there is a need to address the topic and optimize biological performance through intelligent design and create safe environments for individuals who are electromagnetically hypersensitive (EHS) or otherwise adversely affected by EM pollution.

In both building and home environments, the ever-increasing use of electrical devices means that occupants may be inadvertently exposed to EM fields generated within the building and outside, and at levels that some studies indicate may be potentially harmful to human health. Therefore, it is a key reflection for the AEC community to ensure, as much as possible, the health and safety of people in the built environment that they have created. Sufficient evidence now exists to justify taking a precautionary approach in building design to eliminate harmful effects (Maisch et al., 2006). Such a strategy would involve reducing or eliminating unnecessary EM exposures when this can be relatively easily accomplished for a modest cost in the early stages of design.

Artificial EM fields with various frequency bands have largely emerged in the last decades and almost everybody is exposed to rising levels. These fields are produced either intentionally or as by-products. Today, nonionizing fields including ELF fields are used as a platform for electrification (powerlines, substations, and transformers, electrical wiring in buildings, power tools, and appliances), while the higher frequency RF fields are used for wireless communication technologies (broadcasting stations, cellular systems, and other wireless networks), and various biomedical, sensing, and industrial applications (Habash, 2020).

Today's EM fields at all frequencies are one of the most common environmental issues, about which there is growing concern and speculation. They are present everywhere in the environment but are invisible to the human eye. All populations are now exposed to varying degrees of EM fields, and the level of exposure will continue to increase as technological inventions advance. These inventions have become an integral part of modern life. We just need to know they are safe. In several years from now, bioelectromagnetics environmental issues should be openly addressed in the design of buildings, technologies, and environments. By that time it is intended that many buildings will have biologically optimized "low anthropogenic EM field zones," as standard, as will the infrastructures that allow people access to them. An essential component aiding advancement at the edge of digital transformation is the creation of more bioelectromagnetically friendly intelligent and responsive building materials and technologies.

2.10 Safety and security

Dating back to the early times, buildings were designed and built with safety and security in mind and architecture played a role in securing buildings and providing security to occupants. In today's society, the scope of life and safety systems in buildings is growing at a very rapid pace. What was considered a luxury or state-of-the-art technology only a few years ago could easily be a minimum requirement in newer buildings. Also, it is not only the mandate of architecture but also technology in securing the safety and privacy of its owners and occupants. Perhaps most importantly of all, it is vital to provide security professionals and building occupants with solutions they require and need.

With all these advancements in technology, the sense of threat still exists. Such a feeling may trigger a cascade of biological "fight or flight" responses that alter people's physical and psychological functioning (Schneiderman et al., 2005). Psychological security is a state in which an individual perceives that his/her environment is safe and free from threat. While individuals vary in their response, psychological stress can negatively affect immune function with the onset of immune changes occurring in as little as 5 min. Chronically elevated stress hormones suppress immunity which can exacerbate autoimmune diseases and other inflammatory conditions, while elevated blood pressure levels can eventually lead to damaged arteries and plaque formation, putting stressed individuals at greater risk of hypertension and cardiovascular disease (Marsland et al., 2002). Over time, these responses place wear and tear on the body that increases disease susceptibility (For Health, 2017).

The relationship between fear of crime and health may be reciprocal because the fear of crime can negatively impact health and poorer health can heighten anxiety about an individual's increased vulnerability to victimization. People with long-term depressive symptoms, poorer physical functioning, or poorer mental functioning are more likely to report subsequent fear of crime (Jackson and Stafford, 2009). There is growing evidence that particularly for women and older adults, physical inactivity is higher among people who perceive their neighborhood to be unsafe from crimes (Lorenc et al., 2013).

References

ALA, 2020. Populations At Risk. https://www.lung.org/research/sota/key-findings/people-at-risk.

Allen, J.G., MacNaughton, P., Satish, U., Santanam, S., Vallarino, J., Spengler, J.D., 2015. Associations of cognitive function scores with carbon dioxide, ventilation, and volatile organic compound exposures in office workers: a controlled exposure study of green and conventional office environments. Environ. Health Perspect. 124 (6), 2.

Badino, E., Shtrepi, L., Astolfi, A., 2020. Acoustic performance-based design: a brief overview of the opportunities and limits in current practice. Acoustics 2, 246–278. https://doi.org/10.3390/acoustics2020016.

Baker-Laporte, P., 2009. Building Biology 101: What Is Baubiologie? https://www.motherearthliving.com/green-living/building-biology-baubiologie.

CHE, 2020. Built Environment. https://www.healthandenvironment.org/environmental-health/environmental-risks/built-environment.

Cho, Y., Ryu, S.H., Lee, B.R., Kim, K.H., Lee, E., Choi, J., 2015. Effects of artificial light at night on human health: a literature review of observational and experimental studies applied to exposure assessment. Chronobiol. Int. 32 (9), 1294–1310.

Cuce, E., Sher, F., Sadiq, H., Cuce, P.M., Guclu, T., Besir, A.B., 2019. Sustainable ventilation strategies in buildings: CFD research. Sustainable Energy Technol. Assess. 36, 100540. https://doi.org/10.1016/j.seta.2019.100540. ISSN 2213-1388.

DoE, 2015. An assessment of energy technologies and research opportunities. In: Chapter 5: Increasing Efficiency of Building Systems and Technologies. US Department of Energy. https://www.energy.gov/sites/prod/files/2017/03/f34/qtr-2015-chapter5.pdf.

Ece, N., 2018. Building Biology: Criteria and Architectural Design. Birkhauser Architecture, Basel, Switzerland.

EPA, 1997. Exposure Factors Handbook. National Center for Environmental Assessment, Washington, DC.

EPA, 2020. An Office Building Occupants Guide to Indoor Air Quality. https://www.epa.gov/indoor-air-quality-iaq/office-building-occupants-guide-indoor-air-quality.

Ferro, A.R., Kopperud, R.J., Hildemann, L.M., 2004. Elevated personal exposure to particulate matter from human activities in a residence. J. Expo. Sci. Environ. Epidemiol. 14, S34–S40.

Guimarães, R.P., Carvalho, M.C.R., Santos, F.A., 2013. The influence of ceiling height in thermal comfort of buildings: a case study in Belo Horizonte. Braz. Int. J. Hous. Sci. 37, 75–85.

Habash, R., 2020. Bioelectromagnetics: Health Effects and Biomedical Applications. CRC Taylor and Francis, Boca Raton.

Hanna-Attisha, M., LaChance, J., Sadler, R.C., Schnepp, A.C., 2015. Elevated blood lead levels in children associated with the Flint drinking water crisis: a spatial analysis of risk and public health response. Am. J. Public Health 106 (2), 283–290.

For Health, 2017. The 9 Foundations of a Healthy Building. https://9foundations.forhealth.org/9_Foundations_of_a_Healthy_Building.February_2017.pdf.

Hosseini, S.M., Mohammadi, M., Guerra-Santin, O., 2019. Interactive kinetic façade: improving visual comfort based on dynamic daylight and occupant's positions by 2D and 3D shape changes. Build. Environ. 165, 106396.

Howieson, S.G., Hogan, M., 2005. Multiple deprivation and excess winter deaths in Scotland. Perspect. Public Health 125 (1), 18–22.

IBN, 2015. Building Biology Evaluation Guidelines for Sleeping Areas. Supplement to the Standard of Building Biology Testing Methods SBM. https://buildingbiology.com/site/wp-content/uploads/richtwerte-2015-englisch.pdf.

IED, 2009. Integrated Energy Design: Some Principles of Low Energy Design. http://www.integrateddesign.eu/downloads/Some_principles_revised_NormalQuality.pdf.

Jackson, J., Stafford, M., 2009. Public health and fear of crime: a prospective cohort study. Br. J. Criminol. 49 (6), 832–847.

Kenisarin, M., Mahkamov, K., 2016. Passive thermal control in residential buildings using phase change materials. Renew. Sust. Energ. Rev. 55, 371–398.

Kensek, K.M., 2014. Integration of environmental sensors with BIM: case studies usingduino, dynamo, and the Revit API. Inf. Constr. 66 (536). https://doi.org/10.3989/ic.13.151, e044.

Kostoffa, R.N., Herouxb, P., Aschnerc, M., Tsatsakisd, A., 2020. Adverse health effects of 5G mobile networking technology under real-life conditions. Toxicol. Lett. 323, 30–35.

Kubba, S., 2017. Handbook of Green Building Design and Construction. Elsevier, Oxford.

Lan, L., Wargocki, P., Wyon, D.P., Lian, Z., 2011. Effects of thermal discomfort in an office on perceived air quality, SBS symptoms, physiological responses, and human performance. Indoor Air 21 (5), 376–390.

Leech, J.A., Nelson, W.C., Burnett, R.T., Aaron, S., Raizenne, M.E., 2002. It's about time: a comparison of Canadian and American time-activity patterns. J. Expo. Anal. Environ. Epidemiol. 12, 427–432.

Logue, J.M., McKone, T.E., Sherman, M.H., Singer, B.C., 2011. Hazard assessment of chemical air contaminants measured in residences. Indoor Air 2, 92–109.

Lorenc, T., Petticrew, M., Whitehead, M., Neary, D., Clayton, S., Wright, K., Thomson, H., Cummins, S., Sowden, A., Renton, A., 2013. Fear of crime and the environment: systematic review of UK qualitative evidence. BMC Public Health 13, 496.

Lotfabadi, P., Hançer, P., 2019. A comparative study of traditional and contemporary building envelope construction techniques in terms of thermal comfort and energy efficiency in hot and humid climates. Sustainability 11 (13), 3582.

Lowen, A.C., Mubareka, S., Steel, J., Palese, P., 2007. Influenza virus transmission is dependent on relative humidity and temperature. PLoS Pathog. 3 (10), e151.

Maisch, D., Podd, J., Rapley, B., 2006. Electromagnetic fields in the built environment—design for minimal radiation exposure. BDP Design Environment Guide 76, 1–4. http://www.emfrf.com/wp-content/uploads/2014/03/Building_Design_EMF_exposure.pdf.

Marsland, A.L., Bachen, E.A., Cohen, S., Rabin, B., Manuck, S.B., 2002. Stress, immune reactivity and susceptibility to infectious disease. Physiol. Behav. 77, 711–716.

Miller, R., Williams, P., O'Neill, M., 2018. The Healthy Workplace Nudge. Wiley, Hobokon.

Miranda, M.L., Kim, D., Jull, A.P., Paul, C.J., Overstreet Galeano, M.A., 2007. Changes in blood lead levels associated with use of chloramines in water treatment systems. Environ. Health Perspect. 115, 221–225.

Münzel, T., Gori, T., Babisch, W., Basner, M., 2014. Cardiovascular effects of environmental noise exposure. Eur. Heart J. 35 (13), 829–836.

Ortiz, M.A., Kurvers, S.R., Bluyssen, P.M., 2017. A review of comfort, health, and energy use: understanding daily energy use and wellbeing for the development of a new approach to study comfort. Energ. Buildings 152 (1), 323–335.

Panagopoulos, D.J., Johansson, O., Carlo, G.L., 2015. Polarization: a key difference between man-made and natural electromagnetic fields, in regard to biological activity. Sci. Rep. 5, 14914.

Pringle, D., 2016. Time for Smart Cities to Prioritize Air Pollution. https://enterpriseiotinsights.com/20160527/channels/news/time-smart-cities-prioritize-air-pollution-tag28.

Raish, J., 2009. Thermal Comfort: Designing for People. https://soa.utexas.edu/sites/default/disk/urban_ecosystems/urban_ecosystems/09_03_fa_ferguson_raish_ml.pdf.

Raja, I.A., Nicol, J.F., McCartney, K.J., Humphreys, M.A., 2001. Thermal comfort: use of controls in naturally ventilated buildings. Energ. Buildings 33, 235–244.

REHVA, 2020. REHVA COVID-19 Guidance Document. https://www.rehva.eu/fileadmin/user_upload/REHVA_COVID-19_guidance_document_ver2_20200403_1.pdf.

Rodes, C.E., Kamens, R.M., Wiener, R.W., 1991. The significance and characteristics of the personal activity cloud on exposure assessment measurements for indoor contaminants. Indoor Air 1 (2), 123–145.

Sakellaris, I.A., Saraga, D.E., Mandin, C., Roda, C., Fossati, S., de Kluizenaar, Y., Carrer, P., Dimitroulopoulou, S., Mihucz, V.G., Szigeti, T., Hänninen, O., de Oliveira, F.E., Bartzis, J. G., Bluyssen, P.M., 2016. Perceived indoor environment and occupants' comfort in European "modern" office buildings: The OFFICAIR study. Int. J. Environ. Res. Public Health 25 13 (5), E444.

Savic, S., 2020. Architecture: Between Weather and Electromagnetic Radiation. http://digicult.it/news/architettura-tra-condizioni-atmosferiche-e-radiazioni-elettromagnetiche-parte-1/.

Schmidt, S., 2019. Brain fog: does air pollution make us less productive? Environ. Health Perspect. 127 (5), 1–7.

Schneiderman, N., Ironson, G., Siegel, S., 2005. Stress and health: psychological, behavioral, and biological determinants. Annu. Rev. Clin. Psychol. 1, 607–678.

Shendell, D.G., Winer, A.M., Weker, R., Colome, S.D., 2004. Evidence of inadequate ventilation in portable classrooms: results of a pilot study in Los Angeles county. Indoor Air 14 (3), 154–158.

Sundell, J., 2004. On the history of indoor air quality and health. Indoor Air 14, 51–58.

Thach, T.-Q., Mahirah, D., Dunleavy, G., Nazeha, N., Zhang, Y., Tan, G.E.H., Roberts, A.G., Chrisotpoulos, G., So, G.K., Car, J., 2019. Prevalence of sick building syndrome and its association with perceived indoor environmental quality in an Asian multi-ethnic working population. Build. Environ. 166, 106420.

Van Kempen, E., Babisch, W., 2012. The quantitative relationship between road traffic noise and hypertension: a meta analysis. J. Hypertens. 30 (6), 1075–1086.
WHO, 2021. Roadmap to Improve and Ensure Good Indoor Ventilation in the Context of COVID-19., ISBN: 978-92-4-002128-0. https://www.who.int/publications/i/item/9789240021280.
Wilson, J., 2020a. The Sound of Sustainability: Acoustics in High-Performance Design. https://www.buildinggreen.com/feature/sound-sustainability-acoustics-high-performance-design.
Wilson, J., 2020b. Lighting Design for Health and Sustainability: A Guide for Architects. https://www.buildinggreen.com/feature/lighting-design-health-and-sustainability-guide-architects.
Wood, A., Salib, R., 2012. Natural Ventilation in High-Rise Buildings. An output of the CTBUH sustainability working group https://store.ctbuh.org/PDF_Previews/Reports/2012_CTBUHNaturalVentilationGuide_Preview.pdf.
Youssef, O., 2021. Therapeutic Architecture Design Index. https://content.aia.org/sites/default/files/2016-04/DH-Therapeutic-architecture-design-index.pdf.

Building as an energy system

3

Riadh Habash
School of Electrical Engineering and Computer Science, University of Ottawa, Ottawa, ON, Canada

Don't design for energy savings; design well, and the result will be incredible energy savings.

Nancy Clanton

3.1 Energy efficiency through a sustainability lens

Energy efficiency is the continuing decrease in energy use while providing the same or better building functions. To understand the challenge of promoting energy efficiency, new pathways are required. From a holistic perspective, an energy-efficient building is the synthesis of sustainability-related to human health and the environment (Ece, 2018). It is a complex combination of energy efficiency with IEQ because of the potential trade-offs and interactions. Some approaches to energy conservation can improve IEQ, and others may sacrifice it. It is best to view both kinds of impacts simultaneously. Therefore, a good approach to thinking about energy efficiency as well as IEQ factors is through a sustainability lens, supported by awareness, policies, expertise, and financial solutions.

3.1.1 The six focus areas of energy use

As a broad guideline, research has shown that energy consumption in buildings accounts for approximately 40% of the world's energy resources and emits approximately one-third of GHG (Pérez-Lombard et al., 2008). This puts the world is in the middle of a climate emergency. The CO_2 concentration in the atmosphere, the highest oxidized state of carbon which is considered as a waste, has increased by almost 50% since the industrial revolution (Vitillo, 2015). Therefore, highly energy-efficient and environmentally friendly practices are essential in every industry.

The building sector is considered the biggest single contributor to world energy consumption and GHG emissions. Therefore, a good understanding of the nature and structure of energy use in buildings is crucial for establishing adequate future energy-efficient and climate-responsive buildings (Allouhi et al., 2015). This process identifies six focus areas of energy use including climate, façade, technical systems, indoor design criteria, building operation and maintenance services, and stakeholders' behavior (human dimension) as shown in Fig. 3.1. While there has been significant progress in the first five focus areas, there is a lack of scientific and robust methods to define and model energy-related stakeholders' behavior in buildings. Stakeholders include building designers, operators, managers, engineers, occupants, vendors, and

Sustainability and Health in Intelligent Buildings. https://doi.org/10.1016/B978-0-323-98826-1.00003-X

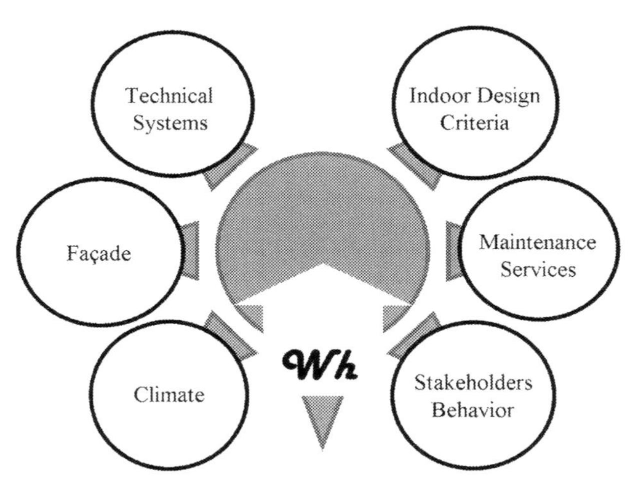

Fig. 3.1 The six focus areas of energy use in buildings.

policymakers, who directly or indirectly influence the development of the built environments, from construction to the urban scale (D'Oca et al., 2017). The human dimension impacts actual energy use as well as indoor IEQ in a building, and it relates to a wide range of stakeholders that together affect all stages in a building's life-cycle. Building managers and operators have the critical task of running buildings effectively while meeting occupant needs that are varied and vigorous. The importance of the human factor in building energy efficiency is evident where the behavior of occupants and operators, in particular, influences energy consumption significantly and is one of the leading sources of uncertainty in predicting building energy use. These behaviors and habits include interactions with lights, operable windows, blinds, thermostats, and various plug-in appliances. Encouraging occupants to change their habits may require lots of creative thinking.

3.1.2 The "Trias Energetica" concept

An opening path for buildings to become energy efficient is by minimizing their energy demand and minimizing waste. This is generally seen as the first step of the "Trias Energetica," a three-step design model, developed by Delft University of Technology, for sustainable design in the building sector as shown in Fig. 3.2. Following the concept, the first step is to create a building envelope that reduces the amount of energy needed to heat or cool it, frequently known as taking a "fabric first" approach. This is realized by adopting bioclimatic architecture which takes into account climate and environmental conditions to help achieve thermal and visual indoor comfort. It should be accomplished by performing insulation solutions, airtightness, and ventilation to ensure IAQ as well as benefiting at maximum from the sun. A highly insulated building envelope is created using high-performance materials, with precisely constructed junctions to reduce air leakage and therefore heat loss.

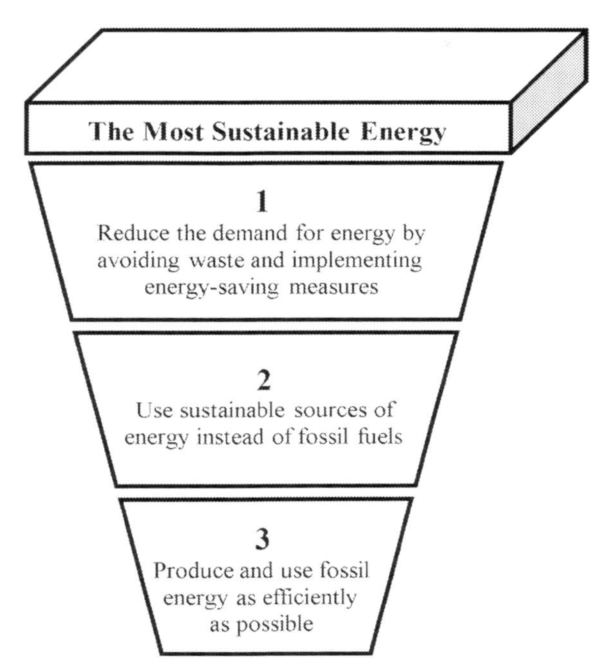

Fig. 3.2 The "Trias Energetica" concept.

The two other steps include the implementation of sustainable sources of energy and the use of fossil fuel as cleanly and efficiently as possible. For example, the use of highly efficient boilers is an approach to turn gas into heat. The Trias Energetica points out that only when a building has been designed to minimize energy loss, then should the attention move to distributed energy resources (DER) and recovery systems. Renewable energy sources can have substantially lower environmental impacts and should be much more sustainable long term if not forever. The concept makes it obvious that energy savings need to come first through ecological and environmental protection.

3.1.3 Energy efficiency strategies

Effective energy efficiency strategies take into consideration several factors to help influence the efficiency of energy planning and through the building design and construction phase. Therefore, an optimized state may be realized by implementing the four strategies, Lean, Clean, Green, and Mean (Fig. 3.3), which complement each other. Lean is a long-term thinking philosophy implemented in Japan particularly in Toyota based on a concept initiated by Henry Ford. It is a process centered on lessening and, at times, eliminating waste within a system. Waste takes many forms and can be found in any place, hidden in procedures, processes, designs, and operations. In buildings, it is a process that incorporates various construction techniques and

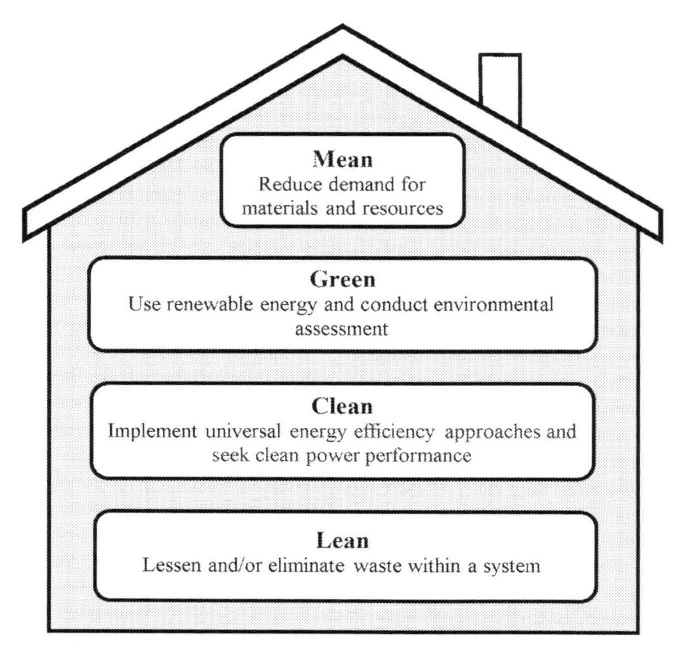

Fig. 3.3 Integrated roadmap through effective energy efficiency strategies.

materials to produce value for a facility (Habash, 2019). Its key factors are quality, cost, and time as well as the ability to adapt to changes. Lean is firmly connected to circularity, however, while lean focuses on locating and eliminating waste, circularity focuses on locating and monetizing waste. Practically, an integrated hybrid design incorporating Lean and circularity enabled by new digital technologies reinforce the energy efficiency strategy and boost sustainability.

The Clean energy strategy aims to reduce energy consumption and GHG emissions, improve air quality, and achieve economic development. Once the energy demand has been minimized, the space heating energy demands should be accurately assessed and demonstrated. This includes a connection to heating networks or onsite combined heat and power (CHP) engines, which are often oversized to meet carbon targets.

The Green strategy calls for intelligent use of renewable energy and technologies. It also identifies and minimizes the operational GHG emissions of the development after each stage of the energy order, including both regulated emissions and those emissions associated with applications not included in building regulations (Hocknell, 2016). This strategy applies environmental assessment at all scales including the design of buildings, new development, and climate change adaptation projects.

Finally, Mean as a supplemental strategy aims at reducing the demand for materials, energy, and other resources. This may be accomplished by developing guidelines for building designers to utilize passive approaches such as natural heating, lighting, ventilation, and external shading.

3.1.4 Exergy analysis

The notion "exergy" represents what in common could be called "useful energy." Other names such as "availability" and "available energy" were in use also (Dewulf et al., 2008). Exergy analysis of an energy form or a substance is a measure of its usefulness or potential to cause change. It is a thermodynamic concept to measure the usefulness and potential to cause change that impacts sustainability in conjunction with economics (exergoeconomics).

Exergy represents the maximum work that can be extracted from a mass or energy stream when it flows from a given thermodynamic state to one in chemical, mechanical, and thermal equilibrium with the environment in a reversible way. Exergy analysis is based on the second law of thermodynamics which shows that different forms of energy have different potentials to produce useful work. The useful work potentially extracted from a system at a given thermodynamic state can be measured by letting the system develop until it reaches a condition of thermal and pressure equilibrium with the surrounding environment (Evola et al., 2018).

Exergy analysis is used to evaluate the performance of a system. Its main advantage is the enhancement of the energy conversion process. It helps identify the main points of exergy destruction, the quantity, and causes of this destruction, as well as show which areas in the system and components have the potential for improvements (Eboh et al., 2017). This perspective is particularly relevant when dealing with buildings and their energy conversion systems, which usually deliver thermal energy at a temperature level that is close to the environmental temperature. This means that the users require low-quality energy; notwithstanding, this energy comes from the depletion of high-quality energy sources, such as fossil fuels and electricity (Evola et al., 2018). Electricity and heat as the two viable sources of energy in buildings are characterized by different transportation and storage possibilities. In basic terms, this may be resumed as electricity is easy to transport but difficult to store, while the opposite applies to heat/cold (Stremke et al., 2011). Accordingly, exergy analysis can help in advancing building energy designs and their energy conversion systems for space heating, cooling, and domestic hot water planning including solar and heat pump (aerothermal, geothermal, aquathermal) systems.

Exergy methods are important since they are useful in improving efficiency. The relations between exergy and both energy and the environment make it clear that exergy is directly related to SD. Therefore, exergy may be considered as the convergence of energy, environment, and SD as reflected in Fig. 3.4. Potential solutions to energy-related environmental concerns have also evolved (Dincer and Rosen, 2021). These include the use of renewable energy technologies and efficient energy conservation, cogeneration and trigeneration, alternative energy forms and sources for transport, smart grid and distribution systems, energy storage technologies, clean energy technologies, AI and machine learning (ML), data management techniques, IoT for monitoring, and evaluation of energy indicators. Importantly, districts, campuses, and cities should implement exergy analysis when planning their path to efficient use of energy. An enormous opportunity exists for researchers desiring to explore this area.

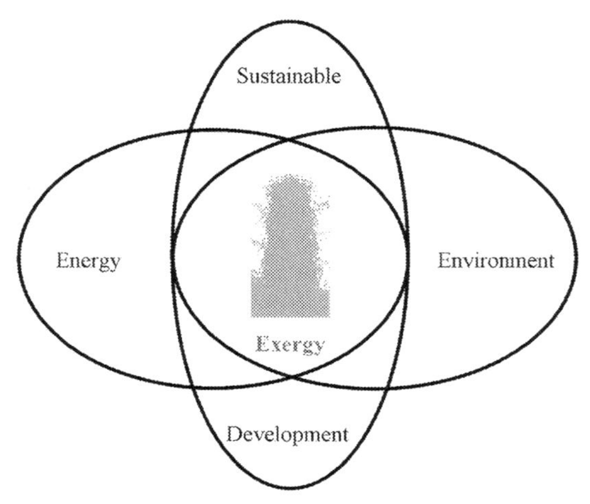

Fig. 3.4 Exergy as an interrelationship between energy, environment, and sustainable development.

3.2 Bioclimatic architecture

The domain of sustainable architecture entails energy and material resource efficiency in the life-cycle of buildings as well as the active involvement of occupants in climate control within the built environment. Many architects and engineers have long been interested in design based on the study of climate, ecology, and biology with urban macroscale and building microscale design to enhance IEQs and decrease energy consumption. Such design has been gaining popularity but is still considered expensive. Bioclimatic architecture is a sector of architecture that is dominated by the principles of ecology and sustainability. It aims to create energy-efficient comfortable interiors while reducing the building's reliance on artificial energy or even fully cover its energy requirements without causing environmental damage.

Historically, bioclimatic architecture is an extension of the experience-based "vernacular architecture" which originated in empiricism, dictated by forebears' rules of a good building. Vernacular architecture stems from long-term development and is part of the traditional widespread culture; therefore, it is considered well adapted to the natural and social conditions of a certain location in which it exists. It arose at a time when there was no technology to answer the needs of air conditioning and lighting, and this led to efficient construction optimized for the building's location within the surrounding environment, using local materials (Cruz et al., 2011). Many vernacular structures were designed and built in a synergetic relationship with their surroundings as repair, mobility, and change were a requirement.

In the bioclimatic approach, energy-saving and a lower environmental impact are consequences of the integration of the design solution to local climatic features to achieve better comfortable conditions and it is not necessarily limited by the building material (Dalhat, 2014). For a long time, minimizing buildings' energy consumption while maximizing comfort conditions has been a serious goal in architecture and urban planning (Lotfabadi, 2015; Lotfabadi and Hançer, 2019). Energy savings inside buildings require a comprehensive design approach to balance bioclimatic strategies with the use of active systems.

Building bioclimatic design (BBCD) recognizes architecture as a filter between outdoor climate and indoor comfort. This way, it encourages the exploitation of freely available climatic resources, before adding any energy system. It aims to mediate external agents both to reduce climate loads and to create a healthy and comfortable indoor environment. It deals with the design and architectural elements, avoiding complete dependence on mechanical systems, which are regarded as support. A good example of this is using natural ventilation or hybrid ventilation (Ness et al., 2019). BBCD incorporates solar radiation for heating and cooling in addition to other kinds of renewable energies including wind or water, and methane generation from biomass.

Energy modeling, lighting models, daylighting studies, computational fluid dynamics are all tools that designers can use to understand how the design best integrates with the local climate and microclimate features specific to the site. For example, several techniques are available to the architect for determining thermal comfort criteria and passive energy-saving strategies that are appropriate for the climate like the software tool "Climate Consultant" developed by the University of California. Another tool to determine the sun's positions and the amount of light coming in is the "Burnett" system which consists of two diagrams. The first shows graphically the position of the sun on a plan at different times of the day and the months of the year, and the second shows the altitude of the sun. Other types of analyses can help solve or avoid, problems affecting occupant experiences in buildings. A glare analysis of the interior space helps to determine the areas and times that will be subject to direct sun glare. It is significant for the areas where occupants spend long hours.

In general, bioclimatic architecture can help reduce the energy consumption of buildings and can be enhanced when coupled with sustainability architecture techniques. The components which affect significantly BBCD are climate and the microclimate, thermal comfort, as well as the utilization of thermal mass.

3.3 Life-cycle energy consumption

Life-cycle energy covers almost every discipline. When designing a building with energy and emission in mind, the AEC community needs to consider both structural embodied energy and operational energy. The main challenge in this regard is to build energy-efficient buildings, which require more materials and services, with low embodied emissions.

3.3.1 Embodied energy

The term embodied energy or embodied carbon refers to the way a building is constructed rather than how it is used. It represents the total energy expended to construct a building during its life-cycle stages as shown in Fig. 3.5. Embodied carbon is an

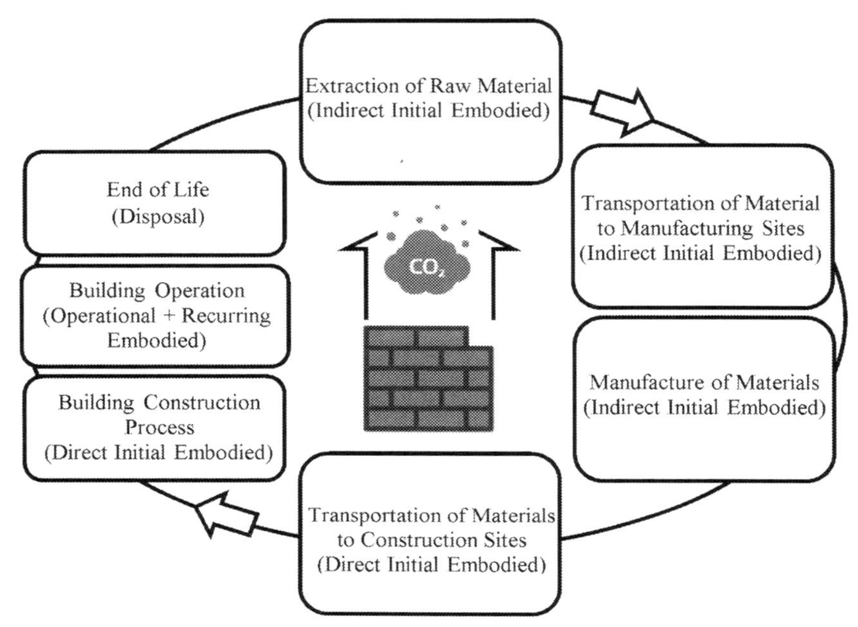

Fig. 3.5 Embodied energy and carbon during the life-cycle stages of a building.

important sustainability initiative as emphasized by the demand for low-carbon solutions and materials to shrink the environmental footprint. Unlike the life-cycle assessment (LCA), which assesses all of the impacts over the whole life of a material or element, embodied energy represents the total energy consumed during the life-cycle of buildings which only reflects the front-end part of the building material impacts.

Considering embodied carbon emissions begins with building design and construction material specifications, heading to policies for low- to carbon-positive building materials. Embodied carbon in materials, in both manufacturing and construction, is a priority to decrease carbon intensity with the use of alternative materials on the journey toward energy efficiency. Buildings use a wide range of materials, and the manufacturing of each material consumes energy and emits CO_2. This has given a complete understanding of the energy consumption and CO_2 emissions as part of building construction, ultimately providing a better consideration regarding the sustainability of buildings over their service life (Taffese and Abegaz, 2019). It is important to keep in mind that there are two types of carbon emissions in buildings, embodied carbon and operational carbon. The latter refers to all CO_2 emitted during the entire life of a building, rather than just its materials, including electricity consumption, HVAC, and others.

The embodied energy of buildings can be categorized into several processes. These include the initial embodied energy which represents the energy consumed in the extraction of raw materials and processing of natural resources to the manufacturing and transport of products to building construction sites. It also includes the energy that is directly associated with the construction activities. The recurrent embodied energy represents the energy required to maintain, repair, and/or refurbish the buildings during their service life (Azari and Abbasabadi, 2018). In addition to the above, is the demolition embodied energy, which is the energy consumed to destroy the building at the end of its life-cycle, recycle and reuse some components, and dispose of others by transporting the debris and waste to landfills or incinerators. It is a largely uncertain component of the embodied energy content due to data unavailability issues (Praseeda et al., 2015).

LCA is a decision support and modeling tool that is used to measure and evaluate the environmental effects of products, construction activities, and the manufacturing process. It is a decision-making tool that includes extraction, processing, and quantifying materials and energy use, manufacturing, distribution, recycling, and final disposal. With its few variants, namely cradle-to-grave, cradle-to-gate, cradle-to-cradle, or gate-to-gate, LCA requires enormous amounts of data which are often hard to find or expensive to purchase (Habash, 2017). An analysis of more than 650 global LCA case studies demonstrates the prospect to design buildings with low life-cycle emissions regardless of the building regulations they have to comply with (Toth and Volt, 2021). This means it is technically feasible to build energy-efficient buildings with low embodied emissions.

Finally, AI, ML, and bog data may be utilized to help the AEC community reduce the amount of embodied emissions in buildings and infrastructure projects. An AI system for predicting embodied emissions begins by collecting embodied carbon data from previous projects to developing ML models, whose algorithms will then analyze this data to spot any patterns and learn from them. As these systems examine through lots of items of data, it will progressively become faster and will eventually be able to come up with material suggestions in a short period.

3.3.2 Operational energy

Operational energy is the energy required during the entire service life of a structure such as building, user, and life safety systems. It is associated with a relatively longer proportion of infrastructure's service life and can constitute 80%–90% of the total energy associated with the structure (Tuladhar and Yin, 2019). Today there has been a significant effort to improve design and technology and reduce the operational energy of the building through various design approaches (Fig. 3.6). Passive design techniques, as the first step to high-performance, is used to reduce the energy demand of buildings to reach comfortable IEQ levels naturally in terms of thermal comfort (heating or cooling), air quality (ventilation), visual comfort (lighting), domestic hot water, and appliances. Next is to use energy-saving equipment across various technical services.

In-use energy consumption of a building is calculated by measuring the operational energy consumed on an annual basis under normal conditions of use, in terms of heating, domestic hot water, ventilation, lighting, etc. These data express final values of energy consumption, measured in kilowatt-hours per square meter (kWh/ $m^2 \times$ year) in kilograms of CO_2 per square meter of housing ($kgCO_2/m^2 \times$ year). In other words, to measure the energy efficiency of a building, the indicators of annual CO_2 emissions and the annual consumption of nonrenewable primary energy recorded by that building are used (Benavente-Peces and Ibadah, 2020).

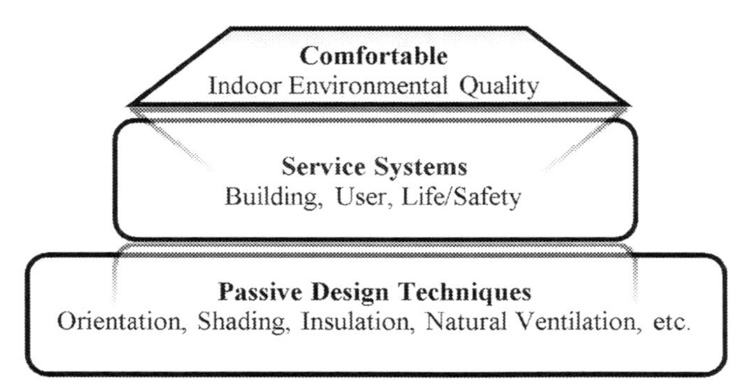

Fig. 3.6 Building design approaches to reduce operational energy.

An important sustainability aspect of the energy chain is its efficiency. At its most basic, energy efficiency refers to a technique of reducing energy consumption by using less energy to attain the same amount of useful output. To reduce energy consumption, the building designers should seek to optimize building passive design, maximize efficiency, employ energy management systems, and prioritize the well-being of occupants. Energy loss can be minimized by improving existing buildings and striving for smart solutions and energy-efficient materials. More efficient energy use leads to reduced primary energy consumption and hence better sustainability scores (Vandyevyvere and Stremke, 2012). To decrease the thermal part of energy consumption in the operational stage, insulation in walls, ceilings, and floors enhance the efficiency of the building.

The most successful choices concerning sustainable design are those are made in the preconstruction and/or during construction stages. So, when designing a building, it is important to use energy-efficient solutions to reduce operational energy, and perhaps even more important to do so without compromising the IEQ. Energy-efficient building design involves building envelopes and technical systems, to work together to be as efficient as possible. The main focus should be on reducing the energy that buildings require for HVAC services. In this regard, innovative ways to cut down energy consumption are necessary.

3.3.3 Energy return of investment (ERoI)

ERoI is an important figure of merit in assessing the viability of energy alternatives. EROI is also referred to as energy return on energy investment. It is a commonly used calculation of how much energy is needed to locate, extract, and refine an output of energy. Too often comparisons of energy systems use efficiency, energy return, energy ratio, or net energy when EROI would be more suitable. Mathematically, ERoI measures the ratio of the net usable energy by a process to the net energy required to extract, refine, and deliver to the process. It is assumed that ERoI >5–7 is required for modern society to function (Habash, 2017).

Using the data from several studies, the EROIs for the most commonly used energy techniques are obtained, each calculated for an unbuffered and buffered scenario.

The power plant's lifetime should be carefully considered since the EROI scales directly with it. It is dominated by the lifetime of the most energy-intense parts. Whereas wind- and solar-based techniques have estimated lifetimes from 20 to 30 years, fossil-fueled power plants can reach 35 years and even more than 60 years.

Weisbach et al. (2013) found EROI values of 4 for solar and 16 for wind, without storage, or 1.6 and 3.9, respectively, with storage. On the other hand, EROI values for hydro and nuclear energy are 49 and 75, respectively. But these power plants require large amounts of water during the process of electricity generation.

3.4 Distributed energy resources (DER)

DERs are denoted as small, geographically distributed generation resources installed and operated on the energy distribution system at voltage levels below 100 kV. Today, renewable power represents the majority of DERs where the installed generation capacity will expand from over 2500 GW today to over 27,700 GW in 2050, more than a 10-fold increase (IRENA, 2021). Renewable energy is often called clean and green because of its minimal environmental impact compared to fossil fuels.

3.4.1 Geothermal heating and cooling

Geothermal energy has been used to heat and air condition buildings for decades where its technologies use heat from the center of the earth. It is considered a renewable resource including the heat retained in shallow ground, hot water, and rock found beneath the earth's surface. Using geothermal heat pumps, this heat can be tapped to provide heating and cooling for homes and buildings. Geothermal heat pumps use the constant temperature of the earth as an exchange medium for heat. They can heat, cool, and supply homes and buildings with hot water. The system consists of a heat pump, an air delivery system (ductwork), and a heat exchanger, a system of pipes buried in shallow ground. In the winter, the heat pump removes heat from the heat exchanger and pumps it into the indoor air delivery system (Hayter and Kandt, 2011). In the summer, the process is reversed, and the heat pump moves heat from the indoor air into the heat exchanger. The heat removed from the indoor air during the summer can also be used to provide a free source of hot water.

There are two major types of geothermal heat pump systems, closed-loop, and open-loop systems. Both approaches can be used for residential and commercial building applications. Fig. 3.7 shows how a heat pump works in heating mode by taking heat from the ground and carrying it to a building and cooling mode, which recirculates water or another fluid from the building and transfers it to the ground. Heat pumps can be a good choice for low exergy systems, as far as their electricity use does not wipe out the advantages of using waste or ambient heat (van den Dobbelsteen et al., 2011). As the fluid passes through the ground loop, it absorbs heat from the warmer soil, rock, or groundwater around it. The heated fluid returns to the building where it is used for useful purposes, such as space or water heating.

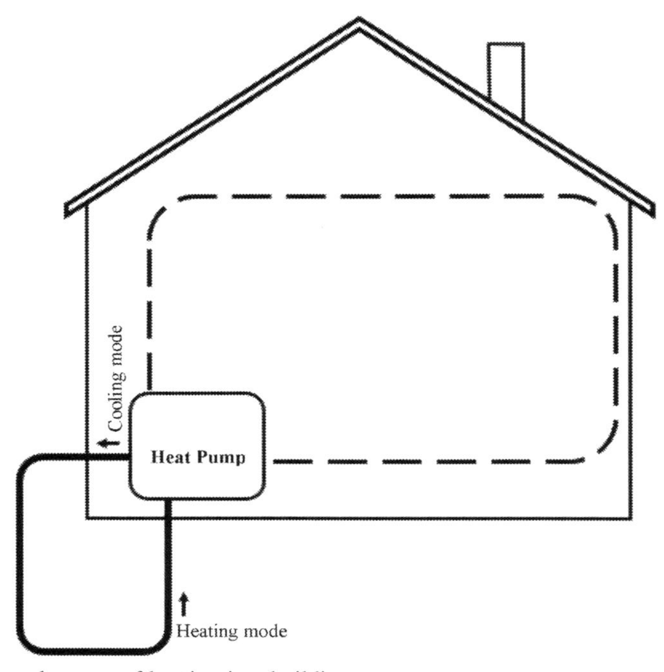

Fig. 3.7 Ground source of heating in a building.

3.4.2 Biomass energy

The concept of biomass energy (liquid or gaseous) is still in its beginning in most parts of the world; however, it does have an important role to play in terms of sustainability in general and buildings in particular. It includes a large variety of materials, including wood, agricultural residues, and animal waste.

Biomass is stored solar energy converted into biofuel through the photosynthesis process. Also, biogas can be obtained from biomass through a fermentation process. Biomass exists in various forms including residual from sewage treatment plants, manure from livestock farming, organic waste from agriculture and food industry, vegetable and garden waste, wood from trees, and other woody waste. They are used for building heating and, sometimes, for electric power generation and CHP. Biomass technologies break down the organic material to release the stored energy and produce biofuels. Ethanol is the most common biofuel. It is an alcohol made from the fermentation of biomass high in carbohydrates (Habash, 2017). Biomass from corn, wheat, soybeans, and wood can also be used to produce chemicals and materials that are usually obtained from petroleum.

Wood has been used to provide heat for thousands of years. This flexibility in materials has resulted in increased use of biomass technologies (Hayter and Kandt, 2011). Wood biomass is commonly used for building heating in various forms including logs, chips, and pellets. Fig. 3.8 shows an example of wood biomass for heat in a building.

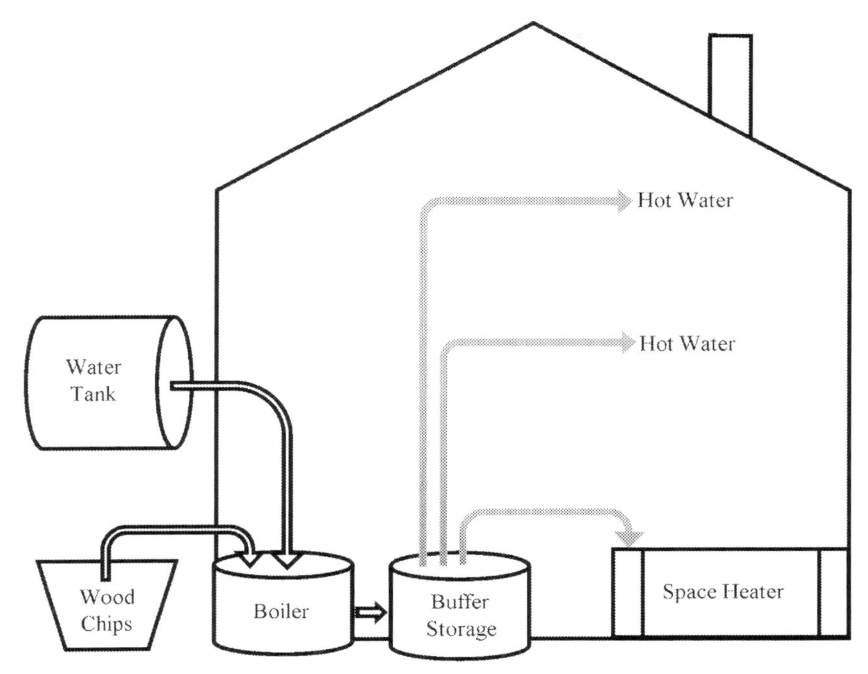

Fig. 3.8 Basic elements of a wood biomass system for heating.

Direct combustion is the most common method of producing heat from biomass. The biomass is burned to generate hot gas, which is fed into a boiler to generate hot water or steam.

Large wood chip systems usually involve additional incorporation with the building, particularly if a chip container is part of the building assembly. If the system is used as a CHP system, the boiler can produce steam to run a turbine and power a generator, and the remaining steam and hot water can then be used for heating (WBDG, 2016).

Additionally, the concept of the biomass-based refinery is an extension to the biomass energy system. It is a facility that integrates biomass conversion processes and equipment to produce fuels, power, and value-added chemicals from biomass. By producing several products, a biorefinery takes advantage of the various components in biomass and their intermediates, therefore maximizing the value derived from the biomass feedstock (Zafar, 2020).

3.4.3 Combined heat and power (CHP)

Sometimes called cogeneration, CHP is not a technology, but an integrated set of technologies to capture waste heat from power plants to produce electricity and useful thermal energy in a single integrated system to supplement building loads like HVAC and water heating (King and Perry, 2017). CHP is often an integral part of district

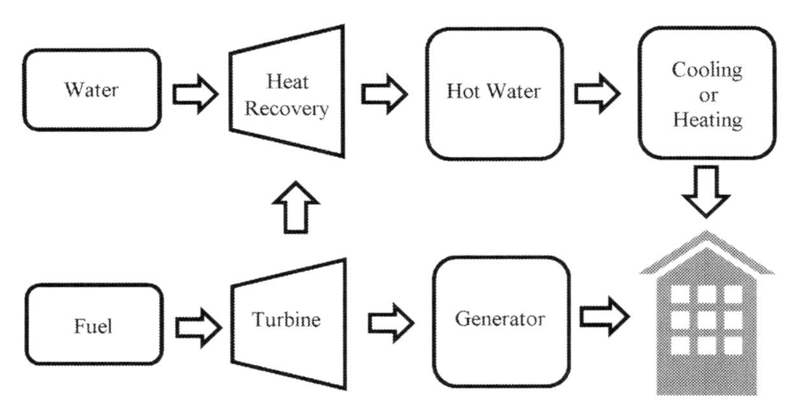

Fig. 3.9 A common CHP configuration.

energy systems. Due to the integration of both power and thermal generation, the CHP system is more efficient than separate generating systems and provides environmental and economic benefits.

Nearly two-thirds of the energy used by conventional electricity generation is wasted in the form of heat discharged into the atmosphere. Additional energy is wasted during the transmission and distribution of electricity to end-users. Fig. 3.9 shows a typical combustion turbine or reciprocating engine CHP system which burns fuels to turn generators to produce electricity and uses heat recovery devices to capture the heat from the turbine or engine. This heat is converted into useful thermal energy, usually in the form of steam or hot water. By capturing and using heat that would otherwise be wasted, and by avoiding distribution losses, CHP can achieve high efficiencies compared to typical technologies. The total system efficiency of the CHP system is the sum of the net useful electric output and net useful thermal output divided by the total fuel input. CHP systems typically accomplish total system efficiencies of 60%–80%.

CHP systems are commonly available in large industries and at institutional campuses. However, smaller systems have begun to exist in commercial buildings and at smaller institutions such as hospitals. They can be located at an individual facility or building, or be district energy or utility resource where there is a need for both electricity and thermal energy.

3.4.4 Urban heating and cooling networks

The concept of heat network has a history dating back to antiquity. The initial network pioneered the use of steam produced by burning coal. Later systems moved from steam to pressurized hot water. It is a large infrastructural heated and cooled water transported via a long underground pipeline system to a large number of buildings. A wider variety of fuel sources of such networks include oil, geothermal, biomass, and waste. Occupants use this energy for space heating, hot water, and cooling. The network generates and distributes heat in the form of hot and cold water

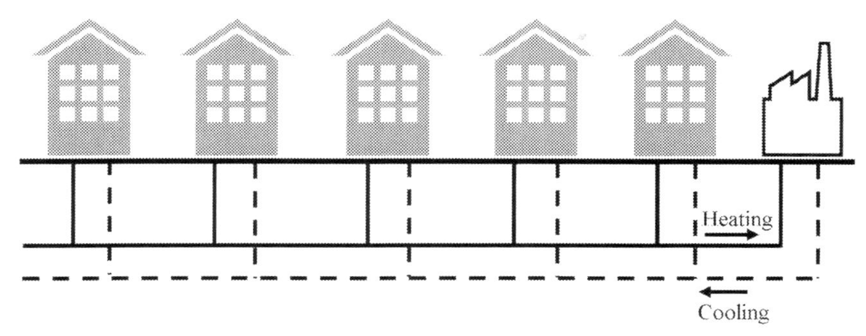

Fig. 3.10 An urban heating and cooling network.

using generating units, while a cooling network is a centralized system that provides chilled water to supply air conditioning systems (Fig. 3.10).

Based on the circular economy principle, urban heating and cooling networks play a key role in integrating the resources of the various available sources such as biomass and geothermal power.

3.4.5 Solar systems

Solar energy (passive and active) is a form of renewable energy available for various loads in buildings. Passive systems use the materials and design elements in the structure itself, while active systems use a technical feature to collect, store, and distribute the sun's energy. Buildings designed to use passive solar energy should have features integrated into their design that absorb and slowly release the sun's heat in materials that absorb and slowly reradiate the solar heat (concrete, stone floor slabs, and masonry partitions). A building using passive solar systems may have large south-facing windows and its designs allow daylight and solar heat to permeate the main living areas (CDCP and DHUD, 2006). In cold climates, the design allows the light and heat of the sun to be stored in the structure while insulating against the cold.

Active solar systems can be installed in new or existing buildings and periodically need to be inspected and maintained. In general, two major types of devices are suitable for buildings to actively produce energy. Solar collectors or solar thermal modules produce heat and PV-cells and modules to generate electricity. The solar collectors consist of collectors, a storage tank, piping or ductwork, fans, motors, and other equipment. These systems can be a cost-competitive way to generate hot water or air and eliminate both the cost of electricity or fossil fuel as well as the associated environmental impacts (Kandt et al., 2011). Most hot water systems use a liquid collector system because it is more efficient than an air-type system. The heated liquid travels through coils in the hot water tank, and the heat is transferred to the water and perhaps the heating system.

Solar PV (utility and rooftop) systems acquire their name from the process of converting light (photons) to electricity (voltage). They are one of the promising renewable energy technologies for producing electricity directly from the sun with minimum maintenance, pollution, and depletion of materials (Victoria et al., 2021) with a global capacity of about 400 GW in 2018. PV systems are made up of modules assembled into arrays that can be mounted on or near a building or other structure. Traditional solar cells made from crystalline silicon either as single or poly-crystalline wafers are usually in the core of the above modules (Hayter and Kandt, 2011). They deliver about 10–12 watts per ft^2 of PV array, under full sun. On the other hand, thin-film solar cells are made from PV-active materials like amorphous silicon or non-silicon materials such as cadmium telluride (Strong, 2016). The major challenge with solar PV technologies is ensuring appropriate siting for maximum electricity production or efficiency. The estimated lifetime of a PV module is about 30 years and performance would be anticipated to continue operating at about 80% of the initial power after 25 years. Fig. 3.11 shows a PV module as part of the power distribution system in a building.

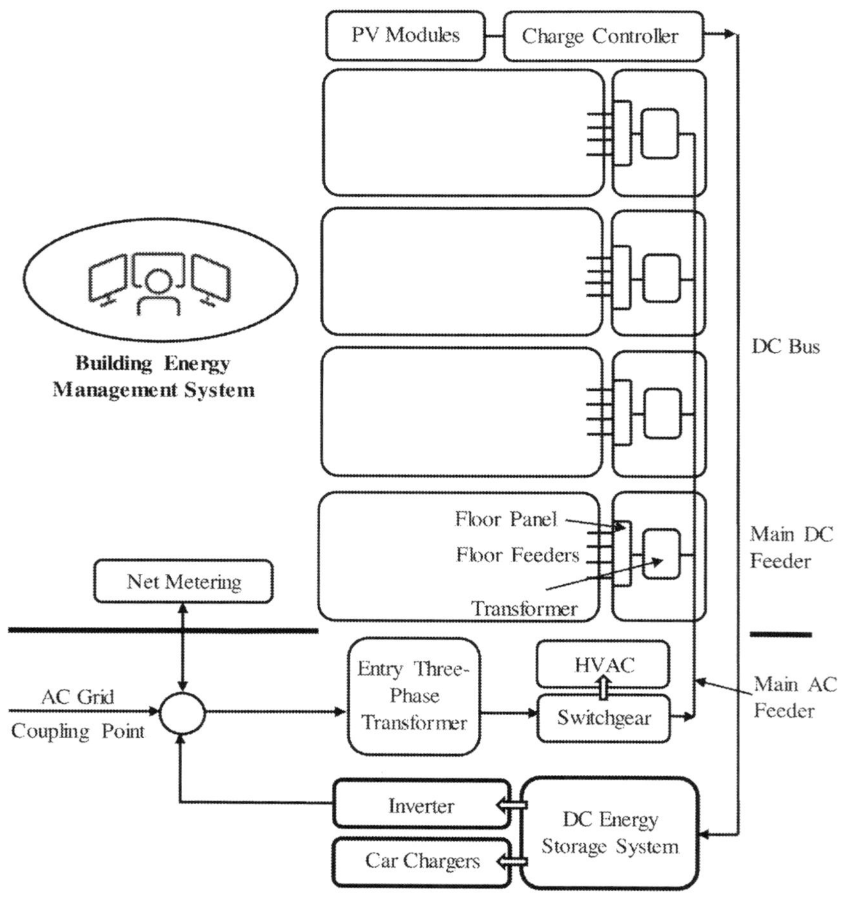

Fig. 3.11 Power distribution in a building including an example of a photovoltaic system.

Solar PV has been consistently advancing in terms of structural design and efficiency. For example, building-integrated PV (BIPV) products comprise parts of the building envelope, such as the exterior cladding or window and skylight glass. By simultaneously serving as building envelope material and power generation, BIPV systems can offer savings in materials and electricity costs, decrease the use of fossil fuels and emission of gases, and importantly add architectural elements to the building. A charge controller is used to regulate the power into and out of the power storage system. The power conversion equipment including an inverter to convert the PV modules' direct current (DC) output to alternating current (AC) compatible with the utility grid. Design considerations for BIPV systems must include the building's usage and electrical loads, its location and orientation, the appropriate building and safety codes, and the applicable utility issues and expenditures.

3.4.6 Wind systems

Wind (onshore and offshore) is a natural energy choice providing an alternative to fossil fuels with a global power capacity of about 590 GW in 2018 (IRENA, 2019) to reach over 8100 GW by 2050 (IRENA, 2021). Historically, a famous vernacular example of a wind system is the wind tower or windcatcher, which is a small tower installed on top of a building. It can supply natural ventilation and passive cooling for the interior spaces of the building (Dehghani-sanij et al., 2015). A windcatcher is a traditional Persian architectural element that existed across the Middle East with use dating back thousands of years. Its name originates from its ability to catch passing winds and push them down into indoor spaces. This passive technology utilizes systems of natural ventilation without energy consumption. These systems may also provide daylight in addition to natural ventilation through a combination of a windcatcher and a natural daylight sun-pipe placed together in a single system (Fig. 3.12). The hot

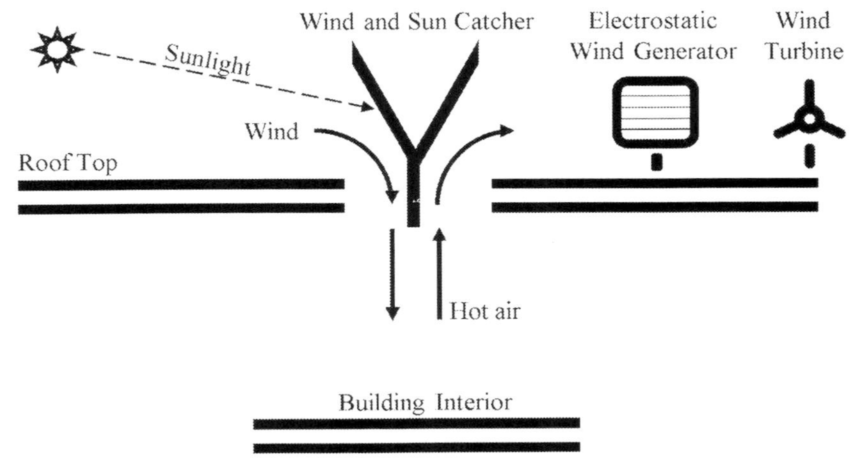

Fig. 3.12 Airflow in a building with wind and suncatcher.

air that returns from the building may be used to heat water for the building, therefore reducing energy consumption.

Wind speed typically increases with height, as it is less affected by trees and surrounding topography. Putting active wind turbines on top of buildings, especially tall buildings should allow them to take advantage of height without an expensive, full-size tower. Unfortunately, the building-integrated wind often does not live up to its potential where noise and vibration from wind turbines are among the utmost problems to incorporating them into buildings (Wilson, 2021). That might explain why different wind turbine prototypes that enclose the blades in a chamber or replace them entirely with a disc-like system are in use.

On the innovation side, researchers in the Netherlands set out to eliminate the need for a mechanical component and created the electrostatic wind converter (EWICON), a bladeless wind turbine with no moving parts that produces electricity using charged water droplets. The design is particularly suited to urban areas (Fincher, 2013). A high-voltage current is passed through a thin wire to ionize the air. The air is ionized and the ionized molecules are dispersed by the wind, this leaves a net positive charge in the ionizing apparatus where a potential difference is built up with the earth. Accordingly, the positively charged ionizer leads to current flow (Nayak, 2020). This concept will diminish many of the drawbacks of conventional wind turbines such as noise, wear, maintenance expenses, and shifting shadows.

The futuristic Dutch Windmill, a 174-m tall, mixed-use building project in the city of Rotterdam aims to implement the EWICON concept. It intends to be a showcase of sustainability by integrating several features including rainwater harvesting, biogas production, and integrated PV systems. The building has been designed for whole life sustainability with the provision for disassembly and reuse at end of life.

3.4.7 Fuel cells and hydrogen

Fuel cells function like a battery or supercapacitor. They are electrochemical devices made up of anode, cathode, and electrolyte that use oxygen and a readily reloaded fuel, such as hydrogen, to generate electricity. In 1838, Sir William Grove invented the fuel cell. But it was not until a century later that Francis Thomas Bacon invented the hydrogen fuel cell which became the cornerstone for the commercialized fuel cells which fall somewhere 40%–60% efficient. However, batteries are 80%–90% efficient. Fuel cell systems have a wide range of applications because they are fuel adaptable and can be sized to adjust energy needs. To maximize the efficiency of a fuel cell system, utilizing the heat generated as well as the electricity is important. Depending on fuel cell type, access to a hydrogen distribution network is required. Today, the leveled cost of energy in $/MWh for fuel cells is between $100 and 150. Meanwhile, solar and wind are dramatically cheaper starting at around $40 and $30, respectively.

Fuel cells can produce the highest proportion of electricity of any CHP technology. They are a flexible, modular technology that can easily be scaled up from serving individual homes to large office blocks and industrial complexes (Dodds et al., 2016). Potential roles for fuel cell and hydrogen products include the replacement of hydrogen for natural gas in some processes and the use of CHP technologies.

The use of hydrogen from renewable resources in buildings and the roll-out of stationary fuel cells (or fuel cell microcogeneration) is an option for newly built houses or in case of renovation. Stationary fuel cells have a significant potential to decarbonize the building sector. Fuel cells offer additional value to the energy systems at times of peak demand. Moreover, fuel cells bring further value in terms of flexibility and demand response to the energy system as a whole. They offer significant benefits to local air quality even when fueled on hydrogen in terms of harmful air pollutants. This becomes very important if these peaks increase through the electrification of heat and transport.

3.5 Energy storage systems

Energy storage is the capture of energy generated at a certain time for later use to reduce the imbalance between energy demand and energy production. It facilitates fixed output sources to handle varying loads, improves the effectiveness of power grids, and enhances the peak power supply capacity without needing to generate more power. Storage systems come in many forms including batteries, thermal, or mechanical systems.

3.5.1 Batteries

Today, the most significant technological challenge is, without any hesitation, in battery technology. The battery did not change since Benjamin Franklin coined the term in 1748, but its materials, technology, and functions have changed. The new battery generation should provide a better ratio between kWh versus dimensions, weight, and cost. Currently, there are various forms of batteries, including lithium-ion, flow, lead-acid, sodium, solid-state, and others designed to meet specific power and duration requirements. Rechargeable lead-acid and lithium-ion batteries support the most common storage systems in the commercial market including consumer electronics, electric vehicle (EV), and grid-scale energy storage. Large lithium-ion batteries are a relatively new invention compared with other building systems, and their safety requirements and fire hazards are still being determined.

Current lithium-ion batteries use electrolyte solution which requires proper management to avoid leakage and fire. Beyond the potential for causing a fire, the liquid electrolytes inside lithium-ion batteries are not the best at endurance. Over time, compounds in the liquid can decompose internal battery components and can cause degradation or solid material build up inside, both of which lead to a degradation of battery overall performance.

For years, solid-state batteries have been promised as the big development in the energy storage sector due to their safety combined with their high-energy density as well as reduced size and weight. They are not a new technology, but their use in the automobile and building sectors is. They have been implemented in consumer devices like pacemakers and wearables. Solid-state batteries use solid electrolytes instead of liquid electrolytes such as oxides, sulfides, and phosphates. Practically, solid-state lithium-metal batteries are considered the best battery technology. It can hold more energy while charging in a segment of the time when compared to standard lithium-ion batteries.

Another type of solid-state battery is the glass battery. It uses glass electrolytes and lithium or sodium metal electrodes. The battery was invented by John Goodenough, a Nobel Prize-winning professor at the University of Texas at Austin. An additional advantage to this battery is that it can charge in minutes rather than hours and perform in very cold (−20°C) and very hot (60°C) temperatures.

3.5.2 Flywheels

The old-school mechanical batteries or flywheels that store energy in a rapidly spinning mechanical rotor are capable of absorbing and releasing high power during short periods. These systems can balance fluctuations in electricity supply and demand where they respond to a control signal adjusted every few seconds (NYC, 2020).

A flywheel contains a dual-function electric motor to store and generate energy so that kinetic energy is stored in the spinning wheel. When it is turned off, the electric motor operates as a generator. Therefore, the flywheel can be considered a large mechanical battery. The first flywheel used for energy storage was built by John Howell in 1883 for military applications. Advanced flywheels are efficient and carbon-free. They can be fully charged very quickly and may reach a speed of 100,000 rpm. Flywheels have a long lifetime of about 20 years; therefore, they have been used in grid applications to provide ancillary services like frequency regulation and power fluctuation support.

Flywheels have a high energy installation cost compared to batteries. For example, the cost of storage for flywheels reaches about $500/MWh, while for lithium-ion batteries is about $210/MWh. interestingly, flywheel technology can be integrated with lithium-ion batteries to develop hybrid storage systems.

3.5.3 Thermal energy

Thermal energy storage is the temporary storage of high- or low-temperature energy for later consumption by using heating and cooling methods to store and release energy. In continental climates, it is possible to store heat from the warm summer for use in winter, while the cold ambient temperatures of winter can charge a cold store to provide cooling in summer. For example, solar thermal multisystem provide space heating and hot water through solar-thermally-charged hot water storage tanks using two energy sources, solar energy and any kind of auxiliary heat. These systems range in size from small-scale systems installed in small buildings to large-scale systems serving large buildings.

Thermal energy storage is a relatively established technology that requires further design innovations. This technology may be used to considerably decrease energy costs by allowing an electrically driven cooling system to be largely operated during off-peak hours when electricity rates are lower. Many commercial chiller models are compatible with ice storage tanks. Chiller units can be used to cool thermal storage at night when the cost of electricity is relatively low. They can be configured to make ice with low-cost off-peak electricity or with surplus output from renewable sources. In such a way, electricity cost is reduced, the efficiency of the chiller is boosted and the peak electricity demand for utilities is decreased. In the future, electricity storage may become more feasible, for example, by using high-capacity batteries as a storage medium within smart grids. Exploiting beneficial interferences between electricity and heat/cold applications is one aspect that gains importance (Clement-Nyns et al., 2011). In addition, compressed air and hydrogen storage are other forms of emerging energy storage that are in different stages of development.

3.6 Building as a charging station

As an innovation in the sustainability industry, it may be necessary to capitalize on the growth in EV use and deliver a resource for tenants, visitors, and the community around buildings. EVs come with environmental, health, and economic benefits, however, they can also result in higher electricity costs for building owners if not properly managed (Fathy and Carmichael, 2019). In 2018, the yearly sale of EVs surpassed 2 million units (IRENA, 2019). Installation of EV charging stations at sites where vehicle parking or storage is one of the primary use of the facility-like buildings is a challenging task in terms of adaptation and usability. Globally, several cities have begun to initiate incentives for vehicle-charging infrastructure in newly built sites.

AC and DC are two concepts of EV charging. The difference lies in the mode of transfer of power into the charging port of the vehicle. In AC (slow) charging mode, the AC power (usually about 20 kW) from the grid is delivered into the car via an AC outlet. The vehicle will manage the rectification (AC/DC conversion) process via an onboard charger. In DC (fast) charging mode (up to 400 kW), the AC/DC conversion is carried out outside the vehicle by an off-board charger. Such a DC facility requires a dedicated high-voltage transformer from the grid. However, DC fast chargers are not suitable for home charging in particular. Several standards govern the operation of EC charging systems. These include the International Electrotechnical Commission (IEC) like IEC 61851 focusing on different topics of EV conductive charging systems, including AC and DC charging up to 1000 and 1500 V respectively. In North America, the standard SAEJ1772 covers AC and DC charging and establishes charging levels.

The last remaining obstacle is the charging time, where slow and fast systems coexist. Rapid chargers are usually DC, handling the rectification process of the AC supply within the charger rather than relying on the EV's onboard rectifier. Another solution could be by connecting the chargers to the DC power storage. Several batteries charge in a dual-mode, initially using a constant current mode until a sizable fraction of their rated voltage is reached, at which state the charger turns to a constant voltage mode while the current is diminished to reach the battery to full charge. With a so-called ultra-fast charging technology, it is already possible today on most vehicles to fully charge the battery in a short period, like about 15 min.

Chargers for use in future buildings must be robust, support fast and adapt automatically to serve multiple different EV types and battery conditions. The addition of such EV charging equipment will modernize buildings and position them more competitively in the future. Whether a building is commercial or residential, as the EV market continues to grow, properties with charging stations can lead the way. The design challenge here is to deliver large amounts of charging power in a way that is safe, robust, and easy to use. The rate at which EV can be charged is becoming a key differentiator in IB development. The reason is that charging an EV's battery is currently a lot slower than refilling a tank with gas to go the same distance. By equipping buildings with charging infrastructure for EVs, it will be taking a major step forward in creating the gas station of the future at suitable charging time which is the ratio between the effective battery capacity in kWh and the average charging power in kW (kWh/kW).

Currently, there are multiple charging systems available in the market, either AC, DC, or combined charging systems which combines AC, three-phase AC, and DC high-speed charging. Homeowners are unlikely to install a combined charging system station because of its high cost and power needs. In this regard, a sustainable transition requires the deployment of fast-charging stations into large buildings, probably a combined charging systems to keep moving with the development of EVs (Fig. 3.13). The station operates between 400 and 1000 V AC, to supply 50 kW and above. Also, it can be combined with the building energy storage system. These EV chargers may exist in a single location or across a range of buildings.

Power can be delivered to vehicles wirelessly. The first generation of wireless power products employed the principles of magnetic induction but with a challenge of power transfer efficiency. Magnetic resonance, also known as highly resonant nonradiative wireless energy transfer, meets those challenges. It allows wireless charging over distance as well as charging through materials (i.e., wood, granite, water) and high efficiency of power transfer. The system is powered by two coils, one at the charging station to produce an oscillating magnetic field, which in turn induces an alternating current that is received by the coil at the vehicle being charged.

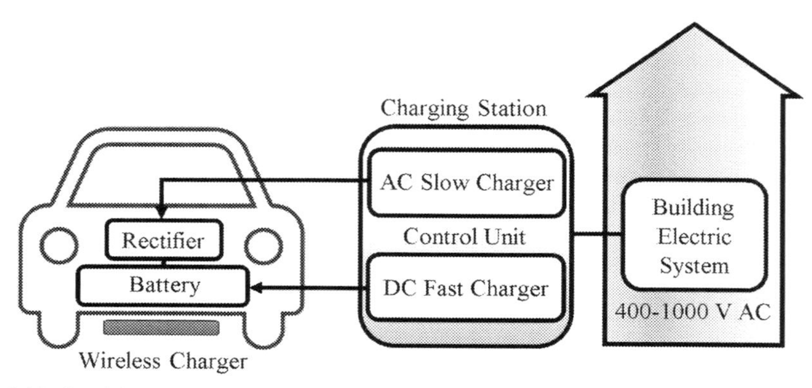

Fig. 3.13 Combined charging station attached to a building.

Finally, employing internet-connected smart chargers allows the implementation of electrical load management (power-sharing or load balancing) across individuals and groups of EV chargers for the optimization of energy consumption, depending on the need and availability of electricity. This underlines adaptability intelligently and flexibly, to both the needs of the user and the building power system. The other related feature is power shaving to avoid exceeding the maximum energy capacity available at the charging point. A combination of power-sharing and power boost capabilities is called dynamic power-sharing, a very relevant to buildings, where strict monitoring of the power demand can prevent surpassing the maximum capacity. When the power demand is higher than the capacity, the accessible power is equally distributed across the loads. On the other hand, if the power demand is lower than the capacity, the remaining available power can be fed back to the power grid.

3.7 Electrification and decarbonization

Electricity would be the main energy carrier of the future with a generation growing threefold from 26,380 TWh in 2018 to close to 78,700 TWh in 2050. The share of renewables would grow to 90% in 2050 from 25% in 2018 (IRENA, 2021). While currently almost all the electrification attention has been focused on vehicles, there is entire another sector that historically uses other sources of energy. The building sector, including everything from homes to building to communities, to cities could benefit from a transition away from traditional fuels to electrification. This sector would see the highest direct electrification rates, reaching 73% in 2050 compared to 32% today.

3.7.1 Electrify everything

Electrification is the practice through which building loads as heating and cooling appliances that are powered by fossil fuels such as natural gas or fuel oil are powered by electricity instead. In buildings, nearly 100% of energy use can be electrified with today's technologies. Examples include gas-powered furnaces and domestic water heaters or boilers that are switched with heat pumps powered by electricity the enabling technology of widespread building electrification (Deason et al., 2018). By 2050, electricity demand in the building sector is projected to increase by 70% worldwide, despite improvements in energy efficiency approaches. However, the limitations of electrification are cost and other practical barriers, rather than lack of deployable technologies. This matter is a serious concern since electrification as a product of the massive electric power grids is just like air and water, which has an enormous positive impact on the QoL and health of people.

All-electric technologies offer key performance advantages including more control and precision than their gas counterparts. Electrification may provide greater flexibility for managing electric loads, opportunities to provide additional supplementary services to the grid, and valuable synergies with electric vehicles, demand response, distributed generation, and energy storage. The primary technological tools of such a transition to electrifying everything are the encouragement of energy efficiency and the use of clean and renewable electricity. In the building sector, as electric storage becomes increasingly inexpensive, the combination of DERs and electric storage may offer a long-term hedge against rising fossil fuel prices as well as greater buoyancy to power outages. A major challenge is the winter peak electricity load and in some cases may switch annual peak load from the summer to winter.

As building sector electrification scales up, market design and demand response programs coupling flexible demand smart technologies with wholesale electricity markets will help to smooth peak loads, reduce renewable limitation, and more broadly bolster the economics of electrification. Additional measures to shift demand such as thermal storage should also be considered (Ruhnau et al., 2019). The eventual obstacles to electrification are not technical, but economic, and this is especially true in existing buildings.

Space and water heating is currently the largest source of nonelectric demand in the building sector, accounting for approximately 70% of nonelectric energy consumption in the residential sector, and 60% in the commercial sector (Deason et al., 2018). While the electrification of end uses such as space and water heating can be relatively cost-effective in new buildings, the up-front capital costs and utility project complexity can be a barrier to the electrification of existing buildings. Coordination and education of key decision-makers in governance, within utilities, and among builders will be critical to ensure a smooth and fast transition.

3.7.2 Decarbonization strategy

Electrification is an opportunity to use carbon-free electricity in place of fossil fuels in end-use applications. This improves the overall efficiency of the energy service supply significantly (IRENA, 2021). Such a path to decarbonization is enhancing the need for a new generation of advanced technologies and materials as well as circular economy measures. Key openings to decarbonization in buildings include enlarged electrification with supply dominated by renewables and more energy efficiency. This includes the use of intelligent efficiency technologies that reduce and/or separate carbon as well as exploration of proper carbon-neutral routes to rethinking construction emissions. It also requires partial employment of carbon capture and storage and carbon dioxide removal technologies. Bioenergy combined with carbon capture and storage systems would play a key role in power plants, cogeneration plants, and in the industry specifically for the cement and chemical sectors, to bring negative emissions in line with a very constrained carbon budget. They would contribute over 52% of the carbon captured over the period to 2050.

The design of a zero-carbon built environment is being driven by several major trends. These include decarbonization of the electric power grid, electrification of building space and water heating, efficiency improvements to reduce energy demand, and digitalization to provide needed flexibility in meeting the needs of building occupants and the energy grid (Nesler, 2020). Direct GHG building emissions are the result of gas, propane, and oil burned onsite, primarily for space and water heating (Ruhnau et al., 2019).

For decarbonizing some heat applications, the total number of heat pumps would exceed 180 million by 2030 and close to 400 million by 2050 compared to around 20 million installed today (IRENA, 2021). Switching from fossil fuels to clean electricity or substituting renewable natural gas for today's fossil gas are two possible approaches to zero carbon emission. However, it is simply not feasible for decarbonizing the buildings sector and when compared to fuel oil, heat pumps are always cleaner. Electric heat pumps will be the main heating technology in the future, as opposed to other direct electrification technologies (electric space heaters or electric resistance heaters), or indirect electrification technologies. They use electricity-powered compressors to move heated air or water into or out of a structure, replacing existing natural gas-fired heaters. And because heat pumps can either expel heat from the indoors during the cooling season or capture heat outdoors from the ground or air and draw it indoors in winter, they offer heating and airconditioning from the same equipment. In general, heat pumps are many times more energy-efficient than their natural-gas counterparts especially those integrated with AI technologies.

The Kyoto Protocol (adopted in Kyoto, Japan in 1997) is an international agreement that called for industrialized nations to reduce their GHG emissions significantly. Other accords, like the Doha Amendment in 2012 and the Paris Climate Agreement in 2015, have also tried to curb the global warming crisis. The European Union aims to be climate-neutral by 2050, demanding a fundamental transformation of the construction and building sectors. Emissions must be radically cut throughout the whole life-cycle of buildings, including all operational and embodied emissions (Toth and Volt, 2021). 2050 is also a target for zero-carbon cities in Japan. Finland aims to become carbon-neutral by 2035. Denmark's new regulation sets whole-life carbon emissions for new buildings, including both operational and embodied emissions, based on LCA. In 2019, the government of Canada announced support to accelerate the transition to zero carbon buildings. Many of the right policies are in place to encourage zero carbon buildings in the US, Mexico, China, India, and other countries.

3.8 Electricity in buildings

Electricity would increasingly become the dominant energy carrier and renewable power would be the provider of the bulk of worldwide power. Technically, electricity is generated by machines that convert energy from different forms into electrical energy. For an electric current (amperes) to run, it must flow from a higher to a lower potential voltage. In an electrical system, the hot wires (black or red) are at a higher potential than are the neutral or ground wires (white or green). Voltage (volts) is a measure of the force at which electricity is provided. It is like to pressure in a water supply system (CDCP and DHUD, 2006). The potential difference or voltage between the hot wires and the ground or neutral wire of a normal residential electrical system is either 110 V in North America or 220 V in the rest of the world. Power measured in watts is equal to volts times amperes.

3.8.1 Electric power distribution in small buildings

Homes and small buildings have a very simple power distribution system. Usually, the power utility will have a transformer, which will sit on a pad outside the building or will be attached to a utility pole. Utility-owned transformers reduce the voltage from a typical utility feeder of 13.8 kV down to 120/208 V (120 V between neutral and any phase and 208 V between any two lines for Wye connection three-phase system in North America) or 240/380 V (rest of the world). The main advantage of a three-phase system is that it allows the operation of heavy electrical equipment such as dryers, cooking ranges, and air conditioners, the majority of which require 220-V circuits. Most of the small buildings like homes, residential, and light commercial units have a single-phase electrical service.

The electricity passes to a meter, which is owned by the utility and keeps a record of power consumption. After leaving the meter, the power is transmitted into the building at which point all wiring, panels, and devices are the property of the building owner. Cables transfer the electricity from the meter to a panel board, which is generally located in the basement of the building. The panel board contains the main service breaker and a series of circuit breakers, which control the flow of power to various circuits in the building. Each branch circuit will serve a cluster of devices and appliances (sometimes to a single heavy load) or several devices like socket outlets and/or lights. Sometimes, there exist shops, manufacturing, and heavier industrial applications that use three-phase or more and have a higher current and voltage service than is ever seen in most residences and offices. The single-phase service enters the meter box at the service entrance from a center-tap transformer through three wires, two hot wires of 120 V each that are 180° out of phase, and a neutral wire that completes the circuit to the transformer. This allows a building to access 120 V power on each hot wire and combines them if 240 V is needed for certain appliances.

The service entrance box also has a grounding bus that is connected to a grounding rod where every building is required to have an independent grounding system in addition to grounded outlets. The grounding system is a link to ground from one of the current-carrying conductors of the electrical system. This system limits overvoltages in the event of a fault and provides personnel safety as well as the means of detecting and isolating ground faults. Therefore, the system ground's main goal is to protect the electrical system as well as the user.

3.8.2 Electric power distribution in large buildings

Large buildings have a much higher electrical load than small buildings; therefore, the electrical equipment must be larger and more robust. As shown in Fig. 3.8, the distribution system can be divided into two types, the vertical supply system (to various floors) and the horizontal supply (distribution at each floor level). In most cases, a high-voltage supply and transformer substation is required, which lowers the voltage to a more usable level (e.g., 480/277 V) (Archtoolbox, 2021). This transformer can be mounted on a pad outside the building, on the ground floor in a transformer room inside the building, or in the basement. The electricity is then transmitted to switchgear. The role of the switchgear is to distribute electricity safely and efficiently to the various electrical panels throughout the building. The equipment has numerous safety features including circuit breakers, which allow power to be disrupted downstream. It should be noted that very large buildings may have multiple transformers, which may feed multiple pieces of switchgear.

The electricity will leave the switchgear and travel along a primary feeder or bus. The bus or feeder is a heavy gauge conductor that is capable of carrying high currents throughout a building safely and efficiently. The bus or feeder is tapped as needed and a conductor is run to an electric closet, which serves a zone or floor of a building. Each electrical closet will have another step-down transformer; in North America, this will drop the power from 480/277 to 120 V for convenience outlets. That transformer will feed a branch panel, which controls a series of branch circuits that cover a portion of the building.

For electric safety, the current flow in any wire is limited to the maximum permitted by using an overcurrent device of a size specified by the code. Several types of overcurrent devices are available including fuses, circuit breakers, ground fault circuit interrupters, and arc fault circuit interrupters. The overcurrent device must be rated at equal or lower capacity than the wire of the circuit it protects.

3.8.3 Wiring system

Today, building construction primarily uses three types of wiring: nonmetallic cable, armored cable, and individual conductors run inside the conduit. The nonmetallic cable is a composite cable consisting of one or more hot conductors, a neutral conductor, and a ground wire. All the conductors are individually insulated and the entire bundle is sheathed in polyvinyl chloride plastic to make a strong package that resists heat and flames. Nonmetallic cable is the most common type of wire used in single-family residential applications. It is protected by two layers of insulation, with each conductor being insulated and the entire bundle also being insulated. This makes insulation failure a rare occurrence and reduces the risk of accidental electrocution and arcing.

The armored cable is sheathed in a flexible metal spiral made either from aluminum or galvanized steel, therefore it is more expensive than nonmetallic cable. The main distinction between the armored and nonmetallic is that the armored can achieve grounding through the outer metal casing, but is heavier than nonmetallic. This casing needs to be attached to metal boxes. It is as good as nonmetallic cable from a magnetic field standpoint but has the added benefit of shielding electric fields. The metal sheathing absorbs the electric fields emitted by the wires and shunts them to the ground. The plastic sheathing of nonmetallic cable does not shield electric fields.

Individual conductors run inside conduits are used primarily in industrial and commercial applications and are rarely used in residential wiring. It consists of individual, insulated conductors pulled from a spool through a conduit, either metal or polyvinyl chloride plastic (HBS, 2013). This is a more difficult and time-consuming wiring method but is more efficient in commercial applications where wiring must be exposed and thus must be inside some form of the conduit.

3.9 Electrical load estimation

Before designing any electric load like a building, realistic preliminary load data must be compiled.

The expected electric power demand on midway substations, and the main electric power supply, should be calculated from the connected load layout by employing suitable factors such as an anticipation of future load growth. A term constantly used in modern building design is "demand factor," the ratio of the maximum demand of a system to the connected load on the system. The load can be measured in either amperes or watts, and it can be calculated based on Ohm's law which relies on a single equation: watts = volts × amps. In most situations, two of the three elements will be known. Typically, electrical loads are measured on a kilowatt (kW) or kilovolt-ampere (kVA) basis while electricity is sold in quantities of kilowatt-hours (kWh).

Preliminary load calculation is based on two major methods including building area or space by space. The building area method is used for buildings that are bigger than that can be handled by the space by space method. The electrical load is estimated based on the activity of the whole building unlike the space by space method which needs to divide the building into multiareas with different activities. The preliminary electrical load is estimated by multiplying the total gross area of the building by its load density (W/m^2) or (W/ft^2). The load density varies with the function of the building (residential, educational, industrial, etc.) and services provided. For example, the load density for an educational building without airconditioning is about 15 W/m^2; however, it is about 90 W/m^2 with airconditioning. ASHRAE and the Institute of Electrical and Electronics Engineers (IEEE) provide load densities for various types of loads.

Space by space methods is based on the estimation of loads in each area of the building taking into consideration the individual loads that exist in that space. These loads would comprise single structures; however, large structures could contain more than one function. Major loads under this category include HVAC systems, lighting, and small power loads. The HVAC system is one of the biggest consumption of energy load in a building. Common electrical loads for HVAC systems include fans, pump motors, chillers, and heaters in addition to variable frequency drives (VFD). An important energy-consuming technology in the HVAC system is electric motors. It is necessary to convert motor horsepower ratings to input kilowatts or kilovolt-amperes before combining them with other loads in the building. However, motors are not the end-consumer of energy. Consequently, they should not always be the initial consideration when seeking energy-saving opportunities. The mechanical loads which are driven by the motor like pumps, fans, compressors often present far more opportunity for energy saving than the motor itself.

Historically, electricity use for lighting usually accounted for the largest share of total annual commercial sector electricity use, but its share has declined over time mainly because of the increasing use of high-efficiency lighting lamps like light-emitting diodes (LED). The lighting system is usually considered secondary to the HVAC system in terms of electric loading. It is defined as all of the components necessary to meet illumination requirements in terms of duration and level. Preliminary estimates of lighting loads may be made by assuming watts per building area. With a demanding effort on sustainability and energy efficiency, recent lighting technologies have produced a wide array of new lamps, control systems, and software programs that are available today for architects and engineers to study, test, and implement various lighting systems.

Small power or plug loads include types of portable and miscellaneous equipment that are served by general-purpose receptacles. As buildings become more energy-efficient, the amount and share of electricity use for computers and office equipment are an increasingly significant source of energy end-use. These loads can represent up to 50% of the electricity use in buildings with high-efficiency systems (Menezes et al., 2014). Considerable improvements in the energy efficiencies of small power loads like computers have been noticed in the last few decades, resulting in reduced energy requirements. In addition to usage patterns, power management capabilities can have a major impact on the energy consumption of loads. In existing buildings, smart small power load controls consist of receptacles and power strips that rely on on-time scheduling, motion sensing, or load detection to completely cut off power to equipment, not in use. Some smart power strips can sense the primary load, such as a computer, and operate peripheral devices accordingly. For centralized control, small power load schedules can be programmed into lighting and management systems.

Finally, the building entry transformer, main panel, and entry cable are sized based on a transformer load level of 70% and power factor (PF) of 0.85. For example, the transformer load in kW = building load in kW/(0.7 × 0.85).

3.10 Clean electricity

Electrification is evolving as a significant solution for reducing GHG emissions but only if combined with clean electricity, which gradually can be obtained at the lowest cost from renewable energy. Electrical pollution, also known as dirty electricity is a term used to describe a type of electrical phenomenon occurring worldwide. It refers to spikes and surges of energy signals circulating throughout buildings, and even between buildings and the grid. This type of EM pollution has recently been recognized as one of the most important and pressing issues facing electrification today. It is risky to electrical equipment as well as unhealthy to biological systems, including humans, insects, plants, and animals.

3.10.1 Power quality

Large buildings and mission-critical facilities like data centers are especially dependent on clean electricity or high power quality. Dirty electricity or poor power quality is a concern in buildings because it wastes energy, reduces electrical capacity, can overheat building and tenant equipment, and may cause a power outage.

Most of the power quality problems in buildings are originating within the same building and mostly on the customer side. The increasing use of nonlinear electronic devices such as VFDs, battery chargers, LED lights, and switched-mode power supplies that replaced conventional power supplies in addition to the proliferation of computers and other devices throughout buildings has raised the need to design electrical systems with an eye on power quality. Power quality problems can be caused by external events also, such as lightning strikes which may impact power quality and cause starting and stopping of heavy equipment, and circuit overloads.

Practically, the vast majority of power quality issues experienced in buildings may be attributed to harmonics, PF, transients, and sags. However, several other power quality conditions can also disrupt processes and equipment, such as swells, undervoltages, overvoltages, interruptions, DC offsets, notching, noise, voltage fluctuations, momentary interruptions, and frequency variations.

Power quality can be measured with power quality meters, analyzed with software, and monitored. However, there are much higher processing and computational requirements to detect, capture, and measure power quality events that may be of high frequency. Power quality mitigation actions may include changing motor starting processes, replacing faulty switches or relays, filtering harmonic producing loads, and altering control of the filters. Once the mitigation processes are in place, further monitoring will verify that the corrective action worked.

Digitization of the electrical distribution helps bring greater visibility, insights, and decision support to help demystify the complex power quality issues that can impact the operations and longevity of critical assets. A common architecture of permanently installed power quality instruments enables continuous monitoring, capture, and reporting of power quality information. This enables an optimal power supply to critical equipment. Analytics of this data ensures a good foundation for interpreting power-related issues and responding accordingly to optimize the reliability of critical equipment.

Finally, the interaction between efficiency and power quality requires a complete approach to design and operation. Building loads and usage change over time, and the usage of energy-efficient devices can alter the power quality status. Installing power quality monitoring equipment at the time of upgrade allows constant inspections on the power quality and efficacy of power quality correction techniques, as well as identifying the source of power quality problems.

3.10.2 Harmonics and power factor

Harmonics are a form of voltage or current waveform distortion which are multiples of the fundamental frequency (50 or 60 Hz). They are integral multiples of the fundamental frequency involving nonlinear loads or control devices, including EM devices (transformers, motors, and lighting ballasts) and various solid-state devices. Harmonics distortion is expressed as the percentage of harmonic voltages or currents compared to the fundamental frequency voltage or current. For example, the third harmonics for North American electrical distribution systems would be at a frequency of 3×60 Hz $=$ 180 Hz.

A common measurement of the cumulative amount of harmonic distortion is total harmonic distortion (THD), which is an important aspect of power systems and should be kept as low as possible. Lower THD in power systems means higher PF, lower peak currents, and higher efficiency. Low THD is such an important feature in power systems that international standards such as IEC 61000-3-2 set limits on the harmonic currents of various classes of power equipment. Harmonics are caused by nonlinear loads in the electrical system, meaning that the loads draw nonsinusoidal current from a sinusoidal voltage source. For example, the main problems arising from HVAC and lighting upgrades are decreased PF and harmonics. PF provides a measure of the efficiency of an electrical system. Importantly, most VFDs available on the market today include PF correction and harmonic reduction circuits. The displacement PF is the ratio of the real power that is used to do work (watts) and the reactive apparent power (VA) that is supplied to the circuit. PF is a dimensionless number that ranges from 0 to 1. A low PF indicates that the reactive load is consuming a larger portion of the total current while a negative PF indicates that the load is generating current.

High levels of harmonic distortion can result in increased transformer, capacitor, motor, or generator heating. While the exact relationship between harmonics and losses is very complex and difficult to generalize. Harmonic filters, passive or active, may be added to electricity systems to suppress harmonic frequencies. Passive filters are tuned to filter a specific frequency or group of frequencies. Although these filters are cost-effective, passive filters are not useful for varying loads. Active filters measure the harmonic currents present in the system and generate opposite harmonics to cancel those produced by the harmonic sources. Harmonics filters are sometimes the most cost-effective in an existing structure where rewiring is difficult or costly. Shielded isolation transformers that are used in buildings are filtering devices that lessen the feed-through of harmonic frequencies from the source or the load.

3.10.3 Transients

Transients are temporary spikes or surges in AC voltage or current that can potentially impact circuits in ways ranging from minor faults to failure. They have a short duration in comparison with normal signal times. A voltage transient can be anywhere from a few millivolts to thousands of volts, and they can last from nanoseconds to hundreds of milliseconds. Some transients are repetitive, such as those caused by inductive ringing in a motor, while other transients are more sporadic, such as electrostatic discharge which is the transfer of electrical charge between two objects, reestablishing the electrical balance in a very short time, often in the shape of a visible spark.

Impulsive transients are frequently referred to as high-energy transients or surges. The most common cause of impulsive transients is lightning strikes (with mean current pulses of about 20,000 A and peak values of up to 150,000 A). For transients to cause damage to a system they must find a way into the system, either directly or via inductive or capacitive coupling. This can result in a large amount of energy transfer with very short rise and decay times. Fast transients can have a 5 ns rise time and have a duration of fewer than 50 ns. An impulsive transient can damage a wide variety of equipment that is not rated for such high-voltage levels, with computing equipment being especially vulnerable (Olikara, 2015). Even if the amplitude of the transient is not very high, the rise and fall rate of voltage can damage solid-state equipment. Good electricity wiring practices and high-frequency filters can also help reduce the impact of impulsive transients.

3.11 Switching between AC and DC

With the greater adoption of solid-state electronics as loads, renewable energy systems, and energy storage systems that supply DC power, there is an increasing need for the potential for DC-based distribution and utilization equipment (Gal et al., 2019). Examples of DC loads are electronic devices like computers, cell phones, tablets, LED lightbulbs, TVs, and a rapidly expanding world of smart devices comprising the IoT.

3.11.1 War of currents

Starting in the late 1800s, the first attempts to distribute energy in the US became a battleground between proponents of DC and AC, known as the "war of currents." DC was championed by Thomas Edison when he invented the DC power system in 1882, whose company later became part of General Electric. However, AC was championed by Nikola Tesla and George Westinghouse of Westinghouse Electric and several European companies that used Tesla's inventions to convert the voltage to higher or lower levels relatively easy using a transformer, making it possible to transmit power over long distances using narrower and inexpensive conductors (Gal et al., 2019). During the early years, DC was the standard in the US. However, Westinghouse had entered the AC business in 1884 and quickly became Edison's competitor. His company acquired a few of former Edison employee Nikola Tesla's AC-based patents and refined the components of the AC power systems known today (such as three-phase power systems, transformers, induction, and synchronous machines), making AC power transmission and distribution a commercially viable alternative to DC. In 1893, the Niagara Falls Power Company decided to award Westinghouse, the contract to generate power from Niagara Falls using the licensed Tesla's AC induction machine.

Edison enthusiastically promoted DC and demonized AC and its proponents, linking AC and Westinghouse to safety concerns, electric chairs, and capital punishment. However, most of the loads those days such as motors and lighting adapted to AC very well and it was more convenient to distribute AC electricity from power plants over long distances. Finally, it was rational to use AC rather than DC as traditionally heavy loads operate on AC and it is more convenient to distribute power from power plants, which are also AC in nature.

AC power won out over DC power as the power distribution of choice, primarily because of the ability to have large power plants in predominant locations and then transmit the power efficiently over power lines to consumers. DC would have required local generators on every neighborhood or even every street or home, which was not feasible nor economically practical at that time. The invention of the transformer made it possible for Westinghouse and others to change AC's voltage so it could be transmitted over long distances at high voltage and convert back to low voltage at its destination. Although AC is more hazardous, particularly early in its history as safety standards were developing, this high-voltage AC transmission was far more efficient and convenient than DC.

Presently, AC distribution is the predominantly utilized method for generating, transmitting, and distributing power. When electricity was new in the late 19th century, AC systems could transmit more power over longer distances than DC. However, techniques are now available for converting DC to higher and lower levels. Since DC is more stable, companies are finding ways of using high-voltage DC to transmit electricity long distances with less electricity loss.

3.11.2 Edison's revenge: A shift back to DC

Today, many power agencies and groups think DC will make a comeback especially with the growth in DC loads from switched-mode power electronics in machine drives, automation, lighting, and computational loads (Los Alamos, 2015). What is interesting about DC today, though, is that the majority of loads in buildings translate AC to DC. Electronic devices, for example, represent a huge load run on DC. With so many electronic devices, lighting devices, and large electricity users like data centers running on DC, the technology can fill a mounting function while chopping energy consumption. Also, the proliferation of renewable generators like PV and wind turbines as promising DC power systems may ease the integration of DC power into the electric grid.

Integrating dedicated DC systems into buildings eliminates the need for these AC-to-DC (rectifier) conversions and may save energy and money, reduce carbon emissions, generate useful data, increase resilience, and add design flexibility. It also has massive potential to improve the connectivity of building systems as part of the growing IoT. Unfortunately, the power coming into buildings from the grid is still AC that has to be converted to DC. Grid-connected renewables require an inverter to convert their DC to AC for the grid, even though it is changed immediately back to DC for customer use. Every time a solid-state device is used, that AC has passed through transformers, power lines (which have around 5% transmission losses), equipment that synchronizes power so the grid remains stable, and other devices before it gets to the device's rectifier. At each of these points, energy is lost.

Though there are a few high-voltage DC power lines in use, AC power infrastructure will be part of the energy future for decades to come. But using more building-wide DC would provide higher efficiency because of the potential to avoid power conversions. DC can lower capital costs because there are fewer parts, requires less complicated controls and less chance of a crash, and has better power quality (Los Alamos, 2015). So the big question goes back to that posted by the giants of the 19th century: AC or DC power? As history has shown, it looks like the two currents will end up working parallel to each other in a sort of hybrid resolution, however, what about the future? Eventually, thanks to the brilliance of both Edison and Tesla.

References

Allouhi, A., El Fouih, Y., Kousksou, T., Jamil, A., Zeraouli, Y., Mourad, Y., 2015. Energy consumption and efficiency in buildings: current status and future trends. J. Clean. Prod. 0959-6526. 109, 118–130. https://doi.org/10.1016/j.jclepro.2015.05.139.

Archtoolbox, 2021. Electrical Power Systems in Buildings. https://www.archtoolbox.com/materials-systems/electrical/electrical-power-systems.html.
Azari, R., Abbasabadi, N., 2018. Embodied energy of buildings: a review of data, methods, challenges, and research trends. Energy Build. 168, 225–235.
Benavente-Peces, C., Ibadah, N., 2020. Buildings energy efficiency analysis and classification using various machine learning technique classifiers. Energies 13, 3497. https://doi.org/10.3390/en13133497.
CDCP and DHUD, 2006. Healthy Housing Reference Manual. https://www.cdc.gov/nceh/publications/books/housing/housing.htm.
Clement-Nyns, K., Haesen, E., Driesen, J., 2011. The impact of vehicle-to-grid on the distribution grid. Electr. Power Syst. Res. 81, 185–192.
Cruz, N.S., Torres, M.I.M., de Silva, J.A.R.M., 2011. Bioclimatic architecture potential in buildings durability and in their thermal and environmental performance. In: International Conference on Durability of Building Materials and Components, Proto, Portugal, April 12-15.
D'Oca, S., Hong, T., Langevin, J., 2017. The Human Dimensions of Energy Use in Buildings: A Review. https://simulationresearch.lbl.gov/sites/all/files/t_hong_-_the_human_dimensions_of_energy_use_in_buildings_a_review.pdf.
Dalhat, A.I., 2014. Application of Bioclimatic Architecture Principles in the Design of Hotels at Katsina Nigeria (M.Sc. Thesis). Department of Architecture, Ahmadu Bello University, Zaria, Nigeria.
Deason, J., Wei, M., Leventis, G., Smith, S., Schwartz, L., 2018. Electrification of Buildings and Industry in the United States Drivers, Barriers, Prospects, and Policy Approaches. https://ipu.msu.edu/wp-content/uploads/2018/04/LBNL-Electrification-of-Buildings-2018.pdf.
Dehghani-sanij, A.R., Soltani, M., Raahemifar, K., 2015. A new design of wind tower for passive ventilation in buildings to reduce energy consumption in windy regions. Renew. Sust. Energ. Rev. 1364-0321. 42, 182–195. https://doi.org/10.1016/j.rser.2014.10.018.
Dewulf, J., Van Langenhove, H., Muys, B., Bruers, S., Bakshi, B.R., Grubb, G.F., Paulus, D.M., Sciubba, E., 2008. Exergy: it's potential and limitations in environmental science and technology. Environ. Sci. Technol. 42 (7), 2221–2232.
Dincer, I., Rosen, M.A., 2021. Exergy, environment, and sustainable development. In: Dincer, I., Rosen, M.A. (Eds.), Exergy, third ed. Elsevier, ISBN: 9780128243725, pp. 61–89, https://doi.org/10.1016/B978-0-12-824372-5.00004-X. Chapter 4.
Dodds, P.E., Staffell, I., Hawkes, A.D., Li, F., Grünewald, P., McDowall, W., Ekins, P., 2016. Hydrogen and fuel cell technologies for heating: a review. Int. J. Hydrog. Energy 40 (5), 2065–2083.
Eboh, F.C., Ahlström, P., Richards, T., 2017. Exergy analysis of solid fuel-fired heat and power plants: a review. Energies 10 (2), 165. https://doi.org/10.3390/en10020165.
Ece, N., 2018. Building Biology: Criteria and Architectural Design. Birkhauser Architecture, Basel, Switzerland.
Evola, G., Marletta, L., Costanzo, V., 2018. Exergy analysis of energy systems in buildings. Buildings 8 (12), 180.
Fathy, A., Carmichael, C., 2019. Why Building Owners Should Take Charge of EV Adoption? https://www.greenbiz.com/article/why-building-owners-should-take-charge-ev-adoption.
Fincher, J., 2013. EWICON Bladeless Wind Turbine Generates Electricity Using Charged Water Droplets. https://newatlas.com/ewicon-bladeless-wind-turbine/26907/.
Gal, I., Lipson, B., Larsen, T., Tsisserev, A., 2019. DC Micrgrids in Buildings. https://www.csagroup.org/wp-content/uploads/CSA-Group-Research-DC-Microgrids-In-Buildings.pdf.

Habash, R., 2017. Green Engineering: Innovation, Entrepreneurship, and Design. CRC Taylor and Francis, Boca Raton.

Habash, R., 2019. Professional Practice in Engineering and Computing. CRC Taylor and Francis, Boca Raton.

Hayter, S.J., Kandt, A., 2011. Renewable Energy Applications for Existing Buildings. https://www.nrel.gov/docs/fy11osti/52172.pdf.

HBS, 2013. Residential Wiring Best Practice. https://healthybuildingscience.com/2013/01/15/residential-wiring-best/.

Hocknell, C., 2016. Effective Energy Strategies Should Be Lean, Clean and Green. https://www.pbctoday.co.uk/news/energy-news/effective-energy-strategies/29308/.

IRENA, 2019. Global Energy Transformation: A Roadmap to 2050. International Renewable Energy Agency, Abu Dhabi, UAE. https://www.irena.org/publications/2019/Apr/Global-energy-transformation-A-roadmap-to-2050-2019Edition.

IRENA, 2021. World Energy Transitions Outlook: 1.5°C Pathway. International Renewable Energy Agency, Abu Dhabi. https://sdgs.un.org/sites/default/files/2021-05/Global%20Energy%20Transformation-%20A%20Roadmap%20To%202050%20%282019%20Edition%29.pdf.

Kandt, A., Walker, E., Hotchkiss, A., Buddenborg, J., Lindberg, J., 2011. Implementing solar PV projects on historic buildings and in historic districts. NREL. Draft Technical Report. NREL/TP-7A40-51297. NREL. https://www.nrel.gov/docs/fy11osti/51297.pdf.

King, J., Perry, C., 2017. Smart Buildings: Using Smart Technology to Save Energy in Existing Buildings. American Council for an Energy-Efficient Economy. https://www.aceee.org/sites/default/files/publications/researchreports/a1701.pdf.

Los Alamos, 2015. DC Microgrids Scoping Study-Estimate of Technical and Economic Benefits. Los Alamos National Laboratory. https://www.energy.gov/sites/prod/files/2015/03/f20/DC_Microgrid_Scoping_Study_LosAlamos-Mar2015.pdf.

Lotfabadi, P., 2015. Solar considerations in high-rise buildings. Energy Build. 89, 183–195.

Lotfabadi, P., Hançer, P., 2019. A comparative study of traditional and contemporary building envelope construction techniques in terms of thermal comfort and energy efficiency in hot and humid climates. Sustainability 11 (13), 3582.

Menezes, A.C., Cripps, A., Buswell, R.A., Wright, J., Bouchlaghem, D., 2014. Estimating the energy consumption and power demand of small power equipment in office buildings. Energy Build. 75, 199–209.

Nayak, B., 2020. The Solid-State Wind Energy Converter. https://goodmenproject.com/featured-content/the-solid-state-wind-energy-converter/.

Nesler, C., 2020. Zero-Carbon Buildings Is Possible Following these Four Steps. https://www.weforum.org/agenda/2020/01/zero-carbon-buildings-climate/.

Ness, C., Andresen, I., Kleiven, T., 2019. Building bioclimatic design in cold climate office buildings, 1st nordic conference on zero emission and plus energy buildings. IOP Conf. Ser. Earth Environ. Sci. 352. https://iopscience.iop.org/article/10.1088/1755-1315/352/1/012066/pdf.

NYC, 2020. Types of Energy Storage. https://www.nyserda.ny.gov/All-Programs/Programs/Energy-Storage/Energy-Storage-for-Your-Business/Types-of-Energy-Storage.

Olikara, K., 2015. Power Quality Issues, Impacts, and Mitigation for Industrial Customers. https://literature.rockwellautomation.com/idc/groups/literature/documents/wp/power-wp002_-en-p.pdf.

Pérez-Lombard, L., Ortiz, J., Pout, C., 2008. A review on buildings energy consumption information. Energy Build. 40, 394–398.

Praseeda, K.I., Reddy, B.V.V., Mani, M., 2015. Embodied energy assessment of building materials in India using process and input–output analysis. Energy Build. 86, 677–686.

Ruhnau, O., Bannik, S., Otten, S., Praktiknjo, A., Robinius, M., 2019. Direct or indirect electrification? A review of heat generation and road transport decarbonisation scenarios for Germany 2050. Energy 166, 989–999.

Stremke, S., van den Dobbelsteen, A., Koh, J., 2011. Exergy landscapes: exploration of second-law thinking towards sustainable landscape design. Int. J. Exergy 8, 148–174.

Strong, S., 2016. Building Integrated Phorovoltaics (BIPV). https://www.wbdg.org/resources/building-integrated-photovoltaics-bipv.

Taffese, W.Z., Abegaz, K.A., 2019. Embodied energy and CO_2 emissions of widely used building materials: the Ethiopian context. Buildings 9, 136. https://doi.org/10.3390/buildings9060136.

Toth, Z., Volt, J., 2021. Whole-Life Carbon: Challenges and Solutions for Highly-Efficient and Climate Neutral Buildings. https://www.bpie.eu/wp-content/uploads/2021/05/BPIE_WLC_Summary-report_final.pdf.

Tuladhar, R., Yin, S., 2019. 21-Sustainability of using recycled plastic fiber in concrete. In: Pacheco-Torgal, F., Khatib, J., Colangelo, F., Uladhar, R. (Eds.), Woodhead Publishing Series in Civil and Structural Engineering, Use of Recycled Plastics in Eco-efficient Concrete. Woodhead Publishing, pp. 441–460.

van den Dobbelsteen, A., Broersma, S., Stremke, S., 2011. Energy potential mapping for energy-producing neighborhoods. Int. J. Sustain. Build. Technol. Urban Dev. 2, 170–176.

Vandyevyvere, H., Stremke, S., 2012. Urban planning for a renewable energy future: methodological challenges and opportunities from a design perspective. Sustainability 4 (6), 1309–1328.

Victoria, M., Haegel, N., Peters, I.M., Sinton, R., Jäger-Waldau, A., del Canizo, C., Breyer, C., Stocks, M., Blakers, A., Kaizuka, I., Komoto, K., Smets, A., 2021. Solar photovoltaics is ready to power a sustainable future. Joule 5 (5), 1041–1056.

Vitillo, J.G., 2015. Magnesium-based systems for carbon dioxide capture, storage and recycling: from leaves to synthetic nanostructured materials. RSC Adv. 5 (46), 36192–36239.

WBDG, 2016. Biomass for Heat. https://www.wbdg.org/resources/biomass-heat.

Weisbach, D., Ruprechta, G., Huke, A., Czerskia, K., Gottlieba, S., Hussein, A., 2013. Energy intensities, EROIs, and energy payback times of electricity generating power plants. Energy 52, 210–221.

Wilson, A., 2021. The Folly of Building-Integrated Wind. https://www.buildinggreen.com/feature/folly-building-integrated-wind.

Zafar, S., 2020. The Concept of Biorefinery. https://www.bioenergyconsult.com/biorefinery/.

Building as a smart system

4

Riadh Habash
School of Electrical Engineering and Computer Science, University of Ottawa, Ottawa, ON, Canada

I didn't fail 1000 times. The light bulb was an invention with 1000 steps.

Thomas A. Edison

4.1 Sustainability through smartness and intelligence

"Smartness" rightfully implies a form of "intelligence"; therefore, the recognized need for smart capabilities likely represents a reflection on the relative lack of intelligence in buildings. These new capabilities rely on physical and digital infrastructure whose potential is only beginning to be realized through the evolution of communication networks and the through the lens of sustainability.

Two major driving forces sustained the paradigm shift toward IBs. First, the rising knowledge and innovation economy that sustains contemporary development worldwide. Second, the spread of the Internet as a major technological innovation. In these buildings, embedded computers and wireless sensor and actuator networks (WSAN) are applied to gather and control various physical processes, mostly with feedback loops where physical processes and computations influence each other. These buildings are regarded as CPS which require close harmonization and integration between the computational (virtual) and the physical worlds. A CPS interface machinery and embedded systems completely with the cyber world through ICT networks and empowers remote control of the entire systems.

Smart building technology is the digital extension of architectural and engineering evolution. It may be initially seen as one version of the smart grid concept at the level of the building. Integration to improve energy efficiency as well as to enhance the occupant's health and wellness is the anticipated requirement from today's building. As society now lives through the age of technology, buildings are starting to adjust to that integration to better serve their occupants. As the design and infrastructure of the built environment are experiencing substantial transformation, broadband wireless communication is becoming as important and universal as the establishment of various public facilities. A smart-ready infrastructure with several technologies such as 5G, WSANs, smart grid, smart industry, ongoing urbanization, and growing population will allow for smart systems to be implemented and upgraded as necessary over the lifetime of the building to address applications involving indoor climate monitoring and control, energy management, occupants' health, comfort, and productivity.

Technology continues to make life convenient. Besides convenience, it has the potential to reduce the impact on the environment. For example, ICTs offer opportunities to further improve the intelligence of a building, enhance the quality of its

Sustainability and Health in Intelligent Buildings. https://doi.org/10.1016/B978-0-323-98826-1.00004-1

occupancy, and also save on energy consumption. The capability of intelligent behavior in buildings presents challenges to software engineering, more specifically with the difficulty of developing building system monitoring, management systems, and their interoperability (Kevitt, 2010). While the AEC community has been exploring IB design for several years now, there is still no majority approval of these new IoT technologies. However, the mindset is changing gradually. With limited examples of solid empirical data demonstrating a value beyond energy efficiency, the challenge of bringing new and untested technologies into existing and often unrelated and fractionalized organizational management structures (e.g., facilities, operations, and IT departments) remains a barrier to adoption and acceptance (Bonin, 2019). On the other side, regardless of the cities and buildings being called smart or intelligent, there is a great need for robust commitment toward circularity via sustainability and health, ensuring more effective involvement of smart ICT and intelligence in all phases (Fig. 4.1). Having a roadmap in mind, with various stakeholders involved in decision-making, the governance of such an environment is not only restricted to implementation, but it also has to be shaped based on interactions among all the corresponding elements.

The demand for buildings with intelligent and smart solutions is on the rise as technological and business-related transformations shape the evolutions in the. It is thought that the intelligent-smart examination will remain but using the present alliance allows strengthening the existing contexts and conceptual models to shed more light on the real essence and future directions of the built environment. This environment needs to be much more than only intelligent and smart, where the desired goal should be overlaying the path toward building better living environments for people regardless of what they would be named intelligent, smart, or circular. However, one

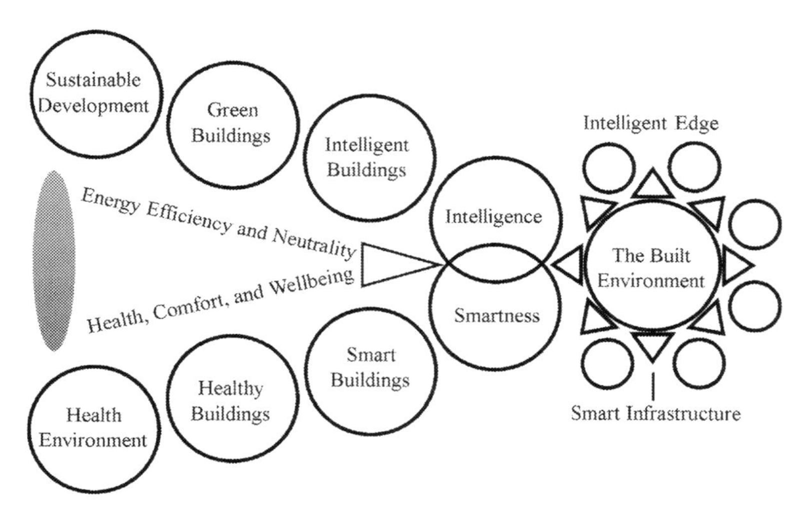

Fig. 4.1 The built environment through the lens of the triadic relationship between sustainability, health, and intelligence.

of the prevalent issues that emerge from the building sector is the need to have transdisciplinary experts involved in the building design team early in the process so that they have input in the various infrastructure needs.

4.2 Electrical grid

The electrical grid is a magnificent paradigm of human creativity and engineering. It covers a hundred of thousands of kilometers. It is more than just generation, transmission, and distribution infrastructure. It is an ecosystem of asset owners, manufacturers, service providers, and regulators, all working together to operate one of the most reliable electrical grids in the world. The electric infrastructure is aging and modernizing the grid to make it smarter and more resilient through the use of cutting-edge technologies, equipment, and controls that communicate and work together to deliver electricity more reliably and efficiently can greatly reduce the frequency and duration of power outages, reduce storm impacts, and restore service faster when outages occur. Three modernizing concepts, namely smart grid, microgrid, and virtual power plant (VPP) (Fig. 4.2) often become interchanged, are attracting a variety of different understandings as the industry is transitioning to electrification and decarbonization.

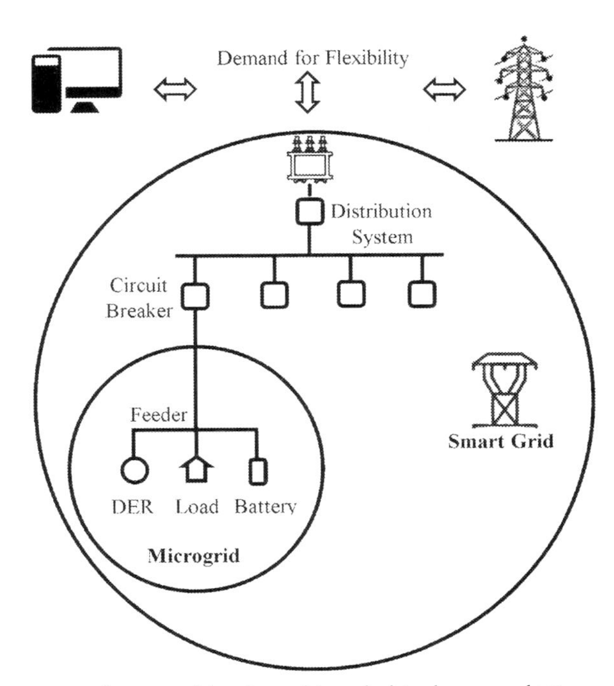

Fig. 4.2 The concepts of smart grid, microgrid, and virtual power plant.

4.2.1 Smart grid

Today, electrical engineering shines with a new era of smart buildings through the concept of smart grid, which should allow, among many things, collecting and using smartly intermittent renewable energies through an energy network in which the fluxes of energy will be multidirectional and massively orchestrated (Wurtz and Delinchant, 2017). This results in a new scheme and new architecture for the stability and success of both buildings and the smart grid.

A smart grid involves two major elements: the power grid, including the generators and consumers; and an associated sensing and control system. Consumers can better manage their energy consumption and costs because they have easier access to their data. Utilities also benefit from a modernized grid, including improved security, reduced peak loads, increased integration of renewables, and lower operational costs.

The smart grid represents an unprecedented opportunity to move the energy industry into a new era of reliability, availability, and efficiency that will contribute to economic, health, and environmental health. In addition, the smart grid is a way to address an aging energy infrastructure that needs to be upgraded or replaced. It is a way to address energy efficiency, to bring increased awareness to consumers about the connection between electricity use and the environment. It is an intelligent and integrated system of interregional connected electric utilities, consumers, and DERs.

Like the Internet, the smart grid is an intelligent and integrated system of interregionally connected electric utilities, consumers, and DERs. This evolving form of electrical generation, transmission, and utilization uses advanced metering, monitoring, management, automation, and ICTs to provide reliable two-way delivery and consumption of electric power and respond digitally to the quickly changing energy demands of users (Chen et al., 2013). This development is meeting the electricity demand throughout the world, causing high power system loading resulting in exaggerated system equipment. Balancing the supply and demand is the main objective of the smart grid. Therefore, smart grid provides an outstanding opportunity for the support and development of smart buildings and communities. Like the Internet, the smart grid consists of controls, computers, automation, new technologies, smart buildings, and equipment working together, but in this case, these technologies will work with the electrical grid to respond digitally to the quickly changing energy demands of the users (Kolokotsa, 2016). This can create a revolution in the building sector.

In the end, smart grids can play an important role in the built environment because they can create physical proximity between consumers and energy distribution that may help grow occupants' awareness toward reasonable consumption of energy in buildings. Moreover, smart grids can offer new openings for the reduction of gas emissions by developing technical conditions that enhance the connection of devices and DERs at low voltage levels.

4.2.2 Microgrid

A microgrid is not a new idea but comes very much from typical models of campuses and remote communities that thrived throughout the mid-20th century, which include a small network of electricity users with access to a local source of energy to sustain a portion of electricity use. The notion of microgrid was initially introduced as a solution for the reliable integration of DERs, including energy storage systems and controllable loads. It is a smart grid on a small scale, in terms of electrical load and generation. According to the US Department of Energy (DoE), a microgrid is "a group of interconnected loads and DERs within clearly defined electrical boundaries that act as a single controllable entity to the grid. It can connect and disconnect from the grid to enable it to operate in both grid-connected or island mode."

Microgrids are modern, small-scale versions of the centralized electricity system. They contain all the elements of complex energy systems. A microgrid is an electrical system that includes single or multiple loads, as well as one or several distributed energy sources, that can be operated in parallel with the broader utility grid. It is a cluster of distributed loads (e.g., houses, EVs), storage devices (e.g., batteries), and DERs (e.g., thermal, solar PV, wind). However, the intermittent nature of DERs makes the operation of the microgrid more difficult. Energy storage devices are necessary to smooth the power generation of renewable resources. They achieve specific local goals, such as reliability, carbon emission reduction, diversification of energy sources, and cost reduction, established by the community being served. They maintain the balance between generation and consumption, and they can operate on and/or off-grid. They are ideal for supplying power to remote or poorly developed regions with no connection to a public network. In addition, more and more industrial operators are using micrograds to produce the electricity they need cost-effectively, sustainably, and reliably.

Microgrid usually is managed by software that will use information like energy market constraints (energy price), weather forecast, and other external and internal constraints. The microgrid structure assumes an aggregation of loads and microsources operating as a single system providing both power and heat. This control flexibility allows the microgrid to present itself to the bulk power system as a single controlled unit, have plug-and-play simplicity for each microsource, and meet the customers' local needs. The ability of the microgrid to operate connected to the grid as well as with a smooth transition to and from the island mode is another important function (Davis, 2003). A microgrid can connect and disconnect from the grid to enable it to operate in both grid-connected and island mode.

Islanding is a major challenge to the development of microgrids because it is time-consuming and difficult to evaluate. Islanding is the intentional isolation of a part of the power system during external widespread grid disturbance in which a DERs remains energized in a localized area (island) like a campus while the remainder of the electric power system loses power, a situation that may happen due to unbalance in generation and load. The mismatch between the two will result in a change in frequency from 50/60 Hz where an under-frequency relay is employed to sense the failure. The main advantage of islanding is that the power supply is not interrupted on the island even during the grid disturbance.

For forthcoming intelligent microgrids, the need of deploying smart devices in the controlling and monitoring activities is important because, the system requires real-time synchronization, stability, and quality power, and secure ICT. IBs can improve the operation of a microgrid by which they are served. As load centers in a given locality, buildings that are technologically enabled to monitor their energy consumption can be further enabled to reschedule certain power usage to off-peak hours, improving the overall efficiency of a microgrid. These IBs also can monitor and adjust building performance to reduce load and bolster cost savings.

4.2.3 Virtual power plants (VPP)

A VPP is a network of decentralized, medium-scale power generating systems such as hydropower plants, solar parks, wind farms, and CHP generators, as well as flexible power consumers including energy-efficient buildings and storage systems. The idea is to propose virtual multisource, multiload, and multistorage systems, all of them massively coordinated by ICTs. This notion should also be based on the emerging evolution of AI combined with other high-performance digital and wireless technologies which are quickly transforming the way of designing, operating, and thinking about living and working built environments. The outcome is a multifaceted functioning environment where generation is transferring from mass, centrally controlled power plants to distributed and weather-dependent resources. This requires new technologies, regulations, and new business models, as well as the smart grid to perform additional tasks.

VPP aims to be able to manage the customers' energy demand collectively, supply targeted areas with renewable energy, and challenge potential network failures. It consists of remote software and data systems to regulate certain energy consumption by connecting, coordinating, and monitoring storage and controlled energy generators. Unlike conventional power plants, VPPs do not own the energy properties themselves. Instead, they own the data that is transferred between each asset in the network. The VPP focuses on optimizing how each property is linked in the network. The purpose of the VPP is to ease the load on the grid by smartly allocating the power generated by the individual units during periods of peak load and fast-responding power services to the grid. Moreover, the combined power generation and power consumption of the networked units in the VPP are traded on the energy exchange.

Using intelligent algorithms, the control system can produce individual schedules for steerable plants. This allows production to meet demand with higher incomes for the plant operators. Based on price information, flexible power generators can be ramped up and down precisely to the needed time slot. Variable power consumers such as buildings can also be managed on optimized price schedules by consuming their electricity when it is inexpensive and demand is low.

4.3 Smart building loads

While conventional buildings have systems functioning independently, smart buildings use ICT to join building systems to optimize operations and performance. Although the greatest penetration of smart technologies in current buildings has been in offices, their application is increasing gradually in all building categories.

4.3.1 Smart HVAC system

To resolve energy-saving and IAQ issues, interest in a new generation of smart HVAC systems, including demand-controlled ventilation, has been growing since the 1980s. Many ventilation international standards and national regulations have progressively integrated an allowance for smart ventilation strategies in buildings. Each country uses different IAQ indicators, calculated through different procedures, and compared to different thresholds (Guyot et al., 2017).

Typically, HVAC systems consume about 30% of building energy and using controls to properly manage their operations is significant in saving energy. However, building operators frequently manage these systems through trial-and-error adjustments in reaction to occupant comfort feedback. Smart HVAC systems have the potential to reduce energy consumption while maintaining or even improving occupant comfort (King and Perry, 2017). They can also constantly adjust the operation to fit the demand and detect needs for maintenance. Smart software interprets information from a variety of HVAC sensor points and maintains that information in real-time, usually in a cloud-based system that is remotely accessible. These advanced controls can limit HVAC consumption in unoccupied building zones, detect and diagnose faults, and reduce HVAC usage during times of peak energy demand.

Smart HVAC is based on sensors and the use of control strategies that adapt to the technology of the system, by modulating temperature, humidity, presence of occupants, air velocities, air, and water flow rates, and pressures. The most popular are probably connected thermostats which are programmable devices intended to help residential building occupants to know and manage their heating and cooling loads, especially from a smartphone. A duct static pressure sensor that measures the amount of resistance against the air flowing through a duct is another example. HVAC fans must work harder to overcome greater resistance in ducts, so reducing static pressure saves energy. A duct static pressure sensor contains a sensing element that reacts to physical changes such as static pressure (Dunn, 2015). Connected IAQ sensors are other types that are designed to inform about indoor pollutant concentrations and could be used to control ventilation airflow rates. IAQ is a major area of concern in buildings that are affected by ventilation.

One of the largest energy efficiency benefits of smart building HVAC controls is found through optimizing the amount of conditioned (i.e., heated or cooled) air supplied throughout a building. For example, a whole-building ventilation control system, with smart capability, senses the amount of CO_2 in occupied areas of the building and can modulate the amount of airflow in one area without starving or over ventilating another (King and Perry, 2017). This can save considerable energy in heating and cooling and ventilation fan operation. In addition to controlling HVAC operation based on CO_2 levels, smart controls can optimize airflow using data provided by occupancy, temperature, humidity, duct static pressure, and air quality sensors.

Adjusting the use of conditioned air is one of the most effective applications of smart building equipment, especially in multizone systems. For example, a multizone variable air volume system with several boxes could use smart controls to more effectively condition each of the zones. With sensors installed in each office area, each variable air volume box can be programmed to cycle back or shut off completely when the corresponding space is vacant. In addition to the above, smart buildings can limit HVAC use during periods of peak demand through demand management and response.

ASHRAE has published Guideline 36, "High-Performance Sequences of Operations for HVAC Systems." This new idea is to influence these new recommended sequences of operation and the capabilities of the HVAC-Cx tool to develop and validate a library of automated commissioning tests linked to the standard sequences of operation. This validation work can become the basis for new industry guidelines or standard methods of test for building system commissioning tools (NIST, 2021).

4.3.2 Smart lighting

Smart lighting consists of advanced controls that incorporate daylighting and advanced occupancy and dimming functions. Demand-response programs are incentivizing steps and continuous dimming control. Smart lighting systems can be controlled wirelessly and scheduled into lighting management systems. Wireless controls facilitate easier retrofits while lighting management platforms allow users access through web-based dashboards.

Smart lighting controls aim to improve on issues related to performance, energy efficiency, and comfort as well as in increasing positive user experience. It consists of networked LED and linear fluorescent luminaires with advanced sensing and control capability. Advanced controllability consists of dual technology (infrared plus ultrasonic) occupancy sensing, vacancy sensing, daylight harvesting, continuous dimming, and task tuning (King and Perry, 2017).

Daylight harvesting controls use photosensors to measure indoor ambient light levels and reduce the amount of artificial lighting needed to meet design requirements. They can operate single or multiple zones and can dim the lighting instantaneously through step changes that reduce light levels either by the percentage of max output or gradually over 1–30-min increments. Some types of smart lighting solutions use luminaires that are manufactured with embedded wireless microcontrollers and sensors.

Networked sensors installed throughout a building can monitor multiple locations and collect minute-by-minute room conditions. Smart lighting solutions with wireless sensors and controls can be centrally managed through a web-based lighting management platform. The lighting management platform can utilize real-time data to autoconfigure and auto commission its operations. Luminaires can be programmed to operate individually or in zones and respond to other networked devices such as window shading and daylighting sensors. Facility personnel and even occupants can also control their lights by interfacing with the dashboard through their personal computers, smartphones, or tablets. Voice-activated control is now an option in some smartphones and tablets. Power over Ethernet is also gaining momentum for advanced lighting applications in new and existing commercial buildings.

4.3.3 Smart plug loads

A plug load, also known as receptacle load, refers to the various types of portable and miscellaneous equipment in buildings, but not related to HVAC, lighting, and water heating. It is the energy used by a device or appliance that typically plugs into an electrical AC outlet. Most plug loads can be physically powered down with built-in power buttons and shutdown procedures. The turning on/off of home loads depends on customer behaviors. However, plug load control may be accomplished in two basic forms of energy savings which are achieved when the device is either transitioned to a low-power state or is deenergized to eliminate the power draw. Both can be executed either manually or automatically. A low-power state is between a deenergized state and a ready-to-use state. This includes standby, sleep, and hibernate modes as well as any "off" state that has a parasitic power draw.

Because plugged-in devices are not for base building use, but rather for occupant- or business-specific tasks, these devices are generally portable and brought into the building by tenants. As a result, plug load energy use is closely tied to occupant behavior. An effective controls strategy for plug loads should begin by identifying the loads most common to the space type and addressing the highest energy consumer among these loads first (King and Perry, 2017).

Smart buildings address plug loads strategically by controlling devices at the outlet. Emerging technologies such as smart plugs, circuit breakers, and advanced power strips are gearing hopes of achieving these energy savings. Smart plugs (automatically controlled receptacles) easily replace existing receptacles and communicate with a controller, such as a timer or an occupancy switch. Plug load monitoring and management tools remotely turn off receptacles based on feedback from occupancy sensors located in tenant spaces. Smart plugs make it possible to turn on/off and control electrical devices from anywhere by using a smartphone or tablet, for example.

4.4 Grid-interactive efficient buildings

Energy efficiency has become a considerable challenge for the advancement of the world due to the increasing electricity demand. Thus, grid-interactive efficient buildings which can provide energy-efficient to consumers have been investigated by many researchers as an extension of the smart grid. It is a building that integrates energy efficiency and demand flexibility enabled by ICT and is characterized by the use of DERs and energy storage. Demand flexibility or load flexibility is the capability provided by DERs to reduce, shed, shift, or generate electricity. It reduces energy and carbon waste while offering flexible building loads to the grid. As the grid becomes increasingly complex, demand flexibility can play an important role in helping maintain grid reliability, improving energy affordability, and integrating a variety of generation sources. Clients, utilities, and system operators can gain important advantages from buildings optimized for energy efficiency and grid integration. Electrical loads in many buildings are flexible and through advanced controls can be managed to operate at specific times and different output levels. Advanced controls and communications enable buildings to adjust power consumption to meet grid needs through a variety of control strategies applied to existing equipment. The grid-interactive efficient buildings have a holistically optimized blend of energy efficiency, energy storage, DERs, load flexible technologies, and smart controls (Fig. 4.3). With the quick advance of renewable and DERs, the value of buildings as flexible, responsive assets grows. This integration delivers a resilient and productive building, reduces operating costs, and provides access to new revenue. At the community scale, additional strategies such as microgrids and district energy systems may also be advantageous.

As electricity demand increases, integrating buildings and the grid utility is a significant step to increasing energy efficiency. The goal of a building-integrated microgrid is to propose a power balancing strategy with smart grid interaction, aiming at reducing grid peak consumption and avoiding undesirable grid power injection (Sechilariu et al., 2013). Intermittent and variable generation sources, such as PV and wind systems, as well as new load sources, such as electric vehicles, are being connected on the grid in growing numbers and at more distributed locations. Effective integration of buildings and their systems with the grid also requires interoperable data interchange. However, the implementation and integration of newer control and automation technologies into buildings can be challenging with older legacy HVAC and building control systems. Flattening peak demand during periods of the day would allow the grid utility to rely less on fossil-fuel power plants to fill the gap. In addition, by extending the storage capacity across the grid, integration could make whole communities more resilient to failures and outages.

From an electrical and physical engineering point of view, a smart building can thus be seen as an energy microgrid connecting microturbines, PV panels, wind turbines, fuel cells, energy storage capabilities, EVs, and building loads (Wurtz and

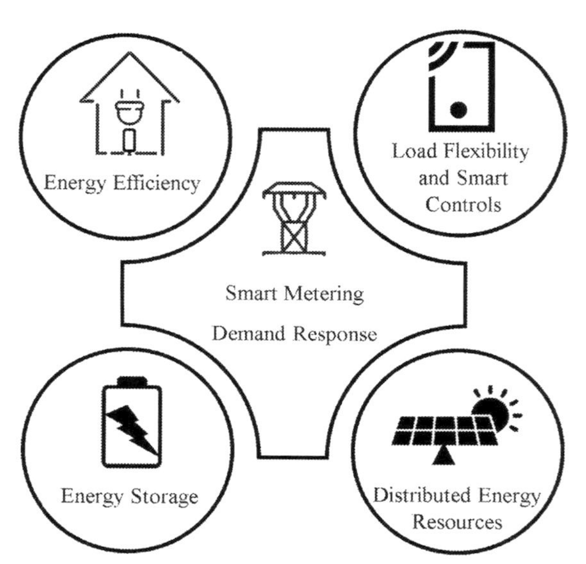

Fig. 4.3 Components of the grid-interactive efficient building.

Delinchant, 2017). The users all are directly connected to the main grid, but during outages, the entire small network can disconnect itself from the main grid, or operate in "island mode," to continue operation. Balancing supply-and-demand resources including thermal and electrical loads-within its defined boundaries, a microgrid system provides resiliency. Therefore, smart grids create an exceptional opportunity for the support of the development of smart zero energy buildings and communities and offer the step toward the IoT for the energy and building industry.

Some buildings can generate electricity for on-site consumption and even dispatch electricity to the grid in response to a signal from the grid. Sometimes, batteries are included in this process, as they improve the process of dispatching such generated power.

4.5 Smart load management strategies

As part of the global drive toward synchronizing the production and consumption of electricity, energy distribution utilities are integrating new strategies into the next-generation grids depending on the consumption levels and operating requirements or even generating energy on-site. For utilities, it is no more just a matter of purchasing and selling electricity, but rather one of harmonizing a multifaceted grid by realizing the basics of utility load management. Today, several methods of smart load management can be implemented, such as consumption forecasting, load shedding, demand response, as well as installing energy-efficient processes, energy storage devices, and power factor correction.

4.5.1 *Smart metering*

The next-generation electricity meters commonly called smart meters record the consumption of electricity, water, and natural gas and transmit information wirelessly to the utility company to record electricity usage and billing purposes. Several different wireless technologies can be used, including cellular and WiFi (Habash, 2020). The design of smart meters depends on the requirements of the utilities as well as the customers. The initial step in implementing a smart grid is to implement an advanced metering infrastructure that describes the entire power infrastructure, from smart meters and two-way communications to the control equipment and all the applications that collect and transmit energy usage information in near real-time. Metering technology is the base of a smart utility and some applications are related to consumer privacy and confidentiality, creating cybersecurity issues.

Smart metering is a prerequisite and starting point for the effective implementation of smart grids and the NZEBs. For electricity providers to deliver intelligent services for customers, bidirectional metering interfaces should be used to obtain customers' energy demand information. Data collected from smart meters, building management systems, and weather stations can be used by advanced AI techniques and machine learning algorithms to infer the complex relationships between energy consumption and various variables such as temperature, solar radiation, time of day, and occupancy (Kolokotsa, 2016).

4.5.2 Energy consumption forecasting

Energy forecasting for buildings is the process of building statistical models to make predictions about consumption levels and corresponding utility costs in both the short term and long term. Underpinning many approaches for enhancing energy usage are accurate and reliable forecasts. These approaches take into consideration historical trends, weather inputs, tariff structures, and occupancy schedules to make predictions. Such forecasting has an enormous impact on energy efficiency and sustainability practices of buildings and campuses. The forecasts can be used by building operators to implement operational strategies to stabilize factors that lead to increased energy consumption. Energy forecasts can make help in plans to optimize the operations of HVAC and lighting systems. For new buildings, where past recorded data is unavailable, computer simulation techniques are used for energy analysis and forecasting future scenarios. However, for existing buildings with historically recorded time series energy data, statistical and machine learning techniques have proved to be more accurate and quick.

Energy prediction can be broadly classified into engineering, AI-based, and hybrid approaches. While engineering techniques use physics-related thermodynamic equations to estimate energy consumption, the mathematics-based approach uses historical data to predict future energy use under constraints. In between, the hybrid techniques use physical and statistical techniques. Owing to the ease of use and adaptability to seek optimal solutions rapidly, the AI-based approach has gained popularity in recent years (Wang and Srinivasan, 2017). The mathematics-based approaches include multiple regression, genetic programming, artificial neural network (ANN), and support vector machine. The above prediction approaches are developed based on several years of observed data of several different parameters such as solar radiation, temperature, wind speed, humidity, and weekday index. The weekday index is an important parameter introduced to differentiate between working and nonworking days. The early data is used for training the models and to obtain prediction data for last year (Amber et al., 2018). At the end, the predicted electricity consumption is compared with the actual consumption of the last year.

4.5.3 Load modulation

Load modulation represents approaches that can rapidly respond to poor power quality or short-term grid disturbances, with duration and response times typically in the seconds-to-subseconds time frame. Power quality focuses on the control of active and reactive power and back-to-back converter. These approaches are integrated into larger smart grid initiatives with specific resilience, islanding, and distribution system control needs. One example of a common load modulation technology is smart PV inverters that can manage volt-ampere reactive (VAR) output as a function of incoming grid voltage and can prioritize either power output or VAR output. In addition, smart PV inverters remain connected during brief voltage sags or swells from the incoming grid power, all of which serve as a form of subsecond load modulation.

4.5.4 Demand response

One of the services offered by the smart grid is called demand-side response where the appliances operate according to the real-time pricing signal. It is the modification of consumer's demand for energy consumption. It provides major services including energy management, demand response, and storage. These services are used by electric utilities to compel their users to reduce or shift energy consumption from high-cost hours to low-cost hours of the day (Hussain et al., 2020). They are tools that enhance flexibility by better aligning the supply from DERs with electricity demand patterns by focusing on the intelligent management of energy resources and loads.

Demand response know-hows are considered a revolution in the operation of the smart grid. It typically entails voluntary and compensated programs that facilitate a power system to encourage or control load reduction as required to maintain grid stability. This happens by actively engage customers in modifying their consumption in response to pricing signals. Demand response increases energy market efficiency and security of supply, which will ultimately benefit customers by way of options for managing their electricity costs and lead to reduced environmental impact (Bradley et al., 2013; Bartusch and Alvehag, 2014). Demand flexibility is gaining traction, with evolving trends and opportunities for DERs, buildings, and electricity services.

The already available demand response programs are generally categorized into incentive- and price-based. Incentive-based programs provide economic incentives for customers to reduce demand at times of capacity shortage or exceptionally high electricity prices, whereas price-based programs involve dynamic tariff rates that promote general changes in patterns of electricity use (Kolokotsa, 2016). The user's loads are switched to an ON/OFF state by sending a short message to the building. When a high peak is detected, the appliance is switched to an off state and vice versa. In the price-based program, the user is motivated to use their loads during low price hours or off-peak hours (Yang et al., 2014). This is generally done in response to electricity price signals.

Knowledge of future electrical demand is essential for power systems and electrical distribution networks operating in the grid edge. This leads to lower energy demand during peak hours or during periods that the electricity grid's reliability is put at risk. Several new technologies are facilitating the integration of energy efficiency and demand response programs. They include smart thermostats and WiFi-enabled devices such as water heaters, refrigerators, clothes dryers, and air conditioners. This allows utilities to simultaneously enroll customers in the programs and provide energy efficiency incentives such as rebates for efficient devices. In general, buildings can implement different demand response strategies in HVAC systems, lighting systems, and other services.

4.5.5 Load shedding and shifting

Load shedding is an effective technique for the achievement of demand response programs aiming at harmonizing demands and supplies for various purposes. It is the ability to reduce electricity use for a short period and on short notice, especially during peak demand periods and/or emergencies. Load shedding is a control approach that can be performed either by the utility or by the consumer in a bid to function an optimum number of devices at a given time interval. It is a way to distribute demand for electrical power across multiple power sources where a building reduces demand for a short period during peak demand or emergency events. It is also a possible strategy to reduce electricity costs as well as to relieve stress on a primary energy source when electricity demand is greater than the supply. For example, a building dims the lighting system by a preset amount in response to grid signals while maintaining occupant visual comfort.

Intelligent load shedding is an integrated model-driven controller with real-time operational digital twins to optimize, predict, and manage load shedding for physically dispersed systems such as buildings. It is typically dispatched during peak demand periods and emergencies. Most buildings acquire electrical power from power utility providers. To lower the cost of power, while also ensuring uninterrupted delivery, a building operator may negotiate an agreement with the power provider to voluntarily load shedding on a prescheduled or on-demand basis. Such buildings change the timing of electricity use to minimize peak demand and take advantage of DERs. During load shedding events, the building draws power from its secondary source(s) rather than from the utility.

Load shifting is the ability to change the timing of electricity consumption. For some technologies, there are times when a load shed can lead to some level of load shifting. Load shifting through effective thermal energy storage or electricity storage is a major part of smart grids and is already exploited by various studies. For some technologies, there are times when load shed can lead to some level of load shifting. A combination of load shifting and smart charging is used to control the chargers so that when a certain load is plugged in, the load will not start charging when the electricity is expensive, but will be delayed until after midnight when the price is lower.

4.6 5G digital ecosystem

The next revolution is digital connectivity, enabling people and devices to send and receive data in the building environment. Wireless plays a significant role in connectivity development, but the fiber is still dominant in terms of keeping different technologies connected to the web. Wireless is an enormous convenience but is still demonstrating its limitations with interference and reliability.

5G, as its name suggests, is the fifth generation of wireless networks which facilitate communications and large data transfer at very high speeds and short latency. These networks are going to include billions of sensors, intelligent management across advanced networks, and leading-edge learning through real-time data analytics. They will enhance environmental sustainability and help to uphold public health. However, they will lead to increase EM field emissions, with many of those devices intended to be used close to the occupants inside buildings.

5G is a transformative ecosystem that includes a heterogeneous network that integrates 4G, WiFi, millimeter-wave (MMW), and other wireless access technologies. It integrates cloud infrastructure, virtualized network cores, intelligent edge services, and distributed computing models that derive insights from the data generated by billions of sensors. The above progress is made achievable by the use of additional higher frequency bands. 5G will be similar to 4G systems that are already in use (Habash, 2020).

The core of 5G technology is a new wireless spectrum networking architecture utilizing MMW bands capable of both high capacity (bandwidth) and high speed (up to 1000 times faster than the existing 4G). One of the objectives of 5G is to advance mobile and fixed Internet access at broadband speeds of the order of 10 Gbps, about a hundred times faster than theoretically possible with the 4G LTE. Additionally, being completely backward compatible with existing wireless networks, 5G creates a ubiquitous wireless broadband "blanket" around the areas where it is deployed (Habash, 2020). MMWs refer to EM fields ranging from about 30 to 300 GHz in terms of frequency, from 1 to 10 mm in terms of wavelength in free space, and from about 10^{-4} to 10^{-4} eV in terms of photon energy. They constitute the extremely high-frequency portion of the RF band. They are considered nonionizing radiation because the photon energy is not nearly sufficient to remove an electron from an atom or a molecule. They remain several orders of magnitude below the level required to ionize biological molecules (typically 12 eV is required) (Brenner et al., 2003). Compared to other wireless technologies, MMWs offer several advantages, including faster data rates (over 2 Gbps).

As 5G systems progress, frequency spectrum availability is a critical matter. Resource management and sustainability initiatives require sufficient bandwidth to power digital applications. In terms of spectrum, 5G technology makes use of a wide range of licensed frequency bands around the world. 5G deployment proposes to add frequencies in the low- (0.6–3.7 GHz), mid-(3.7–24 GHz), and high-band frequencies (24 GHz and higher) for faster communications. As these higher frequencies do not travel far and are blocked by buildings, 5G will have to use a dense network of fixed antennae outdoors as well as indoor systems. While operating at MMW frequencies has its advantages, a significant disadvantage is system path loss.

A differentiating characteristic of 5G is a much denser network with more cellular towers and the employment of smart antennas that can transmit numerous beams (up to 64 with present designs, or even more) that can be independently steered to individual subscribers. The future 5G network is envisioned to be soft, fast, and energy-efficient. It will operate within several frequency bands of which the lower frequencies are being intended for the first phase of the 5G networks. Energy efficiency is the number of bits that can be transmitted per joule of energy, and it is measured in b/J. 5G technology should target higher energy efficiency against the increased equipment and network energy consumption needed on wireless communications. Much higher frequencies are also intended to be employed at later stages of the 5G evolution. The total amount of power transmitted from a 5G cell site may exceed that from a 4G site of an otherwise similar size (microcell or small cell) (Habash, 2020). Based on the above, a change in the exposure to EM fields of humans and the environment is anticipated.

5G will be a network of networks, a set-up with numerous technologies sustaining a global infrastructure of the satellite, small cells, WiFi, typical mobile wireless networks, and enormous machine-type communications, among many others. An area where 5G networking can be very helpful is big data and cloud computing where data is flooding in at a very high rate. 5G can keep up with consumer and enterprise data demand while lowering communication operating expenses. The core networks of 5G will move from copper and fiber to MWW connections, allowing rapid deployment and mesh-like connectivity with cooperation between base stations. As a result, the 5G network will need to accommodate a huge diversity in types of traffic and it will need to be able to accommodate each one with great efficiency and effectiveness. Often it is thought that a "one type suits all" approach does not give the best performance in any application, but this is what is needed for the 5G network. Fig. 4.4 gives an idea about the structure of the 5G network.

Radio access network has been the air interface in application since the beginning of cellular technology and has evolved through the generations (1G through 5G). Small low-powered cells such as RAN nodes having a range of few meters to a few hundred meters in diameter will play an essential role in major applications of 5G. Small cells compromise three types namely Femto (~0.1 km), pico (~1 km), and microcells (~2 km). However, a macrocell is used in the cellular network to offer radio coverage to a wide area of mobile network access (~2 km). By using small cells, the network can increase area spectrum efficiency by reusing a higher frequency. Elements of the access network comprise a base station that connects to sector antennas that cover a small region depending on their capacity and can handle the communication within this small sector only. Future access networks require the ability to manage with one-way latencies of only 200 μs.

To address lower latency and high data rates, researchers are developing cutting-edge technologies for the envisioned 6G wireless communication standards. It is expected to drive the Internet of Everything, with 10^7 connections per km^2. To achieve this, 6G will leverage on subterahertz and Terahertz spectrum (300 GHz–10 THz), which provides a higher frequency spectrum as against the MMW spectrum (30–300 GHz) adopted in 5G (Imoize et al., 2021).

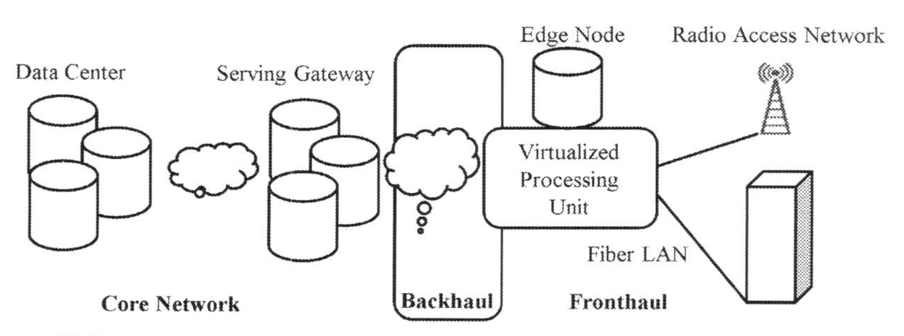

Fig. 4.4 5G future network.

4.7 High-performance wireless

5G technologies will be the strategic pathway to enable the extensive implementation of IoTs in smart buildings. If the cluster of 5G and IoT provides all the perceived potential, then the AEC community is heading toward an age of innovation where applications, systems, and services will be realized soon.

4.7.1 *In-building wireless communication*

In-building wireless refers to communication solutions that sustain connectivity. These solutions are often used by service providers that seek a quality of service (QoS) to their clients in the demanding wireless environment. 5G's arrival coincides with a lot of other developments around in-building infrastructures, such as WiFi 6, optical local area networks, digital remote powering, smart buildings, and edge computing. Taken together, ideally converged in planning and financing, there will be great emphasis on designing "holistic" in-building networks, with deeper use of fiber and RF propagation intelligence (iBwave, 2020). Fig. 4.5 shows a diagram that reflects the enabling infrastructure for an in-building communication network. The base station interfaces the core network with the building wired system. The access points represent the connection point between the wired and wireless networks. These points enable communication with wireless clients within the building and are operated by surveying software to verify signal strength at the user's location. Indoor coverage setup consists of hub repeaters and multiband antennas. The main focus of all the components is to ensure that coverage is perfect.

As the frequency increases, the RF signal will be impeded by the concrete walls. Therefore, 5G distribution signal and/or boosters will be an essential part of delivering next-generation services inside buildings. There have been multiple options for building owners and carriers for delivering in-building coverage that fits requirements. Active and passive distributed antenna systems and in-building wireless networks

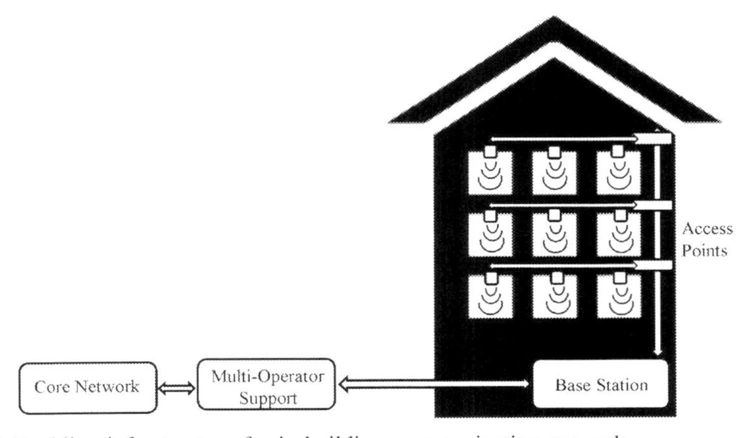

Fig. 4.5 Enabling infrastructure for in-building communication network.

are examples. Active systems need the power to transport the RF signal, while passive systems do not use power to amplify or convert the RF signal and require only cables and antennas to operate. In the era of 2G and 3G, the primary purpose of indoor networks was to extend the voice and short message coverage to large public venues like stadiums and convention centers. For this purpose, passive antenna systems was a good choice as it allowed the existing base stations and radio equipment to remain unchanged. This passive distribution of RF was cost-effective for single transmit/receive streams within a limited spectrum range. As networks evolved to advanced 4G and 5G technology, the need for higher performance started to stress traditional DAS architectures, resulting in the emergence of the active antennas for increasing the capacity for bandwidth-seeking clients. These antennas replace much of the coaxial RF cables with structured fiber and copper cabling. Rather than taking the RF signal directly from the radio and directly propagating it, the active antenna captures the signals, transmitted them over a digital network within the building, then rebuilding the signals closer to the antenna points. Today, in-building wireless connections are readily available in airports, railway stations, and other public places.

Indoor MMW 5G is still a largely unexplored domain. In the short term, it may only be used outdoors, for high-density mobile broadband needs. However, it will be needed indoors as well, especially if multigigabit concepts like the TI come to fruition. Furthermore, in-building network design for MMW brings new modeling considerations in terms of wall penetrations, reflection loss, and signal scattering due to varying material surfaces. One thing seems certain, however, whatever in-building systems eventually support new 5G spectrum bands, deeper fiber infrastructure is likely to be important, as well as availability and effectiveness of the antennas.

4.7.2 Wireless power transfer (WPT)

WPT offers a convenient noncontacting charging solution for everyday equipment, from low-power consumer electronics to implanted medical devices to high-power industrial applications such as EVs and public transport. The main function of WPT is to allow electrical devices to be continuously charged and avoid the constraint of a wiring. WPT was initially practiced by Tesla in 1894 to wirelessly light up phosphorescent and incandescent lamps; however, efficiency and safety have prohibited WPT from real-life applications. Compared with conventional plugged-in charging, WPT brings a number of obvious benefits, including safer charging, enclosed design, a better user experience (UX), and avoidance of messy charging cables (Song et al., 2020) and without the hassle of replacing or disposing of batteries. Based on various EM field operating principles, WPT can be realized by nonradiative (inductive and capacitive) and radiative (RF, Infrared, laser) power transfer, respectively.

Once WPT advances, the range of applications will grow. The range of the transmitted EM fields will increase, from few meters to entire homes where a single transmitting hub can charge the entire range of home devices. Usually, devices that want to receive power from the hub announce their presence via Bluetooth connection. The hub then uses that connection to direct the wireless power signal to the device. It operates in the same unlicensed frequencies as a typical WiFi system. The Bluetooth connection focuses a tight cone of EM signal on the device, and it can even track the device and recalibrate as required.

From sustainability point of view, WPT has no harmful effect on the environment, but by using wireless technology the environment will be CO_2 close to free. However, safety of possible free flowing EM energy and its health impact on the human body is a matter of concern.

4.7.3 WiFi 6

With IPv6 rollout, the limitation on devices was virtually removed and billions of billions of new IP addresses became available for devices entering worldwide. Although the transition to IPv6 will be a slow process, this opens the door for better WiFi as a short-range premises signaling method and protocol which acts as a corresponding technology to cellular coverage in a building. IPv6 is perfect for IoT environment, the fact that improves communications within buildings.

WiFi 6, or IEEE802.11ax, is a short-range wireless access technology at frequency bands of 2.4, 5, and 6 GHz. Unlike its predecessor, WiFi 5 (IEEE802.11 ac), the standard can support up to 12 simultaneous user streams from a single WiFi access point, 8×8 multiuser MIMO for both uplink and downlink. It promises faster speeds (up to 10 Gbps), provides better coverage, reduces congestion in WiFi networks, reduces congestion in WiFi networks, and provides more efficient use of bandwidth (better traffic management and more connected devices). WiFi 6 routers employ several new technologies that are designed to boost overall performance by offering increased throughput speeds; theoretically, compared with max speeds of around 3 Gbps. Also, it aims to relieve network congestion, provide greater client capacity, and reduce client power consumption.

Today, most in-building wireless is designed around the requirements of WiFi for laptops and smartphones. Depending on the building, the locations are predictable as well. In essence, capacity is most needed in places where people walk, sit, or gather. With 5G and related developments, this assumption has to change significantly. Network designers have to consider many more user types, applications, devices, and usage locations. Many of the efficiency enhancements for WiFi 6 maintain great promise for future IoT devices where it sustains a high density of users and devices, with an emphasis on a quality user experience. Examples include high-density outdoor environments, outdoor mobility, and building-to-building connectivity.

4.8 Tactile internet (TI)

TI has been defined by the IEEE P1918.1 TI ad-hoc definition group as "a network or network of networks for remotely accessing, perceiving, manipulating or controlling real or virtual objects or processes in perceived real-time by humans or machines." It refers to communication networks that are capable of delivering real-time control, touch, and sensing/actuation information through adequately reliable, responsive, and intelligent connectivity. It is the next evolution of the IoT, incorporating human-to-machine (H2M) and machine-to-machine (M2M) interaction. At the application level, automation, robotics, telepresence, augmented reality (AR), virtual reality (VR), healthcare, education, smart grid, and AI will all exist.

TI will combine multiple technologies, both at the network and application level. At the edges, the TI will be enabled by the IoT and the Internet (Fig. 4.6). The major challenge with TI is having a seamless network and a delay that humans do not perceive. To achieve this, low latency of 1 ms is required- this is the primary requirement for tactile internet. This was projected to be achievable in 5G. However, tests have shown existing 5G networks to have a latency of 10–16 ms. Therefore, it is expected that 6G wireless networks, which are proposed to have a latency of less than 1 ms, will enable TI and haptic communication. Additionally, precision with very high accuracy is desirable and will be facilitated by 6G (Imoize et al., 2021).

Tactile Internet will facilitate real-time interactive systems with several industrial, societal, and business use cases. In principle, all human senses can interact with machines, and technologies to support and improve that interaction will produce the TI. The Tactile Internet will enable haptic interaction with visual feedback involving robotic systems to be controlled with a subtle time-lag (ITU, 2014). In a TI-enabled CPS, machines, and devices can be controlled remotely, which requires high accuracy and quick responses.

The connectivity of vehicles in the TI will enable cooperative traffic modes, where traffic flow will be optimized by heeding local safety constraints as well as parameters such as the overall traffic density in a smart city. TI should enable Internet protocols over the next generation of empowered devices to achieve convergence and end-to-end transparency through IPv6 (Aijaz and Sooriyabandara, 2019). Guided autonomous driving will allow for a continuous traffic flow in which safety and energy efficiency will be significantly improved as compared to today's situation.

In decentralized electricity grids, the TI enables dynamic activation and deactivation of local power generation and consumption, potentially even taking into account the AC phase information to minimize the generation of unusable reactive power. The TI is the technical basis for smart grids, providing for improved energy efficiency and stability in electricity grids.

The technical requirements of the TI place exceptional demands on future networks' latency and reliability, security, system architecture, sensors and actuators, access networks, and mobile edge clouds (ITU, 2014). The main challenge in the

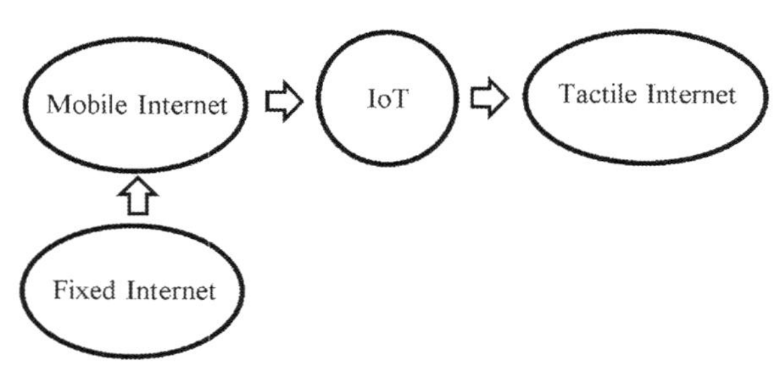

Fig. 4.6 Movement toward Tactile Internet.

realization of the TI is the propagation characteristics that tend to favor "line of sight" connections. This is hard enough outdoors, with trees, vehicles, and even snow in the way. Indoors, this is going to be exceptionally complex, even with new beam-forming/beam-steering techniques, and the potential to reflect off some surfaces.

The most common way of evaluating TI applications is through service quality and the user's experience. Service quality that is termed the QoS is characterized by latency, availability, and reliability (e.g., packet loss). 5G in general and 6G, in particular, provides better QoS and enables novel use cases, such as intelligent transportation systems (ITS) and smart grid in an industrial environment (Aazam et al., 2019). Another anticipated feature of TI is security where blockchain is a proposed enabling technology.

4.9 Internet-of-things (IoT)

Kevin Ashton the then executive director of the Auto-ID Center coined the term IoT in 1999 while working on a presentation for Procter and Gamble. IoT is the notion that every device or object in the environment can be assigned an Internet protocol (IP) address and is directly connected to the Internet. Such devices require only small batteries or no batteries at all because their low electricity needs can be obtained from their environments using motion, light, heat, sound, vibration, wind, etc. from energy transducers.

4.9.1 IoT platforms and gateways

The IoT bridges the gap between the physical and digital ecosystem to improve the quality and productivity of people, businesses, and industries. "Things" are generally considered any device with the ability to communicate information via a network connection, wired or wirelessly. A thing in the IoT should have a unique Internet address connection, a communication device that can send and receive information, built-in computer software that can process information, one or more sensors, and an actuator that can act in the physical environment.

IoT refers to an internetworking world in which various objects are embedded with electronic sensors, actuators, or other digital devices so that they can be networked and connected to collect and exchange data (Xia et al., 2012). In general, the IoT is a new communication and network paradigm, which can offer advanced connectivity of physical objects, systems, and services, enabling object-to-object communication and data sharing. Therefore, an ecosystem of increasingly complex and energy-hungry devices emerging is foreseen. The IoT predicts a future in which physical and digital things or devices.

The IoT is already having an impact on building automation, with the convergence of the IoT and IB via IoT applications which are beginning to be deployed today with 4G long-term evolution (LTE) technology. 5G will build on these LTE capabilities, empowering even higher scalability of device density. This is where 5G is separating itself from 4G LTE. Establishing the stage early in the architectural design process is critical if building the AEC community can help its clients realize the BIoT potential.

The proliferation of IoT related to buildings has caused a dramatic shift from a few building systems to an interconnected system of devices and sensors that can collect and share data within and across buildings. This, in turn, has created a large opportunity to increase the efficiency, security, productivity, occupancy comfort, and profitability of buildings as relevant IoT data can be analyzed remotely and actionable information can be generated.

The IoT helps create dynamic and intelligent cloud-based interoperable networks by connecting electrical, mechanical, and electro-mechanical systems and platforms. By communicating with each other, these systems can help monitor themselves and act when necessary to provide the data and analytics needed for facility managers to intelligently optimize performance and create smarter buildings. The IoT device deployment process typically requires a systems integrator or in-house electrician and IT network professional. The technology to enable this competitive edge is already at hand. The Internet and significant price reductions on IT components such as wireless sensors have made smart building technologies much more affordable, creating a strong business case for owners and investors to invest in more intelligent technologies to increase building performance.

The IoT is based on M2M, which refers to direct communication between physical devices using any means of communication, including wired and wireless communication. In this regard, IoT is the range of solutions that rely on broad, low-cost coverage to carry very small data messages from thousands of devices in a building. For example, integrating HVAC system sensors, occupancy sensors, and smart meters is a relatively typical function found in nearly all modern buildings.

In practice, mechatronics devices will be clustered in homes, buildings, cars, or other facilities and operate as wireless local area networks which may connect to the Internet via a "gateway." IoT gateways are an important middleman element that serves as the messenger and translator between the cloud and clusters of smart devices. They are physical devices or software programs that typically run from the field near the edge sensors and other devices. Large IoT systems might use a multitude of gateways to serve high volumes of edge nodes. They can provide a range of functionality, but most importantly they normalize, connect, and transfer data between the physical device layer and the cloud. All data moving between the cloud and the physical device layer goes through a gateway. IoT gateways are sometimes called "intelligent gateways" or "control tiers."

Today, gateways also support additional computing and peripheral functionality such as telemetry, multiple protocol translation, artificial intelligence, preprocessing and filtering massive raw sensor data sets, provisioning, and device management. It is becoming common practice to implement data encryption and security monitoring on the intelligent gateway to prevent malicious man-in-the-middle attacks against otherwise vulnerable IoT systems.

4.9.2 Building IoT (BIoT) smart structure

The challenge for BIoT is knowing what system data is significant, along with when and how often it should be acquired. Moreover, these individual system databases have historically not been built to work outside of their environment. Fig. 4.7 shows

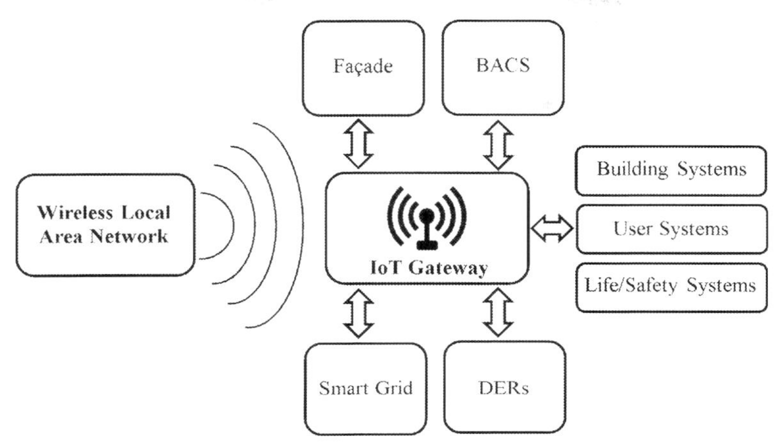

Fig. 4.7 BIoT communication system.

a basic illustration of BIoT smart architecture. BIoT gateways act as the "central nervous system" of the building, securely connecting to a variety of smart sensors that monitor the façade, building systems (HVAC, lighting, water and waste, etc.), user systems (wireless assets, cameras, and control access, elevators, etc.), smart grid and DERs, and environmental detection and protection. Essentially, the interoperability and data integration of the entire solution is facilitated by this IoT gateway. It is easy to implement and still basically demonstrates the communication capabilities of automated façade and systems.

The BIoT brings into focus all the components in a building that could be connected to the Internet to create operational efficiencies, reduce energy consumption, improve occupant experiences, achieve sustainability goals, and effectively optimize financial performance (Badrick, 2019). A concerted data system approach to a BIoT is the key to success for building managers who want to be able to get significant data out of their building façade and systems in a way that can be analyzed.

HVAC is an important system that can evolve through BIoT. Motion and temperature sensors connected through IoT can be placed across the building to ensure that air conditioning and heating are only utilized when occupied and are even capable of integrating weather forecasts to program for maximum efficiency and savings. Critical information can be collected on the building envelope and used to notify managers to build leaks or indicate filters in need of replacement.

A building may be equipped with a lighting upgrade that includes LED fixtures and zone control. This provides an energy-efficient solution with improved light levels and control. But if the system is connected to the network through IoT, the lighting control system can also share valuable data with the electrical distribution, helping to increase each system's functionality and overall performance. Data from electricity usage may provide insight for alteration of regular schedules to replace or fixtures, and detect faulty controls. In addition, intelligent window designs provided with BIoT can distinguish natural lighting and consequentially shut artificial lighting for energy saving.

BIoT can stabilize DERs, water and waste, plumbing systems, environmental and other equipment as buildings constantly struggle with fluctuations in services. Innovative ML tools allow digital utility algorithms to quickly detect faults of any type or size and afford continuous monitoring. BIoT provides enhanced intelligence from building behaviors and systems and learns from external factors, such as weather forecasts and occupancy levels. Once different systems of a building are supplied with sensors and begin to deliver data, the incoming information can be integrated into BAS. This will allow the building to react to real-time scenarios and conditions as per the given control criteria. Challenging features of BIoT involve suitable and scalable sensor networks and devices with robust and secure architectures, open data setups and interfaces, and dependencies that are feasible in the rough and dynamic built environment.

4.10 Wireless sensor and actuator networks (WSAN)

A WSAN is a collection of sensors that accumulate information about the environment and actuators, such as motors, that interact with them. They consist of networks comprised of devices with sensing, processing, storage, and wireless communication capabilities. All components communicate wirelessly and the interaction can be autonomous or human-controlled. Fig. 4.8 shows WSANs as a part of the field level of the building automation and control system (BACS). Each node of the network may

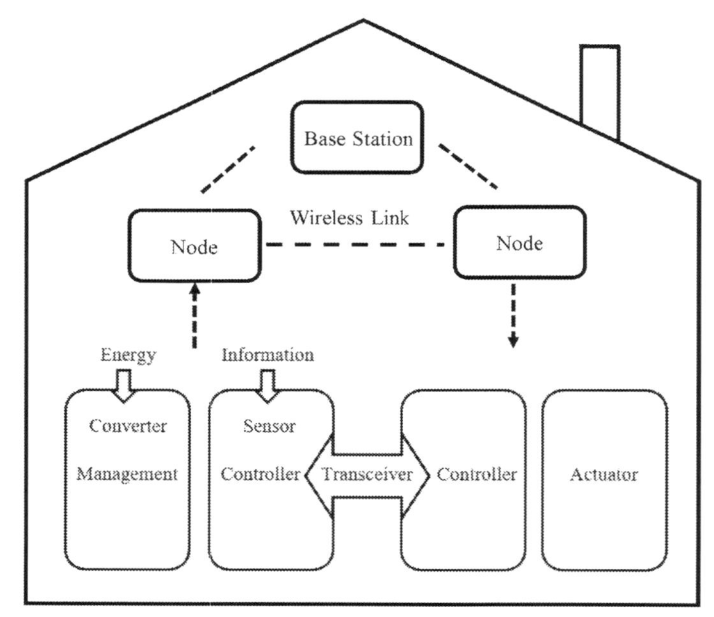

Fig. 4.8 Wireless sensor and actuator networks as a part of the field level of the building automation and control system.

have several sensing units, which can measure physical variables, such as temperature, humidity, and vibration to record and react to an event or a phenomenon. The nodes in a WSAN have limited computing resources and are usually powered by batteries or can be self-powered.

There is presently a wide range of applications for WSANs, ranging from environmental monitoring to building envelop damage detection. WSANs are used in places where exact measurement and control of an environment are necessary. In such a situation, autonomous processing of data, response, and control allow for actors to input change instantly. WSANs should be able to self-manage failures and dynamically self-adapt to the environment. WSANs are an element of the increasing development of the IoT with the ability to transmit data over the network.

A sensor may be broken down into three elements. Sensing element, which is a component that undergoes a measurable change, such as voltage or electrical resistance, in response to a change in the variable to be measured. Second, the transducer, which is an active device that produces an electrical signal which is a function of the change in the sensing element. And third, a transmitter, which is a device that produces an electrical signal that is a standardized function of the change in the physical variable and which can be used as an input to the control module (Wang, 2010). The functions of the transducer and transmitter are often combined, which may be referred to as signal conditioning. This includes filtering to remove noise, averaging over time, or linearization.

Intelligent sensors represent a rapidly expanding area of research. They are incorporated with distributed signal processing functions in the lower layer of the sensing system hierarchy. These sensors are distinguished from other sensors by their computational processing potential. They can convert analog signals into digitally encoded signals for direct communication over a network for onward transmission to other intelligent devices for control and measurement purposes. They can process, analyze, and transmit this data to locations across a network for use in various applications. The function of sensor intelligence can vary between "think for itself," or be part of a "think as a group" methodology in which sensors and a central data analyzer perform together.

An actuator responds to the output signal from a controller and provides the mechanical action to operate the final control device, which is typically a valve, damper, or switch. A wide range of actuators is available and the chosen actuator must address the following concerns: matching the mechanical requirements of the controlled device; matching the characteristics of the control system, especially the output signal of the controller; being suitable for its operating environment.

Underlying the IoT is WSANs, where sensors facilitate the collection of information about surrounding physical environments over extended periods, while actuators take decisions and then perform appropriate actions upon the environment, which allows remote, automated interaction with the environment. A significant component of a thing in the IoT is a microcontroller or microprocessor that can perform software instructions. Another key component is the IPv6 protocol that has a central role in managing all the things that are connected to the Internet. Today, the geometric expansion of the network is based on the adoption of a new IP version 6 (IPv6), which is replacing IPv4, and boosts the Internet packet address from a large (32-bit) address space of 232 bits or 4,294,967,296 unique addresses to 3.4 × 1038, or 340 trillion trillion trillion, unique addresses in the future. When linking sensors and actuators in a system, the most common principle to be employed is feedback. In a feedback loop, the result of an actuator is measured, and the sensor reading is used to evaluate the effect of the actuator. Based on the outcome, the actuator operation is adjusted.

4.11 Internet of energy (IoE)

The IoE is the result of the implementation of IoT technology with distributed energy systems. Its purpose is to optimize the efficiency of the generation, transmission, and utilization of electricity. It involves the integration of WSANs, smart meters, integrated computing systems, and other components of the power grid together with ICT. The main goal of the IoE is to build hardware, software, and middleware for seamless, secure connectivity and interoperability achieved by connecting the Internet with smart energy grids, automated vehicles, and other utilities. IoE uses the bidirectional flow of energy and information within the smart grid to gain deep insights on power usage and predicts future actions to increase energy efficiency and low overall cost. Therefore, the applications of IoE are increasingly being used in a variety of fields such as BAS, EVs, DERs, and some other applications. The architecture of the IoE is depicted in Fig. 4.9.

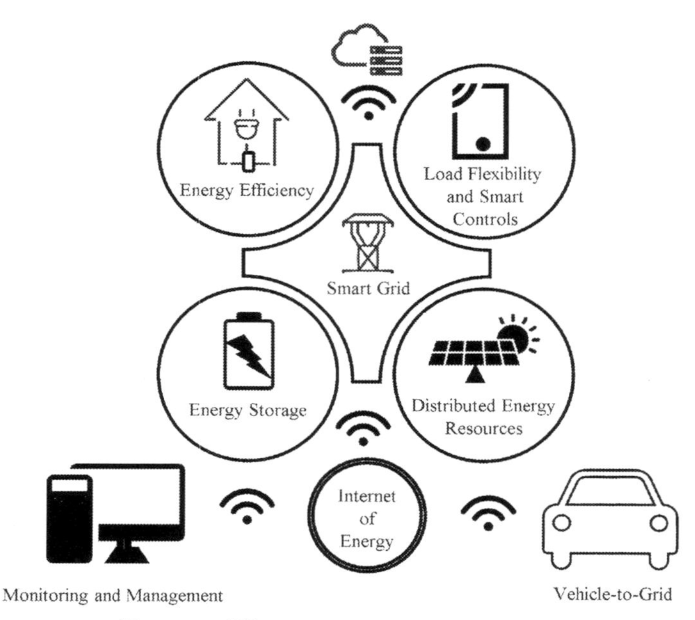

Fig. 4.9 Components of Internet of Energy.

IoE introduces the energy industry to the world of digitization, decarbonization, optimization, reliability, and scalability, bringing in immense possibilities for a greener and sustainable environment. It will enable units of energy to be transported across the globe (Joshi, 2019). An enormous amount of data is gathered from these various connected intelligent energy devices, revealing patterns and insights to steer the cycle of power production, distribution, and consumption. In particular, data about energy consumption in all sectors including building, automation, and telecommunication sectors is smartly used to reduce through the IoE management system, which oversees line loss, leakage, and shutdown.

The IoE breathes light into smart architecture. It showcases its potential to control energy consumption levels. From automatic lighting and temperature controls to HVAC system and air ventilation balance, the IoE not only helps to reduce energy consumption but also plays an active role in financial savings (Joshi, 2019). The IoE features distributed energy generation, demand-side response, and interconnected storage networks. With a growing focus on decarbonization, renewable energy would be an integral component of the IoE, thereby leading to distributed energy suppliers. For stable operation of the power grid, it is essential to balance total generation and total demand. Therefore, it becomes particularly important to have dynamic control over distributed energy suppliers. Moreover, synchronization of suppliers is necessary as out-of-phase power injection results in reactive power, which cannot be used (Aijaz and Sooriyabandara, 2019). Such synchronous cophasing of distributed energy suppliers requires an end-to-end latency on the order of a single-digit millisecond, which can be ensured through the ultra-responsive connectivity provided by the Tactile Internet.

BAS using IoE helps to reduce the building energy consumption by providing information on energy demand and supply. Technologies like AI, ML, and blockchain should be leveraged for acquiring the data that will be generated. Such a collaborative effort will help utilities to build the entire infrastructure needed for making the IoE a perfect experience.

4.12 Smart urban mobility

Urban mobility is about the question of how people and goods can move around effectively and sustainably in densely populated areas in the built environment. As populations enlarge, current and evolving cities and communities face the challenge of meeting rising demands for efficient mobility within limited physical infrastructure capability. Urban population growth will continue to strain urban mobility; however, the new mobility services, enabled by developments in digital technology, are rapidly expanding across many of the world's largest cities. Such smart solutions are needed, not only to minimize the traffic but also to limit environmental pollution and maintain air quality.

Cities and communities everywhere are concerned about environmental sustainability, and transportation is a major contributing factor. Fueled by population growth, urbanization, and a shortfall of investment in public infrastructure, congestion has a deleterious impact on urban life. While most cities remain relatively small, more than 500 of them are home to at least 1 million people. By 2030, there could be 41 "mega-cities" with populations of more than 10 million, there are already 11 such cities in China and India alone (Corwin and Pankratz, 2019).

Road transportation is a leading cause of air pollution. An estimated 22% of global CO_2 emissions are attributed to the transportation sector, which will rise as cities grow and individual citizens adopt carbon-burning transportation. Exhaust particulates are capable of penetrating deep into lung passageways and entering the bloodstream causing cardiovascular and respiratory impacts. In major US urban areas, the annual health costs of congestion exceed US$30 billion (Levy et al., 2010).

The concept of smart urban mobility has been introduced in recent years as a part of the digital ecosystem to describe various technological developments and produce an efficient and more sustainable infrastructure. To harness emerging technologies to solve the most challenging problems, cities would need a comprehensive and integrated system that transcends existing infrastructure, drives standardization, and enables value creation by key parties, and cultivates technological advancements. This is an integrated platform that brings together physical engineered infrastructure (roads, rails, parking, transit hubs), modes of transport (cars, public transit, bike sharing, etc.), ITS, digital infrastructure, IT, and transportation service providers (Fig. 4.10). Data is collected and grouped by public and private systems engaged in data processing analytics, modeling, simulation, and software.

Integrated mobility systems utilize advances in ICT and the proliferation of 5G technology to present multimodal transportation options to users via a single online platform. A simple incarnation of these systems includes online journey planners that can plan commuter journeys across a city using a range of transport options under the

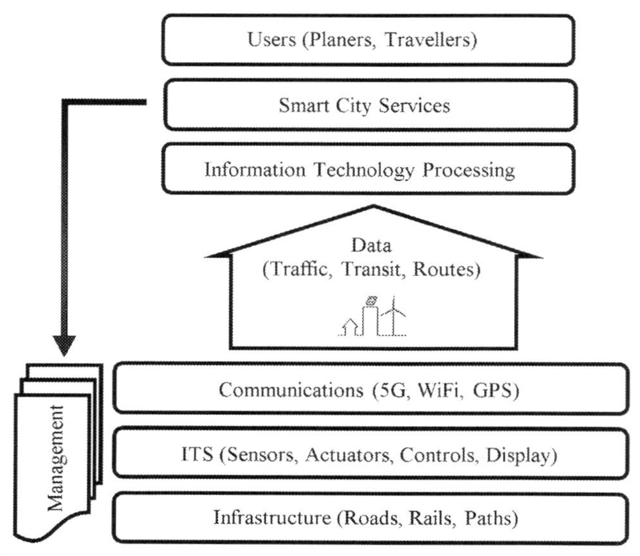

Fig. 4.10 Major components of an urban mobility scenario.

control of multiple mobility providers (Golub et al., 2019). The growing availability of real-time traffic data and the full-scale embracing of smartphones are establishing new opportunities for seamless multimodal urban mobility services including connecting shared mobility, public transportation, autonomous cars, and bikes.

In particular, ITS utilizes the latest computing and improves the overall efficiency and safety of the mobility system. The primary opportunities of such innovation relate to improvements in in-vehicle assistance systems that can remove driver distraction and reduce human error. Improved energy efficiency resulting from better traffic flow, alternative fuel engines, and the use of integrated systems to promote sustainable transport is another opportunity.

From the perspective of urban mobility, smart implies a connected, technology-enabled environment to manage city services and offer a range of benefits to advance urban sustainability goals. Sustainability implies social connectivity and economic wealth reasonably and impartially, without causing risks to local or universal environmental quality and resource use (Buscher et al., 2019). Effective mobility has to be safe, accessible, fast, and efficient. Through smart and green urban mobility, all four can be achieved. By enhancing UX public transportation can be made an easier option than private transportation. By mapping out the most efficient paths, traffic congestion can be reduced. With less congestion, the amount of pollution from the emissions will be limited. Such healthy mobility encourages the use of smart vehicles driven by cleaner energy sources such as electric batteries.

4.13 Net-zero energy buildings (NZEB)

Buildings account for 40% of the primary energy consumption, 74% of the electricity consumption, and 39% of the CO_2 emissions in the US (NIST, 2021), and probably a similar range worldwide. Energy use is associated with about $400 B in all kinds of buildings while the energy efficiency and IEQ sectors account for millions of jobs. Therefore, reducing energy consumption and improving IEQ represent a challenge and opportunity by revolving building design around the NZEB initiatives, which are growing worldwide.

4.13.1 The path to net-zero through sustainability

For all buildings in operation, the term "net-zero" represents "zero-energy" and/or "zero-carbon," signifying a balance of energy efficiency with clean energy resources. It is often the result of energy performance calculations. In addition, the whole life asset management emissions of the building should be considered. This means when the carbon emissions associated with a building's embodied and operational energy impacts over the entire life of the building, including its disposal, are zero or negative." Currently, whole-life carbon is not considered as a criterion at present due to limitations in the reporting of carbon from the repair and end-of-life stages of a building's lifecycle.

NZEBs are becoming the mainstream quality in the AEC industry. They are proven to be cost-effective, healthy, and energy-efficient. Rating systems such as LEED, BREAAM, WELL, and other certification and rating systems are the best way to integrate energy efficiency and health aspects for user-centered NZEB (Attia, 2018). NZEB may be achievable through climate-responsive design and reduce loads on the building based on efficient electric power systems, improving power quality, using renewable resources, and employing smart loads (Habash et al., 2014). Another innovative approach is the use of intelligent energy management such as advanced sensors, zone energy control, and monitoring systems.

The objective of NZEBs is not only to minimize the energy consumption of the building's passive design methods but also to design a building that balances energy requirements with active energy efficiency techniques including intelligent ventilation, thermal storage, demand response, continuous monitoring using power measurement devices, and software as well as integrating renewable energy sources. Realizing zero energy is an aspiring yet gradually reachable aim that is gaining thrust across the world. The best way to achieve the above is to design the building to operate as efficiently as possible before even thinking about adding an on-site renewable energy source. While NZEBs have a higher initial cost than standard construction, they make up for it through energy conservation and lower (or no) utility bills over the lifetime of the facility.

The less energy the building needs to operate, the less renewable energy it will need to produce on-site. This may be achieved by choosing efficient HVAC, lighting, and plug loads as well as maintaining effective passive design approaches. On-site renewable energy may be exported through transmission means other than the electricity grid such as charging electric vehicles used outside the building. Delivered energy to the building includes grid electricity, district heat, and cooling, renewable and non-renewable fuels (DOE, 2015).

Besides economic and environmental advantages, NZEB entails a positive impact on society as well. Most of the benefits are related to the health of the occupants of the building. Therefore, NZEB not only focuses on the environmental perspective but also aims to provide a happy and healthy lifestyle. Buildings of the future, whether being NZEB or not, should be designed and implemented with both active and passive elements and technologies to save energy, enhance IEQ, advance interaction with the smart grid utility, and promote sustainability.

Since the beginning of 2021, all new buildings constructed within the EU must be nearly zero-energy buildings, according to Article 9 of the EU Energy Performance of Buildings Directive 2010/31/EU (EPBD).

4.13.2 System structure and key elements

An NZEB is regarded as an integrated solution to address problems of energy-saving, environmental protection, and CO_2 emission reduction in the building section (Deng et al., 2014). It is typically an energy-balanced grid-connected building where the annual energy production is almost equal to the annual energy consumption. Therefore, the building is its energy source and does not rely on the power grid although it is connected to the grid. The NZEB is a system of building service loads and on-site

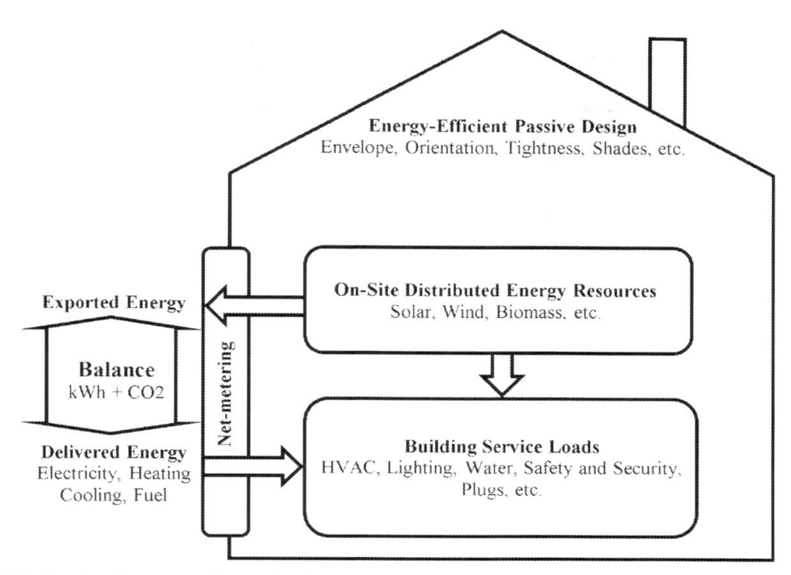

Fig. 4.11 Basic elements of a typical net-zero energy building.

DERs that works in synergy with the utility grid, avoiding the addition of stress on the power infrastructure. ICT and smart grid technologies are the keys to achieve the aforementioned zero energy goals (Kolokotsa, 2016). Fig. 4.11 shows the basic elements of a typical NZEB.

The initial consideration toward NZEB addresses the building as a whole system, evaluating IEQ, water quality, energy consumption, cost-effectiveness, and overall sustainability. Building envelope load reduction includes heating, cooling, and outdoor air ventilation is the second consideration (NIST, 2021). The volume of energy consumption is decreased by appropriate energy-efficient passive design approaches including insulation, thermal mass, orientation, tightness, shading, and space conditioning are needed to manage heat loss or gain through the building envelope and to control the concentrations of airborne contaminants within the building. Controlling outdoor ventilation in buildings is important as it impacts indoor IAQ and energy. Once building loads are reduced, the next is through the use of energy-efficient equipment and on-site DER generation. The gain is especially true when energy efficiency and integration of DERs are selected early in the design process. Using energy modeling software can help eradicate human error by recognizing the least expensive factors to build energy-efficient buildings.

4.13.3 Performance indicators

NZEB is an ambitious goal, to achieve it, there is a need to understand the actual performance of buildings considering that comfort and energy are tightly linked. Performance measures are being developed and integrated into both government policies and in the design of buildings, and the specifications of systems and materials.

A performance measure for an NZEB may be the real-time performance indicator that could be measured using a smart metering solution inside the building as well as being smart in how to operate and manage the building (Hubbard, 2018). NZEB relies on occupants' understanding of energy consumption patterns and knowing how to reduce energy use to meet the overarching goal. The psychology of designing an NZEB to promote sustainability is a significant factor that is based on user experience (UX) principles.

An NZEB balances its energy use so that the exported energy to the grid or other energy network is equal to or greater than the delivered energy to the building on an annual basis. To make a clear balance calculation for the net-zero goal, a boundary needs to be clarified for building loads with on-site renewable. The building systems consume delivered energy, from the utility grid and on-site DERs, and output energy back to the grid when the DERs generate excess electricity. The NZEB implies that the energy demand for electrical power is reduced, and this reduced demand is met on an annual basis from renewable energy supply which can be either integrated into the building design or provided, for example, as part of a community renewable energy supply system (Habash et al., 2014).

While energy consumption is a key operating cost, it is vital to also ensure comfortable and healthy indoor environments for occupants. This can be considered as a precondition indicator for the NZEB evaluation. It includes all elements of IEQ and quality of life (Deng et al., 2014). This requires direct human involvement in the operations process to meet living comfort goals.

ICTs and smart grid implementation are the keys to achieve the aforementioned zero energy goals. ICTs for energy management in buildings have evolved considerably over the last decades leading to a better understanding of IBs. Advances in the design, operation optimization, and control of energy-influencing building elements (e.g., HVAC, solar, fuel cells, CHP, shading, natural ventilation, etc.) unleashed the potential for the realization of significant energy savings and efficiencies in the operation of both new and existing dwellings worldwide (Kolokotsa, 2016).

LCA is not a performance indicator but is a research method used for the quantitative assessment of material used, energy flows, and environmental impacts of products. It has been widely applied in the building industry to determine top design priorities and quantitatively inform sustainable design decision-making for various buildings (Deng et al., 2014). A mathematical equation of LCA NZEB definition is

$$\begin{aligned}\text{Net Energy} &= \text{Output} - \text{Input} - \text{Embodied Energy} \\ &= \text{Net Operational Energy} - \text{Embodied Energy}\end{aligned} \tag{4.1}$$

According to the US DoE (DOE, 2015), actual projects seeking to verify zero energy should work to ensure no harm is done in the process of achieving zero energy performance across other, nonenergy–related considerations, such as water protection, optimized comfort for low-load buildings, and comprehensive IAQ. While these considerations do not affect the definition of zero energy, it is important in practice to ensure implement them but not by sacrificing the pursuit of NZEB.

References

Aazam, M., Harras, K.A., Zeadally, S., 2019. Fog computing for 5G tactile industrial internet of things: QoE-aware resource allocation model. IEEE Trans. Ind. Inf. 15 (5), 3085–3092.
Aijaz, A., Sooriyabandara, M., 2019. The tactile internet for industries: a review. Proc. IEEE 107 (2), 414–435.
Amber, K.P., Ahmad, R., Aslam, M.W., Kousar, A., Usman, M., Khan, M.S., 2018. Intelligent techniques for forecasting electricity consumption of buildings. Energy 157, 886–893.
Attia, S., 2018. Net-Zero Energy Buildings (NZEB): Concepts, Frameworks and Roadmap for Project Analysis and Implementation. Elsevier, Oxford, UK.
Badrick, C., 2019. What Is Building Internet-of-Things (BIoT). https://www.turn-key technologies.com/blog/article/what-is-the-building-internet-of-things-biot/.
Bartusch, C., Alvehag, K., 2014. Further exploring the potential of residential demand response programs in electricity distribution. Appl. Energy 125, 39–59.
Bonin, J.P., 2019. The New Intelligent Building: IoT and the Promise of 5G Networks. http://thesextantgroup.com/the-new-intelligent-building-iot-and-the-promise-of-5g-networks/.
Bradley, P., Leach, M., Torriti, J., 2013. A review of the costs and benefits of demand response for electricity in the UK. Energy Policy 52, 312–327.
Brenner, D.J., Doll, R., Goodhead, D.T., Hall, E.J., Land, C.E., Little, G.B., Lubin, J.H., Preston, D.L., Preston, R.J., Puskin, J.S., Ron, E., Sachs, R.K., Samet, J.M., Setlow, R.B., Zaider, M., 2003. Cancer risks attributable to low doses of ionizing radiation: assessing what we really know. Proc. Natl. Acad. Sci. 100 (24), 13761–13766.
Buscher, V., Doody, L., Webb, L., Aoun, C., 2019. Urban Mobility in the Smart City Age. https://smartcitiescouncil.com/system/tdf/public_resources/Urban%20mobility.pdf?file=1&type=node&id=1272&force.
Chen, L., Huang, Z.C., Wu, P., Liu, C., 2013. Transmission line conditionitoring based on internet of things in smart grid. Appl. Mech. Mater. 336–338, 2488–2493.
Corwin, S., Pankratz, D., 2019. Toward a Mobility Operating System. https://www2.deloitte.com/us/en/insights/focus/future-of-mobility/urban-transport-mobility-platforms.html.
Davis, G., 2003. Integration of Distributed Energy Resources: The CERTS Microgrid Concept. California Energy Commission. http://bnrg.eecs.berkeley.edu/~randy/Courses/CS294.F09/MicroGrid.pdf.
Deng, S., Wang, R.Z., Dai, Y.J., 2014. How to evaluate performance of net zero energy building – a literature research. Energy 71, 1–16.
DOE, 2015. A Common Definition for Zero Energy Buildings. https://www.energy.gov/sites/prod/files/2015/09/f26/bto_common_definition_zero_energy_buildings_093015.pdf.
Dunn, P., 2015. Measurement and Data Analysis for Engineering and Science. CRC Press, Boca Raton.
Golub, A., Satterfield, V., Serritella, M., Singh, J., Phillips, S., 2019. Assessing the barriers to equity in smart mobility systems: a case study of Portland Oregon. Case Stud. Transp. Policy 7 (4), 689–697.
Guyot, G., Sherman, M., Walker, I., 2017. Performance-based approaches to residential smart ventilation. REHVA J. 3, 9–11.
Habash, R., 2020. Bioelectromagnetics: Human Safety and Biomedical Applications. CRC Press Taylor and Francis, Boca Raton.
Habash, G., Chapotchkine, D., Fisher, P., Rancourt, A., Habash, R., Norris, W., 2014. Sustainable design of a nearly zero energy building facilitated by a smart microgrid. J. Renew. Energy 2014, 1–11. https://doi.org/10.1155/2014/725850.

Hubbard, G., 2018. Five Steps to Becoming a more Responsible Architect in the Age of Climate Change. https://archinect.com/features/article/150099103/five-steps-to-becoming-a-more-responsible-architect-in-the-age-of-climate-change.

Hussain, I., Ullah, M., Ullah, I., Bibi, A., Naeem, M., Singh, M., Singh, D., 2020. Optimizing energy consumption in the home energy management system via a bio-inspired dragonfly algorithm and the genetic algorithm. Electronics 9, 406. https://doi.org/10.3390/electronics9030406.

iBwave, 2020. 5G and In-building Wireless Coverage. https://www.ibwave.com/Marketing/Download/ebook-5G-in-building-wireless-convergence.pdf.

Imoize, A.L., Adedeji, O., Tandiya, N., Shetty, S., 2021. 6G enabled smart infrastructure for sustainable society: opportunities, challenges, and research roadmap. Sensors 21 (5), 1709. https://doi.org/10.3390/s21051709.

ITU, 2014. The Tactile Internet: ITU-T Technology Watch Report. https://www.itu.int/dms_pub/itu-t/oth/23/01/T23010000230001PDFE.pdf.

Joshi, N., 2019. Introducing the Internet of Energy. https://www.bbntimes.com/technology/introducing-the-internet-of-energy.

Kevitt, M., 2010. Distributed Energy Monitoring and Control Application for Intelligent Buildings (M.Sc. Thesis). University of Dublin, Ireland. https://www.scss.tcd.ie/publications/tech-reports/reports.10/TCD-CS-2010-24.pdf.

King, J., Perry, C., 2017. Smart Buildings: Using Smart Technology to Save Energy in Existing Buildings. https://www.aceee.org/sites/default/files/publications/researchreports/a1701.pdf.

Kolokotsa, D., 2016. The role of smart grids in the building sector. Energy Build. 116, 703–708. https://doi.org/10.1016/j.enbuild.2015.12.033. ISSN 0378-7788.

Levy, J.I., Buonocore, J.J., von Stackelberg, K., 2010. Evaluation of the public health impacts of traffic congestion: a health risk assessment. Environ. Health 9 (1), 65.

NIST, 2021. Commissioning Building Systems for Improved Energy Performance. https://www.nist.gov/programs-projects/commissioning-building-systems-improved-energy-performance.

Sechilariu, M., Wang, B., Locment, F., 2013. Building-integrated microgrid: advanced local energy management for forthcoming smart power grid communication. Energy Build. 58, 236–243.

Song, J., Liu, M., Wu, Y., Ma, C., 2020. Application of Wireless Power Transfer (WPT) in Smart Homes and Buildings. https://www.caba.org/wp-content/uploads/2020/11/WP_Application-of-Wireless-Power-Transfer-WPT-in-Smart-Homes-and-Buildings.pdf.

Wang, S., 2010. Intelligent Buildings and Building Automation. Spon Press, London.

Wang, Z., Srinivasan, R.S., 2017. A review of artificial intelligence based building energy use prediction: contrasting the capabilities of single and ensemble prediction models. Renew. Sust. Energ. Rev. 75, 796–808.

Wurtz, F., Delinchant, B., 2017. "Smart buildings" integrated in "smart grids": a key challenge for the energy transition by using physical models and optimization with a "human-in-the-loop" approach. C. R. Phys. 18 (7–8), 428–444.

Xia, F., Yang, L.T., Wang, L., Vinel, A., 2012. Internet of things. Int. J. Commun. Syst. 25 (9), 1101–1102.

Yang, J., Zhang, G., Ma, K., 2014. Matching supply with demand: a power control and real-time pricing approach. Int. J. Electr. Power Energy Syst. 61, 111–117.

Building as a human-cyber-physical system

Riadh Habash
School of Electrical Engineering and Computer Science, University of Ottawa, Ottawa, ON, Canada

A house is a machine for living in.

Le Corbusier

5.1 Digital transformation

Digital transformation optimizes systems and processes to make workflows easier and more efficient. Therefore, infrastructure, services, and solutions are needed to seize and exploit data that is adaptable, intelligent, and responsive to underpin the shift to an era of digital sustainability and circular economy. This culture of transformation requires intelligent efficiency which is all about looking at the whole picture or using a holistic approach to design services through technology-centered for people-centered efficiency. Both approaches consist of a more automated approach with many contributing components, such as sensors, smart software algorithms, monitors, controls, and smart technologies (Elliott et al., 2012). Calling this new wave of developments intelligent does not mean traditional efficiency efforts or measures have less value, rather, intelligence means those efforts or measures are "adaptive, anticipatory, and networked."

The human-cyber-physical system (HCPS) generates two new subsystems: CPS and the human-cyber system (HCS) (Ji et al., 2019). The CPS theory has been employed as a core technology of Industry 4.0 in Germany, which is generally powered by the openings created by current advancements in AI, IoT, and big data analytics. Current trends suggest that next-generation industrial systems will distinguish themselves from their predecessors due to increased use of robotics and automation, ubiquitous wireless connectivity, and integration of virtualization technologies (Aijaz and Sooriyabandara, 2019). This generation will provide greater smartness and intelligence in the building sector. The evolution of the above-mentioned vision is highlighted in Fig. 5.1. The CPS that was summarized in Industry 4.0 has evolved into Industry 5.0 and is fundamentally addressing the relationship between humans and machines. Autonomous and collaborative systems operate vigorously under ICT expanded to include energy conservation networks, smart transportation, drones, smart farming, etc. Society 5.0 recognizes three components that drive social innovation into a super-smart society: data, information, and knowledge. In this regard, a knowledge-intensive society is a significant aspect of Society 5.0.

Sustainability and Health in Intelligent Buildings. https://doi.org/10.1016/B978-0-323-98826-1.00005-3

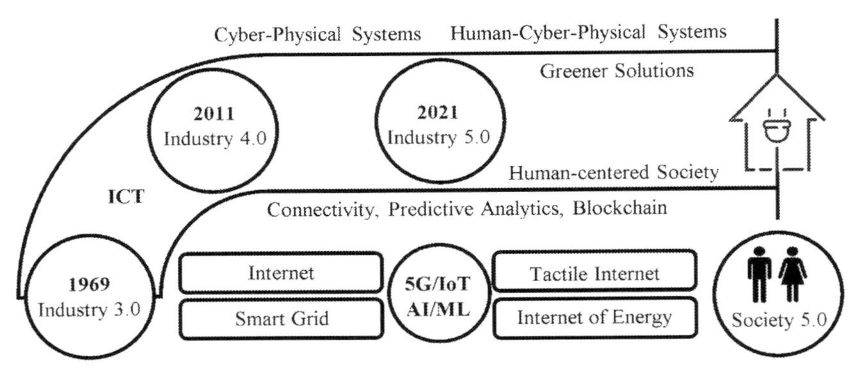

Fig. 5.1 Evolution of technologies toward a smart society.

Building information modeling (BIM) and digital twin are leading examples of digital transformation. They enhance efficiency and exhibit how digital transformation can radically change existing ways of operating. Beyond BIM and the digital twin, the next wave of digital adoption sees the AEC community employing AIoTs. The digital information collected by sensors, machines, and devices making up the IoT platform, is analyzed and contextualized by AI technologies then presented to users in a more meaningful way. Within an entire system, the process requires M2M communication and integration. This can be accomplished through the use of cloud communications technology to make instant calculations, anticipate changes, and adjust controls based on predictions. Through predictive optimization, digital automation, cloud storage, and AIoT solutions, a BAS can use variables such as real-time weather, energy grid, and occupant comfort information, to constantly predict situations and make consistent adjustments.

With the proliferation of AIoT technologies embedded controllers, building control systems are becoming more intelligent, offering a strong opportunity for the AEC community to contribute to creating solutions for occupant comfort and convenience, energy efficiency, and building automation. With a technology-centered approach, human interaction with the management system is minimal, with daily functions being handled automatically after the initial setup of the system. This reduces the burden on a building's operation and maintenance mitigates CPS risks and provides the convenience of use cases, where the common thread sprinting through these cases is the building occupants. This emerging approach to energy efficiency is information-driven with a level of intelligence that increases every day where various building systems will interact autonomously with each other and with other outside systems.

Real intelligence entails the ability to automatically adjust operating parameters interactively between smart elements to optimize building functionality. Optimization will benefit the major elements of the IB including building, user, safety and security, management, and control systems. However, an IB is more than the sum of its smart elements but a design task of resource-efficient buildings with high response to occupant needs. The implementation of IT enables intelligent control of individual components, but also their decentralized interlinking toward a networked system.

The implementation of IT enables control of individual components as intelligent mechatronics evolves with more effective microcomputers and the software that commands the process. Novel innovations in AI and digital technologies will help an era of change that will turn the construction industry into one of the most information-intensive industries, like one that requires extensive exchange of data and information between the project's participants regularly.

5.2 Digital sustainability

Sustainability in construction can be about how to build, and how the building uses and saves energy over time. On the other hand, digital sustainability is a universal phenomenon that integrates digital technology into all professions to transform the way value is transferred to customers and users. It is disrupting the old culture and creating new forms of professional logic. Digital sustainability gives professionals the tools to achieve sustainability goals. The emerging changes can be summed up as a movement toward a digitized networked society. By incorporating a digital mindset, processes can be automated to pull better insight from the data and enable teams to concentrate on priorities. Managing the above transformation involves several domains including customer, competition, data, innovation, and value. These domains may be conceptualized at the intersection of big data and sustainability under three challenges: digital twin, supply chain sustainability, 3D printing, and robotics.

5.2.1 Intelligent digital twin

The idea of using "twins" comes from the National Aeronautics and Space Administration (NASA)'s Apollo space program during the 1960s. The concept was first voiced in 1991, with the publication of Mirror Worlds, by David Gelernter. However, Dr. Michael Grieves, then on faculty at the University of Michigan, is credited with the first application to manufacturing in 2002 and formally announcing the digital twin software concept (Miskinis, 2018). The digital twin concept continued further where Gartner named it as one of the top 10 strategic technological developments in 2017. Since then, the concept has been employed in several industrial applications and practices.

Digital twins are symptomatic of the broader trend toward digitalization that is having a profound effect on businesses and society. The global digital twin market was valued at \$3.8bn in 2019 and is expected to reach \$35.8bn by 2025 (Evans et al., 2020). A digital twin is a virtual environment designed to accurately reflect physical objects such as people, devices, processes, to complex environments, such as buildings that help to make model-driven decisions. As the name indicates the twin means two connected objects, one is physical, the other is its digital replica. The digital twin is a virtual representation of a physical asset in a CPS, capable of reflecting its static and dynamic characteristics. To stem solutions for the physical space, cyberspace must have a formation mirroring that of the real world (Fig. 5.2). The digital twin architecture consists of three layers including physical, cyber, and applications.

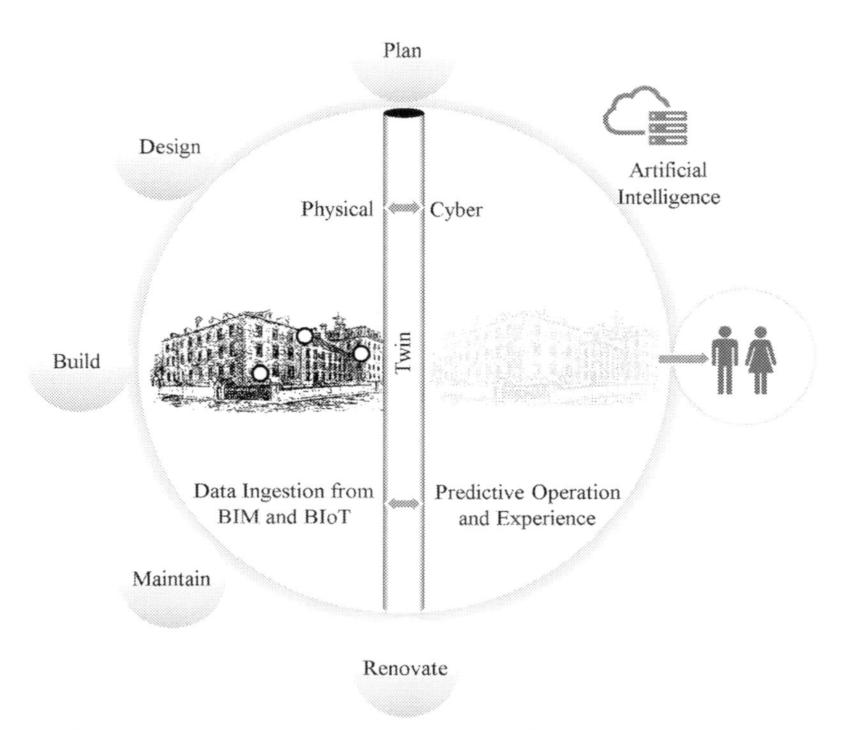

Fig. 5.2 Digital twin within an intelligent human-cyber-physical system.

The cyber layer is fortified with strong intelligence by AI, thereby enabling major capacities like CPS and HCS.

As IoT sensors become smaller and more affordable, the ability to gather, process, and communicate information increases, making the interface between the two worlds invaluable (Evans et al., 2020). The building management software collects together dynamic and static data from multiple sources in 3D models, providing valuable insights and diagnosis into various physical activities. The 3D modeling technology as a digital twin building management solution holds great potential for sustainable operations, helping managers reduce environmental impact (Martynova, 2020). By dynamically integrating data and information throughout the asset lifecycle, they will offer short and long-term efficiency and productivity gains. More than just BIM or a 3D model, the twin is a data (digital thread) resource that can improve the design of new buildings and understanding of existing building conditions, verify the as-built situation, run simulations and scenarios, or provide a digital snapshot for future works.

The difference between digital twin and simulation is largely a matter of scale. While a simulation typically studies one particular process, a digital twin can itself run any number of cost-effective simulations to study multiple processes. It promises more effective asset design, project execution, and operation. In general, the digital twin is the evolution of a model from multiple technologies including 3D simulation like BIM, AI, IoT, and cloud computing. AI including ML and digital twins have a strong relationship where all contribute to each other. They can make the whole process more intelligent by creating more accurate data and creating a single learning system that can offer better predictions and give accurate information to the users. Accordingly, an intelligent digital twin entails all the characteristics of a digital twin as well as AI capability to accomplish an autonomous system (Talkhestani et al., 2019). It can therefore implement ML algorithms on available models and data of the digital twin to optimize conditions as well as paving the way for predictive maintenance.

Digital twins enable the establishment of cyber-physical representation of the building throughout its lifecycle, both before and after construction. They are set to play a crucial role in optimizing and improving energy use and efficiency, personalized comfort, space booking, flexible workplaces, safety, and seamless security. They bring a wealth of useful applications across the construction market and lifecycle of an asset, standing as a bridge between the physical and digital. They enable companies to simulate their entire business to identify optimization opportunities. Today, airplane engines and wind turbines use digital twins to analyze the performance without the need of using the conventional testing protocol so decreasing the expenditures required for testing.

The operational and behavioral knowledge of digital twins will assist the AEC community to achieve better outcomes for people throughout the lifecycles of buildings and infrastructure. Digital twins allow building operators to bring together previously unconnected systems, from security to HVAC to wayfinding and asset tracking, to gain new insights, optimize workflows, and monitor processes remotely. They can be used to give occupants more control over their workspaces and environmental conditions, thereby enhancing the user experience. By optimizing systems and connecting people, owners and operators can use digital twins to reduce costs, increase occupancy rates, and improve overall asset value (Monteith, 2019). The digital twin can predict potential failures in technical systems like HVAC, lighting, access, and safety. This results in saving a lot of time, costs, and expensive shutdowns. In addition, control access to a specific area to influence pedestrian flow, for example, in response to social distancing during a pandemic is another important service offered by the implementation of the digital twins.

5.2.2 3D printing and robotics

3D printing, also referred to as additive manufacturing, is the computer-controlled successive layering of materials to generate geometrically complex structures. It is a physical procedure contrary to screen rendering. The use of 3D printing for construction and architecture is a rapidly developing area. Additive manufacturing, already an innovative electrical design approach, is now crossing over to the field of construction, making it possible to manufacture 3D-printed buildings. It is used not only in the design stages of construction projects but is also bringing remarkable full-scale structures to life. 3D printing is well-known for the ability to make parts with complex geometries quickly and cost-effectively. When combined, robotic additive manufacturing can potentially create large-scale objects. The next innovation step for electrical engineering is the development of embedded devices for the robotics-based construction of 3D-printed IBs.

Although construction tasks are notoriously difficult to automate, the creation of a digital construction platform is motivated by the overall vision of designing robots that excel at repetitive tasks in a controlled environment. Such a vision inspires, for example, combining structures, skin, beams, and windows, in a single production process. Nozzles of the new 3D printing system can be adapted to vary the density of the material being poured, and even to mix different materials as it goes along. The emergence of BIM, in particular, may facilitate greater use of 3D printing. The integration of robotics and 3D printing is also helpful in terms of better material management with a considerable reduction of material waste as the material is deposited only where required. Not only robotic 3D printing presents a sustainable approach to manufacturing, but it also in using organic and eco-friendly materials.

Intelligent industrial robots and additive machines combined with AI could have a much wider range of applications, including 3D printed facade panels, columns, benches, additive manufacturing with metal, and automated repair. These robots could come up with enchanting solutions to the construction of complex problems that architects and engineers had never considered before. Increasingly such robots will be integrated with intelligent tools which are incorporated directly into data infrastructures. The idea is to find insights into data, which may be utilized for several key construction situations, including usage, repair, and safety.

On the sustainability side, however, there does seem to be a strong path to positivity when it comes to intelligent robots and 3D printing. This is via more efficient utilization of material through additive manufacturing practice and fewer emissions from transport and construction equipment as only those materials required should be transported to the site.

5.2.3 Digital building passport

Solutions such as digital building passport and track-and-trace solutions for critical materials are set to shape the closed-loop of resources in the circular economy. Digital building passport is a secure, online repository that digitally stores all postcompletion building information. It is defined as a data set that summarizes the components, materials, and chemical substances or also information on repairability, spare parts, or proper disposal for a product. The data originate from all phases of the building life cycle and are to be used for the optimization of design, production, use, and disposal. The passport also consists of a large amount of data that is collected by different actors in the course of the building's life.

Materials passports are a tool to put the circular economy into practice. Builders and their customers are looking for reliable and convenient data on building designs, pathways, and composition to determine their potential for a circular economy, including optimal productivity, recycling versus downcycling, and optimizing the residual value of materials. Among other things, the digital building passport is intended to provide consistent track and trace information on the origin, composition, repair, and dismantling options of a building, as well as on its handling at the end of its service life. It has the potential to provide consumers and waste management companies with relevant information on a product and thus force decisions toward SD.

5.3 User experience (UX)

Defined by the International Organization for Standardization (ISO13407: 1999), UX is a "person's perceptions and responses resulting from the use and/or anticipated use of a product, system or service." The ISO definition takes into account that "usable systems can provide many benefits, including improved productivity, enhanced user well-being, avoidance of stress, increased accessibility and reduced risk of harm." Another term related to UX is human-centered design, a holistic approach to understand the needs of people, enhance their experiences, and facilitate the system to respond to their feedback.

A popular notion in this regard is "Society 5.0" or a "super-smart society," which is a human-centered society that balances economic and technological advancement to solve society's problems with super-smart AI data systems. The basic idea of Society 5.0 is that data is collected from the real world and processed by IT with the outcomes applied to the real world. In Society 5.0 knowledge and intelligence will come from machines through AI at the service of people. It will create a smarter ecosystem that will provide sustainability at all levels, economic, environmental, social, and political, focusing on the people and the creation of value, aiming for a prosperous human-centered society. Society 5.0 is about a holistic CPS convergence at the level of society. On the other hand, Industry 5.0 works on the basics of collaboration between humans and machines. It strongly focuses on mass customization of a customer experience driven by an enterprise-wide responsive culture and universal collaboration among communities. Although it may sound oversimplification, the advancement from Industry 5.0 to Society 5.0 has some well-known similarities with a journey from big data, AI, and M2M to enterprise AI and H2M.

IBs are digitally connected structures that combine operational automation with intelligent space management to enhance UX, increase productivity, reduce costs, and mitigate physical and cybersecurity risks. Today, the building industry is experiencing a robust change of emphasis from "energy efficient design" to "context and user aware design" and from "creating static objects" (buildings) to "designing interactive systems" with long-term performance in mind. During the design stage, architects and engineers need to fully consider how the interaction of building occupants and operators with the building technologies and its energy systems will impact the final energy use and IEQ outcomes. Accordingly, building designers need data, models, tools, and case studies able to provide an evidence-based understanding of the human dimensions of energy use. During the construction phases of the building life cycle, the value of technologies depends on the building technology manufacturers' and vendors' understanding of how occupants utilize their products. During the operational phase of buildings, occupants require comfortable and healthy spaces to live and work in (D'Oca et al., 2017). Therefore, occupants need to understand the design and operation of building systems such that they may adapt to accomplish optimized personal comfort conditions while reducing energy use.

UX is an exciting area when it comes to the intersection of big data and sustainability. Every interaction that end users have with static and dynamic components of the building as a transaction that could provide a lot of value when collected and analyzed should be considered. Such transactions could reveal unique usage patterns that will allow the designer to re-question the problem and sometimes challenge the initial assumptions. With the IoT in its beginning, the data generated by the sensors that will track occupant interactions with the built environment poses another challenge, visualizing streaming data in a meaningful way that results in action. Therefore, it is important to be able to deliver a user experience that enables an end-user to derive key insights. Design tools that are integrated into data-rich environments and platforms promote user experience, enhance occupant comfort, and increase productivity.

The convergence of UX with the service, often characterized by the quality of experience (QoE) is also of vital significance. QoE is a measure of the user's liking and satisfaction with the service, which depends on the service type, the tools through which the service is being accessed, and the user's context in which the service is utilized (Aazam et al., 2019). The digital transformation on the UX level is not just an issuer of the front-end and user-facing functions; it involves the entire organization and needs to engage the back-end as well. Such engagement at this macro-level could perhaps be described as the merging of Industry 5.0 and Society 5.0 as shown in Fig. 5.3. This relationship may be seen in the concept of dualism of yin yang, the Ancient Chinese philosophy.

Fig. 5.3 User experience and quality of experience at the intersection of Society 5.0 and Industry 5.0.

5.4 Ambient intelligent architecture

Intelligent architecture relates to the design, construction, and maintenance of buildings (Wang, 2010). However, ambient- intelligent architecture is a human-centric architecture where the building identifies and automatically adapts according to the occupants' behavior and preferences, and thereby optimizes comfort, energy use, safety, security, and well-being.

5.4.1 User-focused architecture

Architecture as a profession dates back to ancient times, with a profound impact on the built environment of civilizations all over the world. The evolution of the practice has been relatively slow; while technologies and styles have evolved, the fundamentals today are not all that different than they were historically (Staton, 2020). However, with the evolving technologies and their integration into the profession, the role of an architect is changing faster than it ever has before.

Today, architects are at the forefront of designing the innovative built environment that will envelop living in the 21st century. This entails emerging responsibilities that may include for example new code compliance, environmental considerations, energy efficiency, occupant health, and comfort. The focus of this section is shifting from pure architecture to an environment that is both architectural and user-focused to enhance the occupants' experience while developing innovative ways of using existing buildings and creating new ones.

As early as the 1970s, visionary thinkers like Nicholas Negroponte proposed that advances in AI and the miniaturization of components would soon give rise to buildings capable of intelligently recognizing the activities of their users and responding to their needs, as well as changes in the external and internal environment (Meagher, 2015). This prediction was driven by the value of data which is today one of the biggest byproducts. Over the past decade, a set of emerging technologies have signaled the start of the fourth industrial revolution (Industry 4.0). To secure the opportunities initiated by these technologies, the industry has embarked on a reorientation of its strategic direction. By 2025, the capabilities of machines and algorithms will be more broadly employed than in previous years, and the work hours performed by machines will match the time spent working by human beings (WEF, 2020).

In the field of building science and architecture, data is having a similar impact where an increasing number of sensors in the built environment are bolstering the IoTs and relaying the generated data that they gather. When one examines what is occurring in practice, it becomes evident that data is changing architecture in the following three ways: clients are demanding data from architects; clients are demanding data from buildings; data is changing the process as much as it is changing the output (Davis, 2015). When combining architecture with performance, the association is one of a creative and inclusive approach to achieve results. Such performance-based design requires that architects learn to understand and interpret a new class of information that can impact their decisions. This does not mean architects should become data scientists. Just like understanding building codes and zoning laws does not require architects to become lawyers, the understanding performance of buildings does not require them to become building scientists (Singh, 2013).

Data-driven materials science and materials informatics are umbrella terms for the scientific practice of systematically extracting knowledge from materials datasets. This practice differs from traditional scientific approaches in materials research by the volume of processed data and the more automated way information is extracted, for example through the use of ML (Himanen et al., 2019). Intelligent architecture aims at contributing to the generation and dissemination of knowledge in the field of computational intelligence for design, building, and construction systems. The idea is to bridge and integrate the disciplines of building science, architecture, civil, mechanical, electrical, software engineering with computer science and informatics using research on complex design, building, and construction systems.

Knowledge discovered in data from past projects and buildings in operation can be combined with tacit knowledge for informing future design decision-making. As a result, the huge potential would arise in achieving building design in a sustainable, efficient, and evidence-based manner (Petrova et al., 2019). One of the main research objectives in this regard is to leverage the multiplicity of data sources and types, and thus pave the way to knowledge discovery for evidence-based processes in design and engineering practice.

The BIM process is a commanding paradigm of digital transformation. It improves efficiency and exhibits how digitization can drastically change existing ways of building design. It is a collaborative tool that provides a shared digital data environment where stakeholders communicate tasks for each stage of the project under consideration with high efficiency. In addition to BIM, the next wave of digital adoption sees the AEC community employing IoT. The creation of an IB system with an autonomous intellect that learns from usage trends and which has embedded system interoperability has the potential to improve various building qualities.

Achieving this varying objective requires advanced control, automation, and monitoring systems where buildings rely less on human control and more on the embedded intelligence. The convergence of ICT and data science, particularly ML and big data mining has empowered the retrieval of useful information for building automation. This entails an integration platform that employs IoT, cloud computing, big data, AI technologies, and various proactive and predictive digital capabilities.

5.4.2 Intelligent façades

Today's IBs are electronically enhanced buildings, the forerunners of a new architecture. Intelligent façades are a partial aspect of the wider significance of IBs. Façades, surfaces, and building envelopes are the interfaces between interior space and the exterior environment (Saidam et al., 2017). They have a significant impact on a building's overall performance. Given desired energy efficiency and the high prospects toward occupant comfort, they must provide the highest likely effectiveness.

It is significant to design and develop building facades that are responsive to environmental parameters. These facades, as may be called intelligent facades, can adjust their shape, form, and orientation to automatically respond to environmental parameters including wind, temperature, humidity, etc. Such intelligent features in a building require integration of responsive dynamics capabilities which allow for a change in the building's configurations based on timely stimuli and considering the environmental context, indoor space conditions, and occupants' behavior to improve living conditions, enhance building performance, and reduce energy consumption.

Because of the intelligent façades, IBs may be considered as "living and breathing" structures that predict and respond to the habits of the tenants in ways that benefit their comfort, productivity, and safety while optimizing the building's operations and efficiencies. A breathing facade is recognized by skin texture, it could substantially assist buildings to breathe and have better air circulations. Double facades, air volume with potential natural ventilation, and adaptive sunshade are other standard means to control energy and daylight. Therefore, buildings with breathing facade design, offer a strong linking between the exterior environment and inner spaces. Such a breathing mechanism happens through thousands of holes.

The means driving new developments is the enhanced computing power and affordability of new computer systems that perform thermal and energy modeling. These tools help professionals gain insight into the exchange of energy that occurs between the external environment and the internal spaces of a building over a year or more. By using energy modeling and dynamic thermal models, designers can predict how, why, and when a building façade will consume energy.

Finally, building design that focuses narrowly on aesthetics and buildings as isolated bodies fall far short of the requirements for adaptive development. The advent of IBs has, in part, been a response to this. But much more "out of the box" thinking is needed to fulfill the full potential of the AEC community in challenging conventional notions.

5.5 Computational intelligence

AI and ML have captured a large share of academic and industry attention during recent years, both in terms of new capabilities and the implications to society. The implementation of AI within an IoT solution enhances data analytics, enables ML, and makes the connected devices independent and autonomous, but demands low latency and quick responses. Sensors and related models built on AI and ML are changing the dynamics from static BIM to dynamic digital twins to accelerate the sustainability benefits derived from the integrated approach.

5.5.1 Artificial intelligence

The term 'AI' was coined in 1955 by a group of researchers including John McCarthy, Marvin Minsky, Nathaniel Rochester, and Claude Shannon who organized a famous summer workshop at Dartmouth College on the "Study of Artificial Intelligence" in 1956. This event is widely recognized as the starting point of AI.

Artificial is a concept that refers to something that is made or produced rather than occurring naturally, often replicating something natural (Habash, 2019). Intelligence is a major characteristic of humans. It is a perceiving, understanding, analyzing, and synthesizing procedure based on people's experience of things.

At the heart of Industry 5.0 and Society 5.0 is AI, a disruptive technology, which represents the ability of machines to think, match, and perhaps one day will pass the cognitive ability of human developers. It is scaling across the entire economy and transforming people's life aiming to mimic human intelligence by a computer and machine in solving various problems. All current AI solutions come from a set of computational methods and techniques, instead of from a single method or technique. AI involves data-collecting, analyzing, and decision-making processes based on anything artificial that can be a machine, an algorithm, a computer program, a system, or a device. It represents a combination of various technologies including ML, deep learning, natural language processing, computer vision, speech recognition, context-aware processing, and neural network.

5.5.2 Machine learning (ML)

ML is a field of AI that uses statistical techniques to give computer systems the ability to mine, train, and learn from data to predict future behavior, without being explicitly programmed. It is a scalable process of programming a machine so it can identify patterns based on existing data without human intervention. It focuses on prediction through computers which has an enormous capacity for improvement. Based on the historical data, ML with appropriate software algorithms can learn the nonlinear relationship between the independent variables and target variables. The mining task of ML is to recognize correlations between data, extract meaningful insight from these variables, and transport it to the storage for further analysis.

In buildings, ML brings the ability to automatically identify patterns and detect anomalies in the data that smart sensors and devices generate information such as temperature, pressure, humidity, air quality, vibration, light, and sound (Schatsky et al., 2017). The ultimate goal of IoTs in buildings is the indoor climate data collected from massively deployed sensors and the intelligence mined from the generated data. The powerful combination of AI and IoT technology is helping buildings increase functioning efficiency and enhance the quality of indoor living.

ML can be divided into three major styles: supervised, unsupervised, and reinforced (Fig. 5.4). Supervised learning is the most mature and powerful of these subfields and is used in the majority of ML studies in physical sciences (Butler et al., 2018). It applies in situations where an ML model is trained on input-output pairs from a real process to produce optimal outputs for unseen inputs. Typical applications are predictions of physical properties given the input features of a material or process (e.g., geometry, physical properties, and external conditions) (Himanen et al., 2019).

Unsupervised ML is the process of extracting structure from data or learning how to represent data best. It looks at raw data and spots patterns within it. In unsupervised

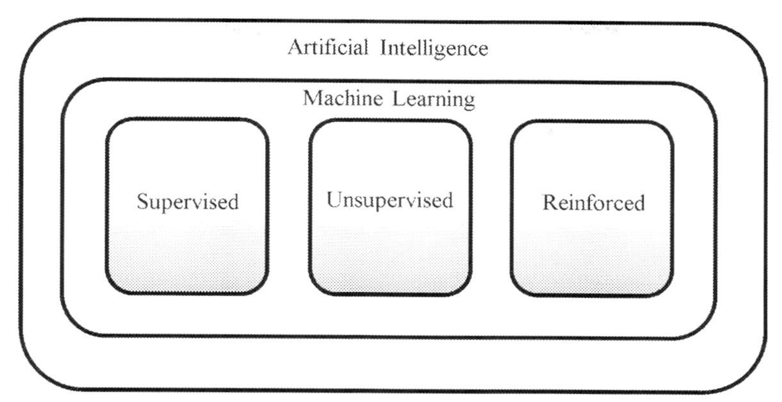

Fig. 5.4 The three styles of machine learning.

learning, only input data is given to a model but no output. The machine is then tasked with a learning objective, for example, to find rankings or patterns for this input.

Reinforcement learning is a teaching model of ML to create a sequence of decision-making. It is a rapidly emerging field with promising applications in tasks that require machine creativity. In reinforcement learning, a model is given the task of choosing a set of actions to optimize a long-term goal (Kelechi et al., 2020). As such, it differs from supervised learning because no correct input-output pairs are presented for individual actions, but the training is a mixture of exploration and exploitation guided by a long-term reward (Bishop, 2006). This type of learning applies to a wide range of decision-making problems such as environmental carbon emissions.

In summary, ML is still very limited when it comes to creativity; however, it is a powerful tool for data analysis and informatics-based tasks. It is a field undergoing very active development, and a surplus of suitable ML methods has been applied to building science and engineering. Quantum ML is going to be the next big step as a result of combining quantum computing and ML.

5.5.3 Human in the loop (HIL) intelligence

HIL is a branch of AI where human and machine intelligence combines to create more accurate AI algorithms. In such systems, humans are involved in every stage of the algorithm's output by creating a feedback loop from training to testing stages resulting in a more accurate model. This technology is a blend of supervised learning (using labeled training data) and active learning (interacting with users for feedback) (Barez, 2019). The HIL approach reframes an automation problem as a human-computer interaction design problem.

Improvements in BAS gravitate around the concepts of bringing the HIL for sensing and control of buildings over the entire building life cycle. The notion of HIL emerges from the CPS domain, as an application that integrates real-time human feedback with the management of complex systems for control optimization purposes.

A variety of existing HIL applications in the energy sector can be classified based on the level of integration of human control over system functioning as active, passive, or hybrid active-passive sensing and control systems (D'Oca et al., 2017). Another challenge is how to bring occupants' comfort, preference, and needs together with the technical aspect of building automation and control into a comprehensive and scientifically accepted modeling framework of HIL interaction. This achievement depends upon understanding the link between energy use and human factors of behavior and operations.

5.6 Big data

Big data is a popular phenomenon that aims to provide an alternative to traditional solutions based on databases and data analysis. It refers to high-volume and high-velocity data that cannot be managed efficiently with the established applications. This data can come from people, computers, machines, sensors, and any other data-generating device or agent. Big data is not just about storage or access to data but it provides solutions to analyze data and exploit their value. AI and ML increase both the possibilities of intelligent data collection and possibilities for effective data analysis.

5.6.1 4 V model

The concept of big data has been defined through the 3 V model, which was defined in 2001 by Laney as "high-volume, high-velocity and high-variety information assets that demand cost-effective, innovative forms of information processing for enhanced insight and decision making." In 2012, Gartner updated the definition as follows: "Big data is high volume, high velocity, and/or high variety information assets that require new forms of processing to enable enhanced decision making, insight discovery, and process optimization." Both definitions refer to the three basic features of big data: volume, variety, and velocity. Other organizations and big data practitioners have extended this 3 V model to a 4 V model by including a new "V": value (Koseleva and Ropaite, 2017). Each "V" presents a challenge, although the volume challenge could be solved by more storage space, and the velocity challenge of faster data generation could be addressed by faster computer processing and enhanced communication techniques. Increased variety is more a challenge for standardization, yet it is also a benefit for AI/ML and materials discovery algorithms. Veracity, however, is the most challenging because it is a softer measure of how to quantify the degree of trust in data (Himanen et al., 2019).

5.6.2 Data centers and cloud

Digital transformation starts in the data center which may be interpreted in different ways. This refers to server hardware on-premises to store and access data through a local network. Second, it may refer to an offsite storage center that consists of servers and other equipment needed to keep the stored data accessible both virtually and physically. Sustaining the security and reliability of data centers is vital to protecting an enterprise's operational stability. Usually, large data centers are located in dedicated buildings. Smaller data centers may be situated in specially designed rooms within buildings made to serve multiple functions. Since data centers consume large amounts of energy, it is significant to ensure the structures that contain them are well-designed and properly shielded to boost temperature controls and energy efficiency. One of the key design challenges related to data centers is that the significant growth of digitalization is stemming from a huge quantity of power and is placing pressure on renewable energy to offset emissions.

On the other hand, cloud data is an online storage system (unlimited capacity) designed to fragment and duplicate data across multiple locations. Cloud is not a physical entity but is a group or network of virtual servers which are arched together to operate as a single entity for an assigned task. Cloud storage minimizes the overall cost of storage because organizations do not need to spend a tremendous amount on hardware installation for data storage. The cloud data can be accessed by anyone with the proper credentials from anywhere with an internet connection. In case of failures, a cloud system ensures that there is always a backup of the backup. The cloud service is by far more cost-effective, especially for small companies.

Cloud computing is a model of computing tools that provides on-demand services to its clients. It provides the computing power, storage space, and security foundation needed to facilitate analytics and AI. Often, construction workers require access to company data to provide timely decision-making and reporting ability while working in the field. The cloud is the application layer that communicates with the gateway, typically over wired or cellular internet. It provides powerful servers and databases that enable robust IoT applications and integrate services such as data storage, big data processing, filtering, analytics, monitoring, and user interfaces.

5.6.3 Edge-intelligence technologies

Edge computing, which deploys the resources, data, and services from the central cloud to the network edge, enjoys increasing popularity. It enables analytics and knowledge generation to occur at or close to users' end. In edge computing, users first offload tasks to the nearby edge cloud, and then the edge cloud executes their tasks and sends outcomes back to users. Compared with cloud computing, edge computing can provide services with low latency, high security, and high mobility but with limited computing capacities.

Edge intelligence refers to the union of edge-computing, advanced connectivity, compact computing power, cloud capacity, data analytics, control, and AI technologies located near devices that use and generate data. The intelligent edge's ability to bring cloud capabilities to remote locations could greatly amplify the construction industry's performance. This reduces latency, lowers cost, reduces threats, and improves reliability.

This rise of the intelligent edge will likely drive the evolution of service architectures to become more location-driven, decentralized, and distributed. Such technology

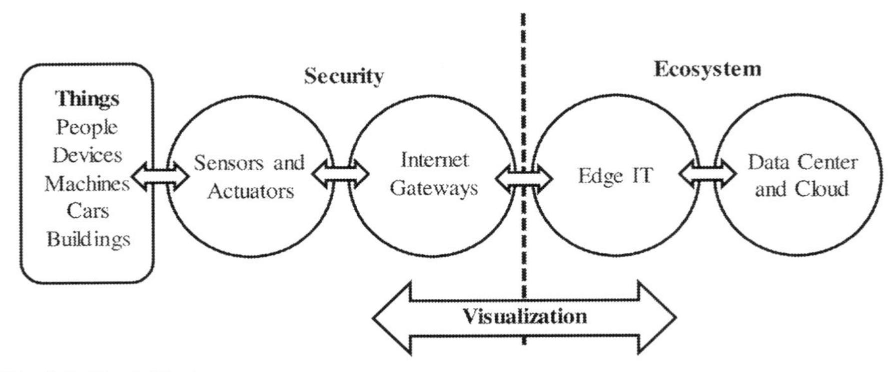

Fig. 5.5 The IoT edge structure.

does not replace the cloud or data centers but, rather, is an element within a holistic cloud-to-edge architecture. Some components of service will run in a centralized cloud, others at the data center, and more yet at the edges on sensor arrays including autonomous vehicles and potentially billions of other machines (Akenberg et al., 2021). Intelligent edge requires 5G solutions including WiFi6 that offer bandwidth slicing, better power management, and support for an increased number of devices, all of which can enable much more robust and dynamic local networks.

There are three broad categories of intelligent edges including operational technology (OT), IoT, and IT. OT edges commonly contain intelligence and controls but have been traditionally limited in connectivity and compute power. OT edge examples include power plants and offshore oil rigs. IT edges are common in the telecommunications and media industries for distributed data transfer and processing, as well as distributed computing in branch offices and campuses. The IoT edge is of great interest today, as the IoT has achieved celebrity status. In many instances, the IoT is a combination of OT and IT. When discussing the IoT, it is instructive to first understand the generalized four-stage IoT solutions architecture shown in Fig. 5.5. Things are connected to sensors for data capture and actuators for control, either wired or wirelessly. These sensors and actuators connect to gateways and data acquisition systems. The IT systems that are at the edge and the remote data center or cloud are situated at the end.

5.6.4 Data analytics

Data analytics is the science of examining raw data to conclude the information. It involves applying various sophisticated software programs to run through several data sets to look for meaningful correlations between each other as well as to discover hidden patterns, associations, and other insights. There are four types of analytics in practice: descriptive analytics (what has happened to the data), diagnostic analytics (why there is a data change), predictive analytics (what will happen next), prescriptive analytics (what actions should be taken to get the desired output).

One of the industries that are reaping the benefits of this technology is the construction industry. IBs are full of data-collection systems. That data is an important source

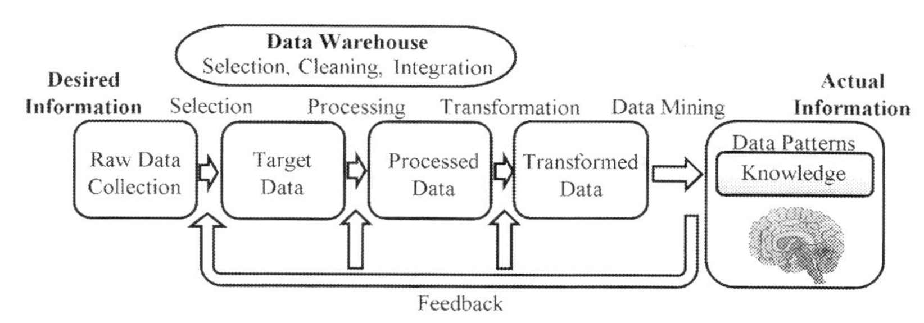

Fig. 5.6 Knowledge discovery in database.

of information to design a resource-efficient and livable built environment for people. Such data is important for energy management and building automation. Two major problems concern the adoption of data tools within building design. First, the translation and systematization of very large and unknown data sets for efficient use. Second, the lack of knowledgeable data experts within design firms who can intelligibly curate diverse data sources and tools according to the project needs. This has led to a growing number of data analytics and software experts making their way into the construction field in recent years to meet the growing demand.

Knowledge is the end product of data analytics, whereas knowledge data discovery represents the overall process of data analytics and the extraction of useful knowledge. Data discovery is a user-oriented process for detecting patterns and outliers by visually navigating data or applying guided advanced analytics. Discovery is an iterative process that does not require extensive upfront model creation. As data is becoming increasingly complex, users entail adaptable roles to be able to efficiently access and prepare their data for analysis. As shown in Fig. 5.6, knowledge data discovery is the process of data selection by capturing the relevant prior knowledge and identifying the desired goal of the process from an end-user perspective; cleaning and enrichment (replace of missing values and removing replicated); transformation into a common format for processing such as numerical using neural networks; data mining for extraction and visualization as patterns for performance prediction; and reporting results with interpretation and evaluation of actionable patterns for knowledge. Data mining is a significant step in the knowledge data discovery process which employs specific algorithms such as clustering, regression, and classification to discover useful and previously unknown patterns in the data. The extraction is generally performed by automatic and semi-automatic (humans and/or AI systems).

5.7 Artificial IoT (AIoT)

Today, IBs, AI, ML, and IoT make an ideal combination to improve the functioning of buildings. AIoT is an example of the above combination of AI and the IoT where AI and IoT are two of the most advanced technologies today. In this hybrid system, IoT collects data while AI processes the data to provide meaning. Individually, these two innovative technologies have their benefits but their true potential can only be realized together. IoT is considered as body and AI is the brain and together they create operating models, which provide intelligent decision-making. It makes the IoT perform intelligence tasks with the help of integrating AI. It helps to connect IoT devices including sensors, actuators, and computing nodes that are integrated with AI capabilities, all of these with no human intervention.

AIoT is transformational and mutually beneficial for both types of technology. IoT advances AI in terms of useful data in real-time from the real world and global connectivity, while AI adds value to IoT through improved security, enhanced analytics, information prediction, and ML capabilities either at the cloud platform or the edge devices to enable decision-making at each component level of the interconnected system. As IoT networks spread throughout major industries, there will be an increasingly large amount of human-oriented and machine-generated unstructured data. AIoT can provide support for data analytics solutions that can create value out of this IoT-generated data. The workflow of AIoT is shown in Fig. 5.7. This combination of powerful technologies is reflected in the intelligent digital twin. It represents a new kind of connected experience that can provide automation solutions that visualize and analyze real-time information for building operation, space utilization, occupant experience, safety, and security. Once the data is collected, it is further transmitted and stored in the cloud due to its high volume. Data processing involves different phases like relevant data extraction from the cloud, data cleaning and making it anomalies-free, data conversion into a standard format, and applying algorithms for deriving insights. After processing the data, the ML algorithms help in predicting future events.

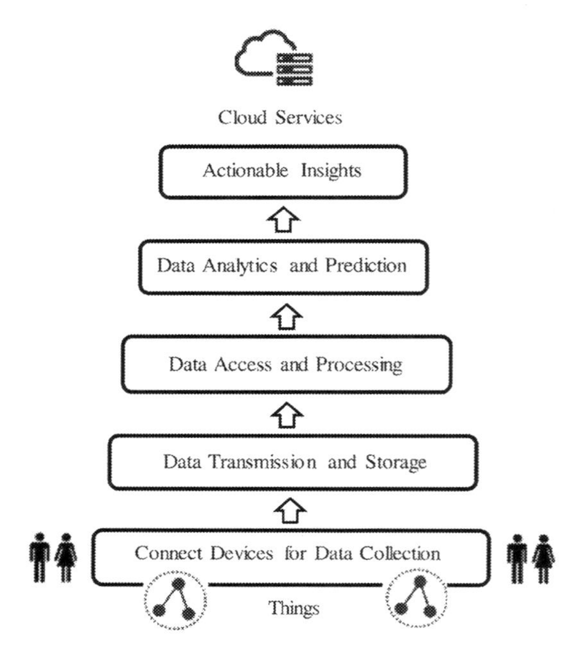

Fig. 5.7 Workflow diagram of artificial intelligence of things.

When the relevant models are generated, it becomes easier to make predictions based on achieved results. After making the predictions, the subsequent step for machines is to take actions as per the generated insights.

Conventional sensing and actuation systems do not offer distributed computing devices for sensing and actuation. To overcome this limitation, WSANs, which represent a viable and more flexible solution to traditional building monitoring and actuating systems, can be adopted (Stankovic, 2008). The paradigm of decentralized control and optimization is a significant component of the intelligent building platform wherein such a scenario, there does not exist a centralized computing node to manage the entire control system (Zhao et al., 2018). Sensed data generated by IoT sensors are transmitted to and processed by the computing nodes. The monitoring node stores the received data and compares the IoT data with the data patterns. If the value lies within the acceptable deviation range (e.g., $\pm 5\%$ of the difference between the received data value and the pattern value), the collection of data will continue. If the value is beyond the normal data pattern range, the monitoring node may request an action (Yoon et al., 2019).

There are many different applications across multiple industries that require AIoT including. Predictive maintenance is the brightest showcase of AI used in AIoT. Predictive or perspective maintenance means that a system powered by ML algorithms can predict a need for maintenance on a building. Today, many software packages are offering AIoT spatial intelligence capabilities that enable modeling benefiting from advanced sensor processing and space topology services. Spatial intelligence, or visual thinking, is the ability to comprehend three-dimensional forms and spaces. It is creating a digital environment for a physical world. For example, in a building environment, spatial intelligence can enable better management of HVAC systems, and occupancy based on how space is used. It will impact how building owners, operators, occupants, and visitors interact with various elements of the built environment.

5.8 Building as a human-cyber-physical platform

An IB is a HCPS, a combination of CPS and HCS, which is realized as an efficient embedded application that engages physical infrastructure with cyber computation to accomplish functional desires including energy efficiency and human-centric values such as health, comfort, and productivity of occupants. The building as an HCPS is about integrating different ICTs and data sources within the building from BIoT to environmental data and technical services while engaging stakeholders (human dimension) in the process. People and CPS functions are fundamental to HCPS, and play various roles in its operation as reflected in Fig. 5.8.

The human dimension is different from the CPS components of the building in being social beings, privacy-sensitive, risk-reluctant, and nonstationary. Their behaviors and activities may be patterned, but their differences remain unpredictable. People are the stakeholders who assume multiple roles in buildings, from occupants who simply enjoy building services to operators who control building services, to managers who work and collaborate with occupants and operators, and sensors who provide

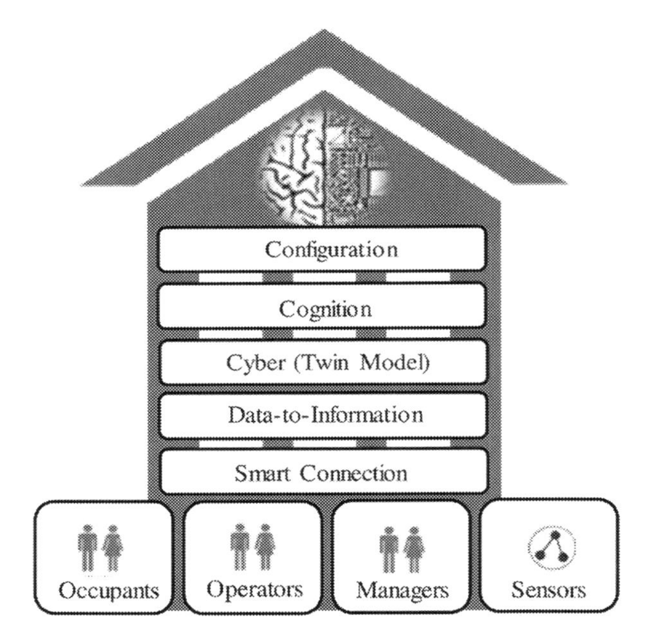

Fig. 5.8 Building as a platform for the human-cyber-physical system.

feedback about current situations (Jin, 2017). Human-centric values such as comfort, productivity, health, and well-being are high priorities for stakeholders.

A Typical CPS structure usually includes five levels, including smart connection, data-to-information conversion, cyber (cyber model), cognition, and configuration (supervisory control and action for resilience). Hints and insights move from cyber twin to physical twin and act as a monitoring system for preventive decisions from operators, machines, or ambient environments. It is a resilient control system in which networked computers and machines interact with the physical world.

CPHS has the potential to adaptively optimize its operational variables toward continuous real-time monitoring and related behaviors. HCPSs continually face these variable operational conditions caused by both internal and external CPS and human factors. The human factor is good at understanding complex situations (contextualization) and making high-level decisions. It has multiple dimensions, delineated by people's interactions with the environment, the system, other people, and themselves. However, CPS is good at the employment of high-level decisions, embedded and remote sensing, fast computation, and information aggregation. Therefore, there is a need to understand the boundaries of responsibilities and assess human–CPS interactions.

An important synonym for CPS is Industry 4.0 Envisioning Industry 4.0 into Industry 5.0 climbs into the distant future exhibits a good deal of uncertainty concerning developmental periods and the detailed way of realization. Such future industrial process control systems require reliable communications between various devices, which is simply M2M. However, HCS and H2M are the essence of Industry 5.0 and Society 5.0

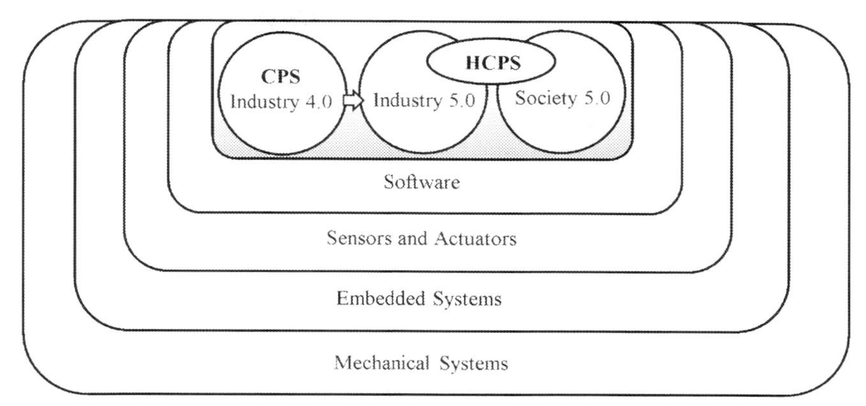

Fig. 5.9 Industry 5.0 and Society 5.0 interaction in a human-cyber-physical system.

which require more human-centric products and services (Fig. 5.9). An HCPS for IBs will likely become functioning and continue to develop in the coming years, contributing to the UN SDGs in affordable and clean energy, innovating for sustainable automation, and combating climate change, which will ultimately benefit the entire world.

5.9 Building information modeling (BIM)

In the construction industry, as in other sectors, big data refers to the vast quantities of information that have been stored in the past and that continue to be obtained continuously. Primary sources of big data in buildings include people who are continuously generating and sharing information, IT-enabled construction devices that gather, share and store data, and holistic systems such as BIM.

BIM is at the foreground of digital transformation in the AEC industry, inspiring cooperation and streamlining data exchange. It is an innovative technology that is considered as the opportunity of the AEC industry to move to the digital era and improve cooperation among the partners of the industry. A BIM is an intelligent digital representation of physical and functional characteristics to efficiently plan and design buildings and infrastructure. It aims to integrate design and construction segments and operation of the building project for decisions during its life cycle from beginning onward. Technically, BIM is a collection of files that can be obtained, replaced, or networked to assist decision-making involving a building or the other built asset. It facilitates the coordination and collaboration between different stakeholders at different phases of the life-cycle of a building project. A true BIM model consists of the virtual equivalents of the actual building parts and pieces used to construct a building before its physical realization through a virtual model.

The existing generation of BIM design tools, including Autodesk Revit and many other tools, grew out of the object-based parametric modeling capabilities. Information is connected via algorithms in a digital parametric structured model so that when a change takes place, components are updated automatically in line with identified

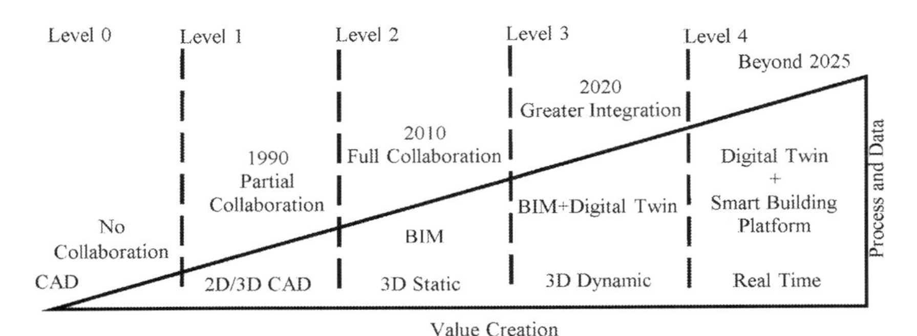

Fig. 5.10 Development levels of building information modeling.

parameters. There are currently four development levels of BIM as shown in Fig. 5.10. At levels 0 and 1 there is either a lack of BIM or an over-reliance on different systems of data. Level 0, is the simplest form, it means no collaboration. It makes use of 2D CAD paper or electronic prints, or a mixture of both, mainly for production information. The majority of the industry is already well ahead of this now.

Level 1 BIM involves using both 3D (width, height, and depth) CAD and 2D (width and height) drafting. While 3D CAD is used for conceptual works, 2D is used for the generation of statutory approval documentation, often managed by the contractor. Level 2 BIM is a managed 3D environment held in separate discipline BIM tools with data attached. This level promotes collaborative working by giving each of the stakeholders its 3D CAD model. Level 3 BIM, on the other hand, is a fully integrated and collaborative process enabled by web services and compliant with emerging standards. It enables all the participants to work on the same model simultaneously which eliminates the chance of conflicting information. Level 3 proposes the use of an integrated solution built around open standards like the industry foundation class where a single server stores all the project data. Level 3 uses 4D where time (construction scheduling information) is added, 5D BIM where cost information is linked to it 5D cost information, and 6D with project lifecycle management information. The BIM model contains also other information, such as material for building components including weight, price, procedures, scale, and size. Level 4 reflects the evolution of BIM extending to the digital twin solutions that enable sustainable construction by incorporating economic efficiency, energy and resource efficiency, and environmental performance in different stages of construction. The key role of the solutions is monitoring, measuring, and managing the processes and operations, in terms of architectural and energy performance, with the ultimate goal of optimizing energy consumption and enabling customers to reach sustainability targets (Royan, 2021).

5.10 BIM-digital twin-driven sustainable design

Sustainability in design entails consideration of numerous factors, such as infrastructure, site plan, architectural design, and the model of the building systems. BIM can enable sustainable design by allowing the AEC professionals access to advanced technology tools than ever before to carefully integrate and analyze quantities like heat gain, solar, ventilation, and energy efficiency in their designs. The three important considerations are the environmental, economic, and community components. The most transformative aspect of BIM is in the modeling and application of sophisticated new technologies such as IoT and the interconnectivity of a wider range of assets that include city infrastructures, utility grids, and transportation networks. This benefit of BIM merges seamlessly with the sustainable construction philosophy, which develops beyond design and construction, into the long-term environmentally conscious maintenance schedules in their finalized project parameters. However, BIM has so far not been directly envisioned for the operation and maintenance of a building, especially not for real-time operations. Therefore, BIM is a small but very useful input into a digital twin, as it provides a great starting point for both an IB and a digital twin.

Sustainable building design requires interaction between multidisciplinary input and implementation of different conditions to affiliate into one high-performing whole. BIM has already brought a profound change in that direction, by allowing the execution of efficient collaborative workflows (Petrova et al., 2019). The significant aspect of BIM that will impact the IB design is the common data environment. This is the single source of information used to collect, manage, and disseminate documentation and the graphical model and nongraphical data for the whole project team. Creating this single source of information facilitates collaboration between project team members and helps avoid duplication and mistakes. However, a digital twin acts as a virtual replica of the building. Enhanced capabilities available for digital twins can enable a truly optimized system, for better indoor living and less energy consumption. It can utilize information and operational technologies such as construction data and building floor plans from BIM, real-time sensor data from the BAS, including data from smart systems and other environmental sensors, as well as data about the building and data of occupancy and movement in it (Wibrand and Saltin, 2020). The digital twin may be considered as an output of a BIM process and is a living edition of the project that BIM processes exist to create. It brings a realization of the physics that determines real-world conditions such as energy use, environmental conditions, and material aspects. Human factors and data democratization are priorities for the digital twin to make information from digital twins accessible to the end-user, without them having to use complex IT technologies. In reality, designing and creating a comprehensive database that stores green building products and materials with their associated costs and integrates it with BIM tools allows architects and engineers to predict the total cost of the proposed building during the design stage. Over the building's entire lifecycle, the digital twin will evolve to provide a way to simulate construction, respond to changes in occupation or energy supply, the need for building upgrades, and assist facility managers in optimizing operations and predicting maintenance.

BIM procedures and digital twin approaches are built on several common principles. Both are concerned with improving process visibility, aligning stakeholders, and supporting planning. But more importantly, they are extremely useful for helping developers look at buildings not as structures, but instead as ongoing projects. The integration of BIM and digital twin, with real-time data from the BIoT networks, offers a strong paradigm for applications to improve construction and operative efficiencies for green and sustainable design. Fig. 5.11 shows the main approach and the overall flow of building design activities using BIM and digital twin.

When AI, ML, data analytics, and the cloud are integrated with the power of simulation, real-time performance optimization of building assets, and the many other benefits that digital twins can offer become a reality. This will help the construction industry toward the goal of significantly decreasing energy use and reaching the zero-carbon targets needed to address the decarbonization agenda and promote a sustainable, healthier built environment. In real-life situations, digital twins can allow operation managers to order services on demand or make maintenance requests instantly, all from their smartphones, for example.

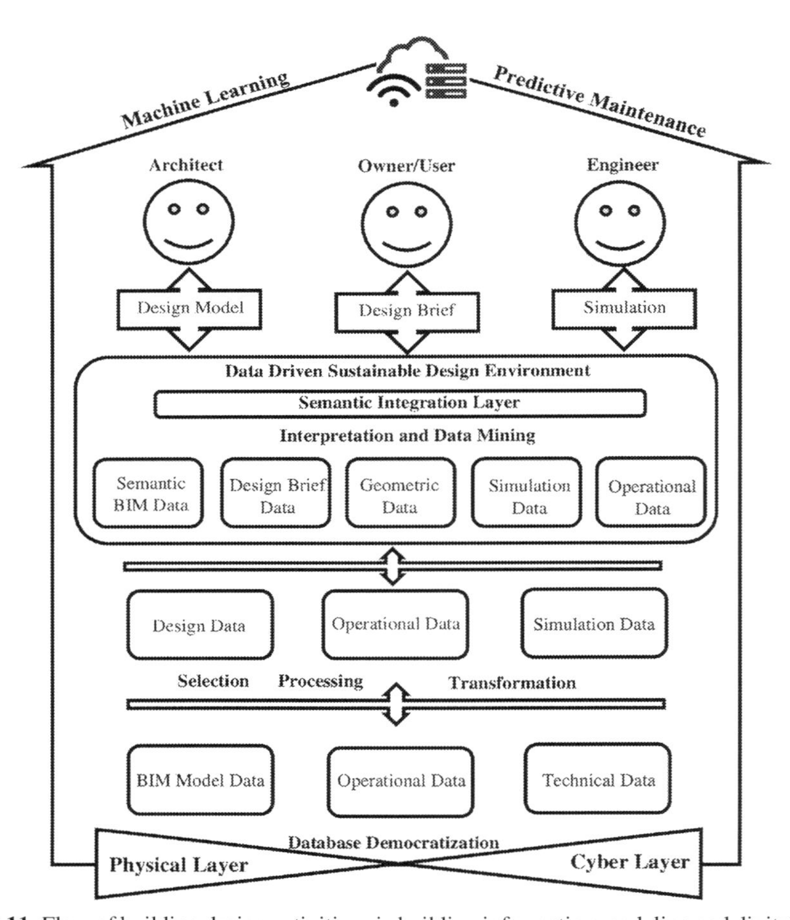

Fig. 5.11 Flow of building design activities via building information modeling and digital twin.

5.11 Surveillance and privacy

Safety is a state in which a place is not in danger or at risk. However, privacy is being free from being observed or disturbed and is a necessity to most people, to some degree. Privacy is the ability and/or right to protect personal secrets, the ability and/or right to prevent invading personal space (Anderson, 2001). It has several well-recognized aspects, e.g., "the right to be let alone," information privacy, privacy as an aspect of personhood, control over information about oneself, and the right to secrecy (Bennett and Raab, 2006). Today, there is a general discussion about privacy and surveillance in IT, which mainly concerns the access to private data and data that is personally identifiable. Surveillance denotes monitoring movements and activities for controlling, managing, and protecting people. Formerly, the video surveillance systems were designed for human operators to remotely observe protected space or concurrently to record video data as an archive for further analysis. Watching surveillance video is a labor-intensive task when a large number of cameras need to be controlled and observed.

Todays, IBs, and communities are about developing a new generation of sustainable and supportive environments which are safe, responsive, and effective to the needs of their users. In this regard, video surveillance systems will be the eyes, accomplishing both a security and management role. Intelligent video surveillance is of great interest in industrial applications due to the increased demand for the reduction of manpower analyzing large-scale video data (Frejlichowski et al., 2015). Although matters surrounding privacy and surveillance are subjects of argument about the extent to which building managements can do monitor and get access to all kinds of content, monitoring the exterior of the building with changing levels of light, wide dynamic range cameras are effective. In an IB, video surveillance solutions can identify potential risks. Surveillance cameras with high-definition capabilities can zoom in to view and capture details such as facial features and report them to an onsite security team, or remotely using smartphones or tablet apps. Video surveillance solutions can also monitor traffic as it happens, and security staff can quickly adjust their positions. A video management system can act as an intelligence center, where security alerts are verified and immediately acted upon. It often serves as the hub for multipurpose, multifaceted integrated systems with the capability to collect, sort, and analyze data generated by multiple systems. AI-based software helps narrow the attention span of the operators so they can focus on crucial areas or unusual behavior. AI technology helps decrease the amount of time spent on surveillance. It enables the operators to focus on what they do best by eliminating the need to continuously display and automate the process of detection. Importantly, AI video surveillance analytics technology will be the main player for security over the next years.

Today a wide range of privacy-enhancing technologies (PET) is designed to extract data value to unleash its full commercial, scientific and social potential, enhancing privacy and protect users' personally identifiable information. PETs that can largely conceal the identity of persons or groups are now a standard staple in data science (Müller, 2020). To protect and improve user privacy, several PETs have been introduced as highlighted in Table 5.1.

Table 5.1 Examples of privacy-enhancing technologies.

PET technology	Description
Homomorphic encryption	Homomorphic (partially, somewhat, fully) encryption is the most widely adopted type of encryption that allows computations to be performed on encrypted data. It builds on the benefits of encryption and helps eliminate vulnerabilities by allowing data to be processed while it remains fully encrypted. Homomorphic encryption allows procedures to be executed on encrypted information, assuring the same output as if the operations were implemented on the clear data. It holds significant promise for use in industries that produce lots of sensitive data, such as the building sector.
Pseudonymization	A data management and de-identification procedure by which fields containing information that enables the identification of an individual (identifiers) are replaced by one or more artificial identifiers or pseudonyms. A single pseudonym for each replaced field or collection of replaced fields makes the data record less identifiable while remaining suitable for data analysis (De la Torre, 2019).
Obfuscation	Obfuscation is a technology that a lot of privacy protection tools, such as virtual private networks use. It refers to the practice of adding distracting or misleading data to a log or profile, which may be especially useful for frustrating precision analytics after data has already been lost or disclosed. Its effectiveness against humans is questioned, but it has greater promise against shallow algorithms (Williams, 2020). A randomly generated noise is added to the underlying data for obfuscation purposes and, as a result, any computations performed on the altered data are only statistically/directionally correct.
Differentially private algorithms	Differential privacy techniques allow the collection and share of aggregate information about users while maintaining the privacy of individual users. It addresses privacy in disclosure rather than in computation, adds noise to the data so that when a statistic is released, information about an individual is not revealed. Three strategies are used in differential privacy: Laplace, Exponential, and Gaussian. The Laplace and Gaussian strategies are utilized for numerical datasets, while exponential techniques are used for nonnumerical datasets (Javed et al., 2021). Differential privacy is believed to be an efficient method to collect combined user information while conserving the privacy of users.
PETchain	The application of blockchain technology is considered a promising solution to preserve user privacy due to its widely recognized accuracy and integrity features. PETchain allows service providers to utilize user data while guaranteeing their privacy. It permits users to securely store their data in a distributed manner and implement their data access policy in an accountable and auditable way (Javed et al., 2021). PETchain can be applied over a blockchain platform that reinforces the implementation of smart contacts.

5.12 Security in human-cyber-physical systems

Security is continuously a question during various stages of the data ingestion process. Attackers inject malware directly into control devices using vulnerabilities in the software. Malware injections can occur through external network connections or internal attacks launched at the device. Keeping attackers away from control systems is the only protection against these kinds of attacks. From a cybersecurity perspective, the greatest risk to a smart building is likely to be from the potential impact on the building control systems, in particular any loss of view or loss of control. Types of cyberattacks can be network-based, system-based (backend), and sensor/equipment-based (frontend). For example, in the case of hacking a building, the HVAC system or other facilities might be accessed and used to gain access to other systems.

The inclusion of HCPS security during the design and construction of control systems will increase the cost of both design and construction; it is more cost-effective to implement these security controls starting at design than to implement them on a designed and installed system (Anixer, 2020). Most industries currently rely on the "security through obscurity" approach to secure engineering, which underlines the confidentiality of the implementation and mechanisms of the cybersecurity system. Therefore, a small leak of information could potentially endanger the entire network (Kshetri, 2017). Security is often among the smart functions of a building, and the capabilities of many physical security systems may contribute to building intelligence.

The rise of the IoT, Industry 4.0 and 5.0, and other sweeping technology initiatives, however, are creating a huge wave of IT adoption at every level of the system architecture. IoT devices, in particular, are a potential weak point for security because they do not generally have the processing power to manage increasingly complex security protocols and encryption schemes, and it would be a challenge to update them even if and when processing power is adequate (Schoechle, 2018). IoT devices are more vulnerable to be attacked than computers or mobile phones, not only because of the surge in the use of IoT devices, but also on account of the complexity, diversity, and inherent mobility of such device application scenarios (Wu et al., 2020). At the same time, IoT has developed rapidly but has not yet matured. Therefore, there is a need to adopt a standardized approach to technology and architecture that leveraged the BIoT and would enable a future path to data analytics and edge intelligence by offering secure mass deployment of IoT technologies that also allow for continuous monitoring and control of user access (O'Brien, 2019). The installation of smart meters in homes and buildings has already created potential access points. Smart utility grid deployment will involve additional potential access points through monitors and sensors that connect into core utility operating systems. The diversity of physically unsecured entry points by itself creates a substantial security risk (OSGF, 2020). Criminals are also increasingly seeking to compromise and gain unauthorized access to building systems.

In general, there are three main issues for buildings seeking to create a safe and secure experience for their occupants: safety and security planning and governance;

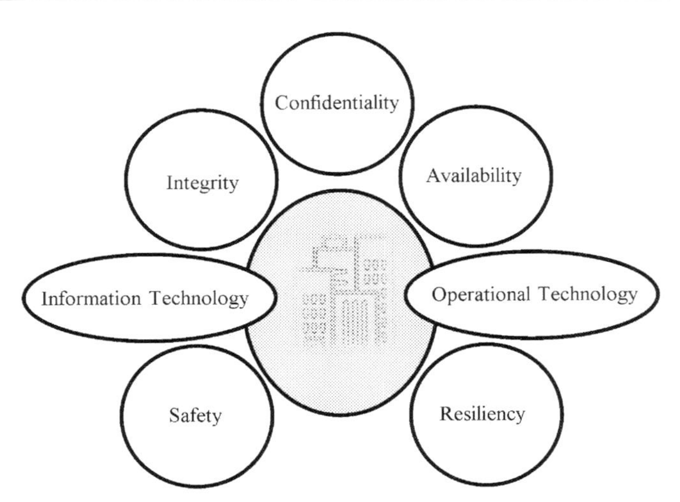

Fig. 5.12 Internet-of-Things as enablement of cyber-physical security.

security practices that can be enhanced by embracing some safety traditional principles; and focus on the HCPS security lessons being learned in the convergence of IT (servers, laptops, and workstations) and OT (HVAC, lighting, elevators, fire alarms, etc.). The building's OT systems reside within the IT infrastructure by IoT and security enablement as shown in Fig. 5.12 where both entities are connected through wired or wireless solutions with means of the ICTs. The security goals of an IB including confidentiality (privacy and authorization of access to data or information), integrity (trustworthiness of the data or information storage), availability (availability of the systems and associated functions when required), safety, and resiliency (predict, absorb, and recover from disturbances) should be grounded on both the objectives of IT as well as those of OT. These security objectives can help building maintain a more secure and resilient operating environment. This holistic approach reflects any cyberattack to a building system and probably on a community that can pose impacts and can pose risks to human safety.

While building automation systems are typically not visible and the vast majority of hackers are not familiar with them, hackers can perform scans of such systems to probe for weakness. Security surveys have concluded that some hackers are sharing such data within their "black hat" communities. A key danger is that compromised building systems could lead to disabled security and access control, which could put lives and critical intellectual property within buildings at risk (O'Neil King, 2016).

Digital twins offer new and exciting possibilities for the built environment by taking a people-centric approach to problem-solving and creating long-term value in terms of security. For buildings that already use IoT systems, digital twins are the next step along unlocking more possibilities to improve efficiencies, optimize processes, detect cyberattacks before they occur, and innovate for the future.

5.13 BIM, digital twin, and blockchain

Professional collaboration among various stakeholders is important for the successful completion of a construction project. However, stakeholders in construction are located in various locations, which in turn impedes responsible information sharing. To address this matter, there is a need for an integrated BIM, digital twin, and a decentralized database framework like the blockchain for traceable data communication (Lee et al., 2021). The digital twin updates BIM in near real-time using IoT, while the blockchain authenticates and adds confidence to all data transactions to the digital twin.

Blockchain is a digitized and decentralized database architecture with built-in security to boost the trust and integrity of transactions. It has four types of architecture including public, private, community, and hybrid. A public blockchain is an open-source platform that allows anyone to sign in and access without any permission. A private blockchain consists of a closed network owned by an organization and is restricted to specific users. A community blockchain allows more than one organization to be involved in the process (Javed et al., 2021). Hybrid blockchain is a combination of public and private blockchain with features of both.

The concept came up in 2008 by Satoshi Nakamoto, who devised the first blockchain database by solving the double-spending problem using the Hashcash method. Blockchain was further adopted as a technique to implement cryptocurrency bitcoin. It differs from a typical database in the way it affords a secure and distributed procedure to record transactions. It stores information where blockchains store data in blocks that are chained together, which accounts for the issues of security and trust in several ways. Each block of information holds transaction data, a timestamp, and an encrypted hash value of the previously linked block. The block is transmitted over the network in a harmonized way that lets each node store a copy of the blockchain through an arrangement between nodes called consensus.

BIM provides all the necessary tools and automation to achieve end-to-end communication, data exchange, and information sharing between project actors. Therefore, blockchain can ensure a secure and controlled collaborative environment around BIM with full governance over the process. The virtual 3D models generated in the context of engaging between BIM and BACS, blockchain can control access to data and information related to a building (Fig. 5.13). The BIoT platform collects data on the behavior of building users, every action that happens to the building, and the performance of every building equipment. All this information is retrieved from BIoT and recorded in a blockchain before it is stored in a private cloud (Skondras and Mavroeidakos, 2019). The management of data could be accessible only by the administrators of the building. Its multilayer encryption using mathematical functions withholds data in a coded string of characters that are hard to crack. Because the information is not stuck on a central server, blockchain can be scaled to fit very large projects.

When blended, BIM, digital twin, and blockchain can work to intensely enhance the efficiency of smart systems. They can work in holding all parties on a project and creates a higher level of intelligibility. Currently, BIM uses peer-to-peer networks for information sharing, but blockchain could make updates in real-time (Ellis, 2020). The combined intelligence of BIM, digital twin, IoT, and blockchain technologies

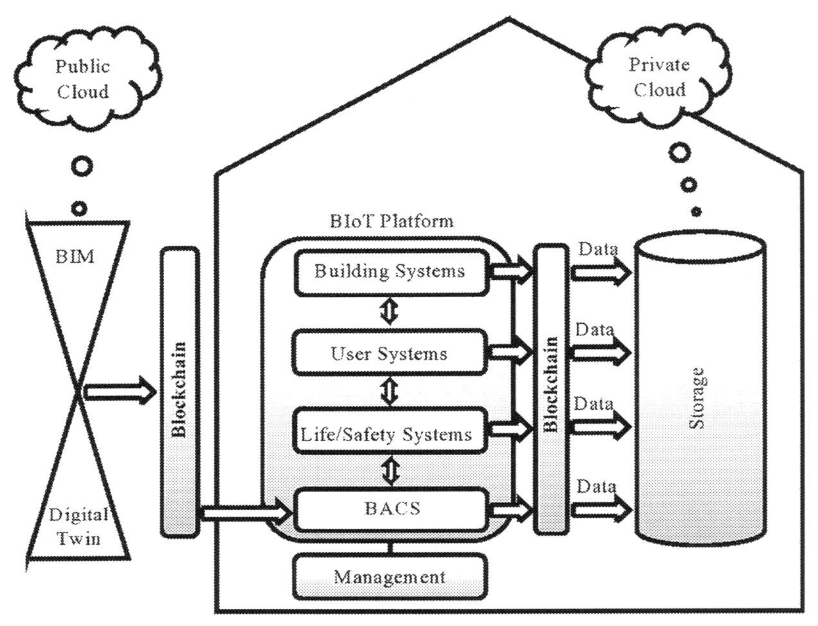

Fig. 5.13 Integration of building information modeling, digital twin, and blockchain in a building scenario.

offer a demonstrable, secure, and robust process to store and manage data generated or managed by connected smart devices. This process will create an innovative framework supporting digital transformation in the building industry, help with intelligibility, lead to better overall communication, and ultimately result in an additional layer of data integrity and security.

References

Aazam, M., Harras, K.A., Zeadally, S., 2019. Fog computing for 5G Tactile industrial Internet of Things: QoE-aware resource allocation model. IEEE Trans. Ind. Inf. 15 (5), 3085–3092.

Aijaz, A., Sooriyabandara, M., 2019. The Tactile Internet for industries: a review. Proc. IEEE 17 (2), 414–435.

Akenberg, C., Nesragi, S.S., Bucaill, A., Littman, D., 2021. Gaining an intelligent edge: edge computing and intelligence could propel tech and telecom growth: TMT predictions. https://www2.deloitte.com/global/en/insights/industry/technology/technology-media-and-telecom-predictions/2021/edge-intelligence-fourth-industrial-revolution.html/#endnote-9.

Anderson, R., 2001. Security Engineering: A Guide to Building Dependable Distributed System. Wiley Computer Publishing, New York.

Anixer, 2020. Smart building infrastructure best practices. https://www.anixter.com/content/dam/anixter/resources/global-technology-briefing/iaap-cb-best-practices-report.pdf.

Barez, F., 2019. Interactive intelligence: human-in-the-loop intelligence. https://www.thedatalab.com/tech-blog/interactive-intelligence-human-in-the-loop-intelligence/.

Bennett, C.J., Raab, C., 2006. The Governance of Privacy: Policy Instruments in Global Perspective, Second Edition. MIT Press, Cambridge.
Bishop, C., 2006. Pattern Recognition and Machine Learning. Springer, New York.
Butler, K.T., Davies, D.W., Cart, W.H., Isayev, O., Walsh, A., 2018. ML for molecular and materials science. Nature 559, 547–555.
D'Oca, S., Hong, T., Langevin, J., 2017. The human dimensions of energy use in buildings: a review. https://simulationresearch.lbl.gov/sites/all/files/t_hong_-_the_human_dimensions_of_energy_use_in_buildings_a_review.pdf.
Davis, D., 2015. How big data is transforming architecture? https://www.architectmagazine.com/technology/how-big-data-is-transforming-architecture_o.
De la Torre, L., 2019. What are privacy-enhancing technologies (PETs)? https://medium.com/golden-data/what-are-privacy-enhancing-technologies-pets-8af6aea9923.
Elliott, N., Molina, M., Trombley, D., 2012. A defining framework for intelligent efficiency. American Council for an Energy-Efficient Economy (ACEEE). Research Report E125 https://www.aceee.org/research-report/e125.
Ellis, G., 2020. Blockchain in construction: 4 ways it could revolutionize the industry. https://constructionblog.autodesk.com/blockchain-in-construction/.
Evans, S., Savian, C., Burns, A., 2020. Cooper C: digital twins for the built environment. https://www.theiet.org/media/4719/digital-twins-for-the-built-environment.pdf.
Frejlichowski, D., Gościewska, K., Forczmański, P., Hofman, R., 2015. Application of foreground object patterns analysis for event detection in an innovative video surveillance system. Pattern. Anal. Applic. 18 (3), 1–12.
Habash, R., 2019. Professional practice in engineering and computing. CRC Press Taylor and Francis, Boca Raton.
Himanen, L., Geurts, A., Foster, A.S., Rinke, P., 2019. Data-driven materials science: status, challenges, and perspectives. Adv. Sci. 16, 1–23.
Javed, I.T., Alharbi, F., Margaria, T., Crespi, N., Qureshi, K.N., 2021. PETchain: a blockchain-based privacy enhancing technology. IEEE Access 9, 41129–41143.
Ji, Z., Yanhong, Z., Baicun, W., Jiyuan, Z., 2019. Human–cyber–physical systems (HCPSs) in the context of new-generation intelligent manufacturing. Engineering 5, 625–636.
Jin, M., 2017. Data-efficient analytics for optimal human-cyber-physical systems. Electrical Engineering and Computer Sciences University of California at Berkeley. Technical Report No. UCB/EECS-2017-228 http://www2.eecs.berkeley.edu/Pubs/TechRpts/2017/EECS-2017-228.html.
Kelechi, A.H., Alsharif, M.H., Bameyi, O.J., Ezra, P.J., Joseph, I.K., Atayero, A.-A., Geem, Z.W., Hong, J., 2020. Artificial intelligence: an energy efficiency tool for enhanced high performance computing. Symmetry 12, 1029. https://doi.org/10.3390/sym12061029.
Koseleva, A., Ropaite, G., 2017. Big data in building energy efficiency: understanding of big data and main challenges. Procedia Eng. 172, 544–549.
Kshetri, N., 2017. Blockchain's roles in strengthening cybersecurity and protecting privacy. Telecommun. Policy 41, 1027–1038.
Lee, D., Lee, S.H., Masoud, N., Krishnan, M.S., Li, V.C., 2021. Integrated digital twin and blockchain framework to support accountable information sharing in construction projects. Autom. Constr. 127, 103688. https://doi.org/10.1016/j.autcon.2021.103688.
Martynova, O., 2020. Digital twins in facility management: the clear path forward the intelligent buildings. https://www.intellias.com/digital-twins-in-facility-management-the-clear-path-forward-for-intelligent-buildings/.
Meagher, M., 2015. Designing for change: the poetic potential of responsive architecture. Front. Archit. Res. 4 (2), 159–165.

Miskinis, C., 2018. Explaining the definition of digital twin and how it works. https://www.challenge.org/insights/what-is-digital-twin/.
Monteith, M., 2019. What is a digital twin? https://www.iotforall.com/what-is-digital-twin-technology.
Müller, V.C., 2020. Ethics of artificial intelligence and robotics. In: The Stanford Encyclopedia of Philosophy. https://plato.stanford.edu/archives/win2020/entries/ethics-ai/.
O'Brien, L., 2019. Cybersecurity for smart buildings. https://www.arcweb.com/blog/cybersecurity-smart-buildings.
O'Neil King, R., 2016. Cyber security for intelligent buildings. Eng. Technol. Ref. 2016, 1–6. https://doi.org/10.1049/etr.2015.0115.
OSGF, 2020. Enabling tomorrow's electricity system: report of the ontario smart grid forum.
Petrova, E., Svidt, K., Pauwels, P., Jensen, R.L., 2019. Towards data-driven sustainable design: decision support based on knowledge discovery in disparate building data. Archit. Eng. Design Manag. 15 (5), 334–356.
Royan, F., 2021. Digital sustainability: the path to net zero for design and manufacturing and architecture, engineering, and construction (AEC) industries. https://damassets.autodesk.net/content/dam/autodesk/www/campaigns/emea/docs/FS_WP_Autodesk_DigitalSustainability.pdf.
Saidam, M.W., Al-Obaidi, K.M., Hussein, H., Ismail, M.A., 2017. The application of smart materials in building facades. Ecol. Environ. Conserv. 23 (Nov, Suppl, Issue), S8–S11.
Schatsky, D., Bumb, S., Kumar, N., 2017. Intelligent IoT bringing the power of AI to the Internet of Things. https://www2.deloitte.com/us/en/insights/focus/signals-for-strategists/intelligent-iot-internet-of-things-artificial-intelligence.html.
Schoechle, T., 2018. Re-inventing wires: the future of landlines and networks. https://electromagnetichealth.org/wp-content/uploads/2018/05/Wires.pdf.
Singh, V., 2013. Why performance based design is the future of architecture. https://sefaira.com/resources/why-is-performance-based-design-the-future-of-architecture/.
Skondras, E., Mavroeidakos, T., 2019. The convergence of blockchain, Internet-of-Things (IoT) and building information modeling (BIM): the smart museum case. In: Wireless Telecommunications Symposium (WTS), April 9–12, New York, USA.
Stankovic, J., 2008. When sensor and actuator cover the world. ETRI J. 30 (5), 627–633.
Staton, B., 2020. What is the future of architecture as a profession? https://www.bdcnetwork.com/blog/what-future-architecture-profession.
Talkhestani, B.A., Jung, T., Lindemann, B., Sahlab, N., Jazdi, N., Schloegl, W., Weyrich, M., 2019. An Architecture of an Intelligent Digital Twin in a Cyber-Physical Production System., https://doi.org/10.1515/auto-2019-0039.
Wang, S., 2010. Intelligent Buildings and Building Automation. Spon Press, London.
WEF, 2020. The future of jobs report. https://www.weforum.org/reports/the-future-of-jobs-report-2020//reports.weforum.org/future-of-jobs-2016/employment-trends/.
Wibrand, J., Saltin, P.-J., 2020. From BIM to digital twins for buildings. https://blog.swegon.com/en/improve-your-building-systems-with-a-digital-twin.
Williams, E.A., 2020. A look inside privacy enhancing technologies. https://www.helpnetsecurity.com/2020/06/16/a-look-inside-privacy-enhancing-technologies/.
Wu, H., Han, H., Wang, X., Sun, S., 2020. Research on artificial intelligence enhancing internet of things security: a survey. IEEE Access 8, 153826–153848.
Yoon, G., Choi, D., Lee, J., Choi, H., 2019. Management of IoT sensor data using a fog computing node. J. Sens. 2019, 1–9. https://www.hindawi.com/journals/js/2019/5107457/.
Zhao, Q., Xia, L., Jiang, Z., 2018. Project report: new generation intelligent building platform techniques. Energy Inf. 1 (2), 1–5. https://doi.org/10.1007/s42162-018-0011-9.

Building as a control system

Riadh Habash
School of Electrical Engineering and Computer Science, University of Ottawa, Ottawa, ON, Canada

Buildings should serve people, not the other way around.

John Portman

6.1 Control systems engineering

The principle of control systems engineering is to understand how the process can be managed by modeling techniques and automation devices to implement such integration into operation. The idea is to decompose a complex system like a building into simpler ones, solve them separately, and then integrate all separate solutions to meet a global objective. This process requires a broad skill set including electrical, mechanical, and computer software engineering.

6.1.1 Model-based control

Model-based control is a mathematical and visual technique of tackling problems associated with designing complex control. One aspect is the control analysis, representing the modeling and simulation aspects of the system including tuning and configuration, and automatic code generation. Each block of the model contains mathematics that allows it to emulate the behavior of the physical system. The other aspect is the system and software, representing continuous validation and verification through testing, and deployment of the system as shown in Fig. 6.1. In this regard, when model-based control is used efficiently, it offers a single design platform to enhance overall system design. Through virtual prototyping, system engineers can easily see whether the whole system working as intended, even before the hardware is manufactured and available for testing.

For control and algorithm designers, the focus is on modeling, which has always been an essential part of the design process. System modeling activities involve creating a mathematical and behavioral representation of the system under consideration. A model is a mathematical representation of a system. It allows simulating and analyzing the system but is never exact. Modeling depends on the goal where a single system may have many models. The main goal of modeling in control engineering is conceptual analysis through various approaches and detailed simulation.

A mathematical model is at best an approximation to the physical world. It specifies the question of which components of the building system should be modeled and

Sustainability and Health in Intelligent Buildings. https://doi.org/10.1016/B978-0-323-98826-1.00006-5

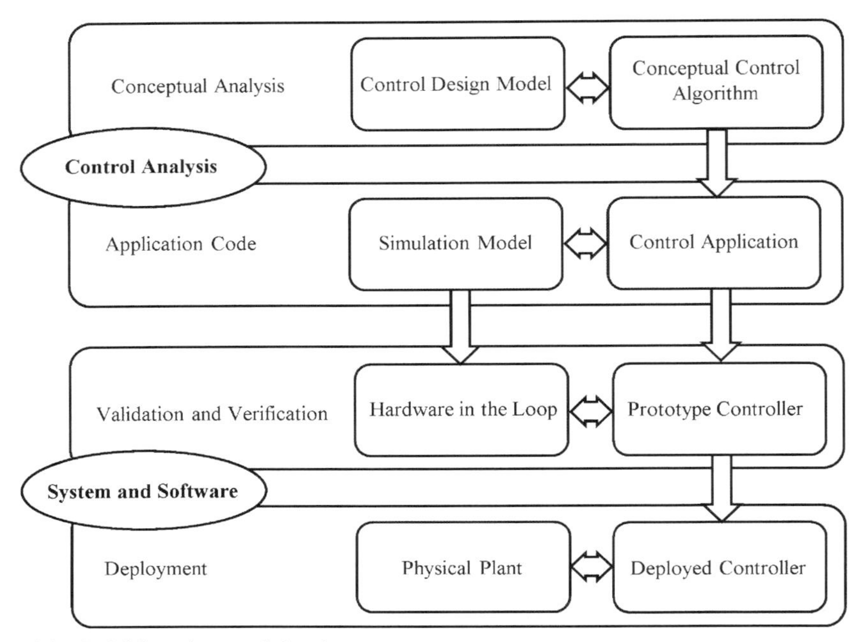

Fig. 6.1 Model-based control development.

then the kind of equation that is used to represent the dynamics of each component. Mathematical models can be classified as linear or nonlinear, steady-state or dynamic, lumped or distributed (Albin, 1997). In finding a mathematical model, it is necessary to make a compromise between the simplicity of the model and the accuracy of the results of the analysis.

Many modeling approaches have been available; however, only two extreme modeling approaches can be generalized. The first one is physical models, which build up models completely based on universal and physical laws. The second approach is called empirical models, which constructs models entirely based on experiments or data. Pure physical or empirical models have both advantages and disadvantages (Lu et al., 2011). Different modeling is required for different computational functions.

Numerous software tools have been developed over the last few decades for the analysis of energy consumption, HVAC design, operation scheduling, lighting information, renewable energies, etc. Regular upgradation is performed on these simulation tools to improve the performance efficiency and decreasing computational cost (Abhinandana et al., 2018). To this end, EnergyPlus is a building energy performance analysis simulation software and console-based program that reads inputs and writes outputs in text files developed based on the building loads analysis and system thermodynamics by the National Renewable Energy Laboratory (NREL) and US DoE Building Technologies Office (BTO).

6.1.2 Control systems

Control systems and emerging communication networks are quickly advancing where occupants and building managers are having more control and interface with the building components while ensuring the minimum comfort levels required by the occupants. Generally, building automation begins with the control of mechanical, electrical, and plumbing systems. For instance, the HVAC system is almost always controlled, including control of its various pieces of equipment such as chillers, boilers, air handling units, roof-top units, fan coil units, and heat pump units. Other systems that are often controlled and/or brought under a complete automation system include lighting system, power monitoring, security close circuit video, card and keypad access, fire alarm system, elevators/escalators, plumbing, and water monitoring.

There are different types of controllers, including mechanical, pneumatic, electronic, digital, and recently intelligent. Early control systems were pneumatic. Due to their large base of installation throughout the 1960s and 1970s, pneumatic control systems are still in place in a majority of existing buildings. Analog electronic control devices became popular throughout the 1980s. They provided a faster response and higher precision than pneumatics. However, it was not until digital control or direct digital control devices came on the scene in the 1990s that a true automation system was possible.

The main feature of a control system is that there should be a clear mathematical relationship between the input and output of the system. When the relationship between input and output cannot be represented by single linear proportionality, rather the input and output are related by some nonlinear relation, the system is referred to as a nonlinear control system. Therefore, the most challenging part in the design of control systems is the accurate modeling of the physical plant or process.

From an academic point of view, the building is a nonlinear multiscale complex system. No single solution exists for such a complex system. Due to the human way of linear thinking, nearly all the theories and methods developed so far are linearly dominated, making it difficult to apply them directly to nonlinear systems. The established linearization approach of analyzing nonlinear systems is based on the assumptions of weak nonlinearities, but these may lead to erroneous results (Kerschen et al., 2006). Therefore, the identification of nonlinear systems is of critical value. Once nonlinear behavior has been detected, model parameters can be estimated using optimization tools such as linear programming, nonlinear programming, and dynamic programming.

6.1.3 Computational intelligence

Analytical expressions are seldom appropriate in practice. Today, computational models and simulation techniques are effective in solving design problems in engineering and other disciplines. They have become affordable and conceivable in both research and industry during the last decades with the fast advancement of the computer industry as well as the vital advancement of computational techniques.

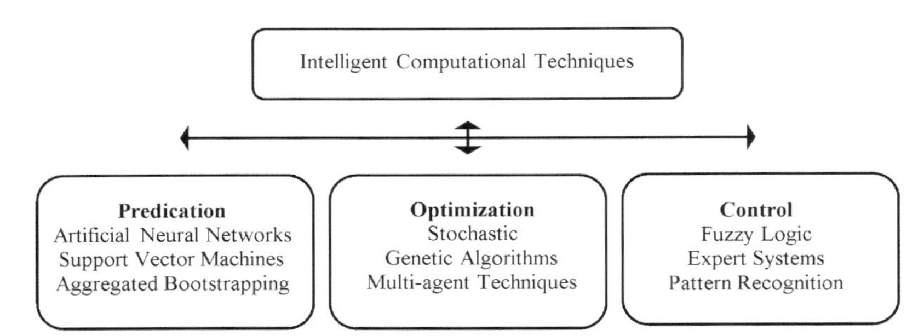

Fig. 6.2 Intelligent computational techniques and their primary area of application.

Intelligent systems underlie what is called "soft computing," sometimes referred to as computational intelligence. Soft computing is an emerging approach to computing that gives the remarkable ability of the human mind to argue and learn in an atmosphere of uncertainty and distrust. It is based on some biological induced methods such as genetics, development, ant behavior, the human nervous system, etc.

Intelligent computational techniques have been successfully applied by researchers to overcome time delay, system uncertainties, and nonlinear features in calculations (Dounis and Caraiscos, 2009), as well as includes HVAC applications such as prediction, optimization, control, and fault detection and diagnosis. These techniques can be classified into three main groups, as shown in Fig. 6.2. ANN and support vector machines have been utilized for prediction, control, and classification purposes. In literature, most of the applications for HVAC systems are mainly based on stochastic-based optimization. For control and fault detection and diagnosis purposes, fuzzy logic (FL)-based controllers and detectors are widely used. Pattern recognition-based methods are mainly applied for fault detection and diagnosis purposes. Multiagent systems can be used for many purposes such as control, monitoring, and detection. For HVAC systems, multiagent systems were mainly used for control purposes.

6.2 Thermal modeling approaches

Thermal modeling is a very powerful tool in IB design and a key to sustainability. Considering thermal phenomena, heat transmission, heat storage, fluid flow, and heat flux represent the fundamental thermal properties of building elements. This requires suitable thermal modeling approaches (Fig. 6.3) depending on how they process information. Comparison between approaches is important since at some stages these may become attractive for energy management.

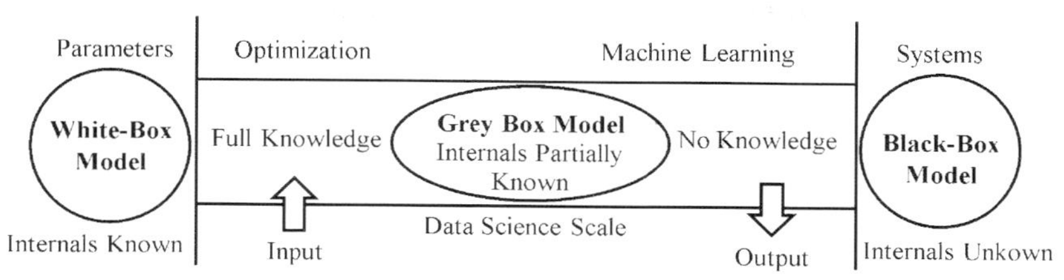

Fig. 6.3 Thermal modeling approaches.

6.2.1 White-box: Modeling with physics

White-box models are thermal dynamics models, which are based on fundamental laws of physics, thermodynamics, and heat transfer. Generally, this approach can be roughly classified into two types: physics-based models using simulation tools such as TRNSYS, EnergyPlus, etc., and physics governing law-based thermal dynamics state-space equation or lumped capacitance model (Salakij et al., 2016).

White-box models are based on static and dynamic models including linear, nonlinear, differentiable, continuous, and noncontinuous. It is a software testing method in which the internal structure/design/implementation of the item being tested is known to the tester. In static models, the output of the system does not depend on time. While in dynamic models, the output is time-varying due to the dynamic heat balance time evolution and is typically represented by differential equations (Amara et al., 2015). Indeed, the white-box models can be constructed from the prior information without the need for any observation.

6.2.2 Black-box: Modeling with data

Black-box models are empirical (statistical) models without significant parameters. This means that black-box models are derived from the inputs-outputs thermal behavior. These are also known as data-driven models which are developed based on statistical models by quantifying historical data parameters and correlating between building performance and data to find an optimal pattern. The parameters are generally adjusted automatically. This automatic adjustment of calibration of black-box parameters provides the greatest benefit over white-box models (Amara et al., 2015). However, a disadvantage is their implicit relationship with physical fundamental principles. The data-driven model approach is often considered less complex with high accuracy and low computational cost. However, the inner process is mostly unknown, leading to reduced control flexibility of the overall process.

The primary requirement of black-box models is precollected data. These can be obtained from; real data collected from the existing building through sensors, smart meters, and other smart systems (Marvuglia et al., 2014); and simulated data collected from the simulation tools such as EnergyPlus, BCVTB, etc. (Macas et al., 2016). In brief, the black-box model is a software testing method in which the internal structure,

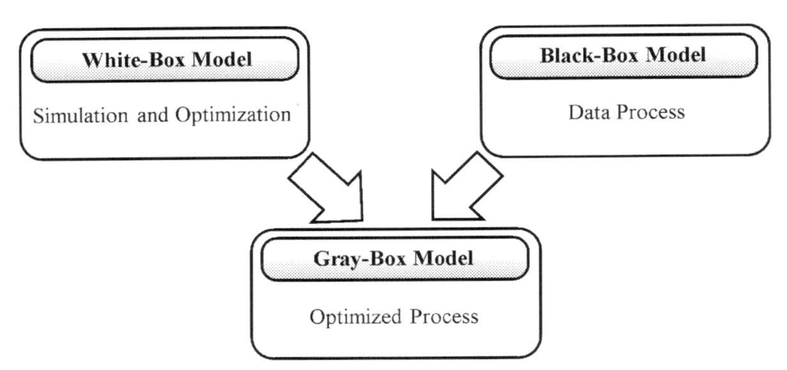

Fig. 6.4 Formation of gray-box models.

design, and implementation of the item being tested are not known to the tester. It has high accuracy, low computational cost, and higher flexibility for building nonlinearities.

6.2.3 Gray box: Hybrid modeling

In thermal modeling of buildings, it is reasonable to combine the relative strengths of the black-box coming from the statistical with the white-box strengths based on physical interpretation to obtain a hybrid model. In that sense, the standard "gray-box" approach is based on both, a statistical method and physical properties that meet the physical fundamental principles (Amara et al., 2015). A gray box model is a balanced model between the white-box model and a black-box model (Fig. 6.4). This model uses a combination of simplified physics and historical data. The process is expressed in a mathematical expression that may be based on the physics and/or thermodynamics laws. It consists of expressions that have a physical explanation (e.g., resistor capacitance network) and a part of the model may be obtained through regression from the available data. This combination ensures that the nonlinearities in the white-box model can be handled using black-box models and the lack of laws of physics reasoning in the black model can be represented through white-box models, but the extra effort is required to design and develop these models (Abhinandana et al., 2018). Compared to the black-box and white-box models, the gray box method has the advantage of using previous knowledge and the information available from existing data.

6.3 Analog control systems

The analog control system, also called continuous-time control or continuous data-flow system, includes elements that only produces or processes signals in continuous time. Whether the system is linear or nonlinear, all variables are continuously present and therefore known at all times. In general, the control system can be classified into

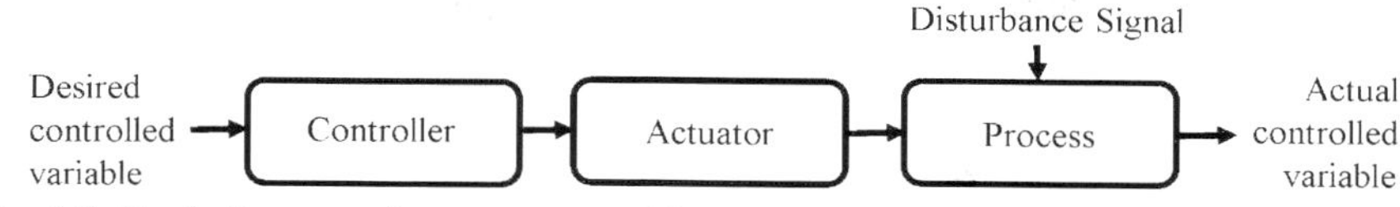

Fig. 6.5 Block diagram of an open control loop.

two fundamental types, open-loop and closed-loop (feedback). The objective of the control system design is to construct a system that has a desirable response to standard inputs. A desirable transient response should be sufficiently fast without excessive oscillations and follows the desired output with sufficient accuracy.

6.3.1 Open-loop systems

An open-control system in which the control action is independent of the output is shown in Fig. 6.5. The process (plant) under control is a continuous-time system (for example, motor, robot, power plant, etc.). In such a system, the process output variable (controlled variable) is determined by the combined effects of the manipulated input and disturbance inputs. The controller, according to the desired value of the controlled variable, generates a control signal for the actuation devices by implementing the control law or control algorithm, which is practically based on the predicted correlation between the process input and output. Open-loop systems, without considering the effects of disturbances can be satisfactory or acceptable if the disturbances are not great and the changes in desired values are not too severe. However, in many practical control systems, the effects of disturbances are great or the errors in the controlled variables caused by such disturbances can be serious.

6.3.2 Closed-loop systems

The closed-loop control is implemented in most controlled processes where feedback is a common and powerful tool when designing a control system (Fig. 6.6). The feedback loop is the tool that considers the system output and enables the system to adjust its performance to meet the desired result of the system. When the feedback signal is negative then the system is called a negative feedback system. For such a system, the error signal is given by the difference between the reference input signal and the

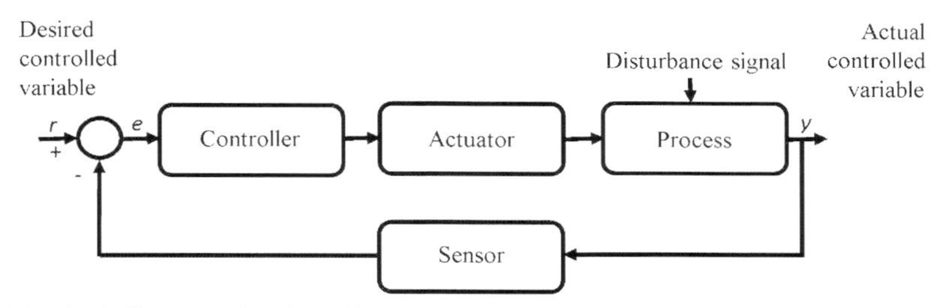

Fig. 6.6 Block diagram of a closed-loop system.

feedback signal. When the sensor reading has changed the operator will then compare the current reading (the controlled variable y) with the required quantity (the set-point r) and compute the difference between them, to obtain the error signal $e = r - y$.

A feedback control system is a control system that tends to maintain a relationship of one system variable to another by comparing functions of these variables and using the difference as a means of control. As the system is becoming more complex, the interrelationship of many controlled variables may be considered in the control scheme. A simple example of the closed-loop control system is a person steering an automobile by looking at its location on the road and making the proper adjustments. Feedback is used in conventional control systems to reject the effect of disturbances upon the controlled variables and to bring them back to their desired values according to a certain performance index.

While the open-loop system is stable, the closed-loop system may not always be stable. The instability problem of the closed-loop control system is often an important and difficult matter in control system design. The addition of feedback to dynamic systems results in several additional problems for the designer. However, in most cases, the advantages far outweigh the disadvantages, and a feedback system is utilized. Therefore, it is necessary to consider the additional complexity and the problem of stability when designing a control. Finally, it is necessary to keep in mind that if an open-loop control, of simple design, for a process can achieve satisfactory accuracy, an open-loop system should be considered first.

6.3.3 Process control

In most engineering applications, it is necessary to control a physical system or plant so that it behaves according to given design specifications. Typically, the plant is analog, the control is piecewise constant, and the control action is updated periodically. The controller (or regulator) is the element that generates the control signal, which is the input signal to the plant. The implementation of the analog regulator requires a system that usually consists of several components and devices. The controller usually provides a transfer function whose input signal is the error, the difference between the reference signal and the feedback signal. The controller computes the error signal (or the difference) and generates an appropriate input signal to the plant to provide the desired output signal.

It has been observed that the control function proportional, integral, derivative (PID) is widely used in various industrial applications including in buildings (Wang, 2010). The controller must operate upon this input in some way to generate a suitable output signal to feed the valve or other final control element. PID controllers are simple in principle and have played a very important role in the progress of automatic control and are used as the most popular functions in process control applied in various forms.

Generally, the gain (P) is an adjustable parameter whose value can be chosen and set by the operator. In theory, the output signal may take any value but in practice, its range is limited by the finite range of controller output or finite range of the final control element: a valve or damper can go only from fully closed to open. Proportional plus integral (PI) control is the most common mode. The reason proportional control suffers from steady-state errors is that a change in desired value or disturbance will require a new value of a manipulated variable to achieve equilibrium at the new operating conditions. Although derivative (PD) functions help solve a variety of control design problems, one of their most important contributions is in system stability improvement. If absolute or relative stability is the problem, a suitable PD control mode is often the answer. If both the transient and steady-state response of the system must be improved, then neither a PI nor a PD controller may meet the desired specifications. With a PID controller, the two functions may be improved.

6.4 Digital control systems

The digital control system (Fig. 6.7), also called a discrete-time system, discrete data-flow system, or sample-data systems, includes components that produce or process discrete signals to one or more parts of the system. The process under control is a continuous-time system. The controller is a digital computer. A digital control system may be considered from different viewpoints including control algorithm, computer program, conversion between analog and digital domains, system performance, sampling process, etc. The process (plant) under control is a continuous-time system.

6.4.1 *Signal converters*

The analog-to-digital converter (ADC) component of the digital control system converts a continuous-time signal into a discrete-time signal specified by a clock. The digital-to-analog converter (DAC), by contrast, converts the discrete-time signal output of the computer into a continuous-time signal to be fed to the plant. The DAC usually contains a sample and hold circuit. The simplest and most popular way of reconstructing a continuous-time signal from a discrete-time signal in control systems is to simply hold the signal constant until a new sample becomes available. Clearly, the smaller the ADC resolution, the better the performance, and therefore, a 32-bit ADC is preferred over lower bit rate systems. However, the cost of the component increases as the word length increases, and the presence of noise might render the presence of a high

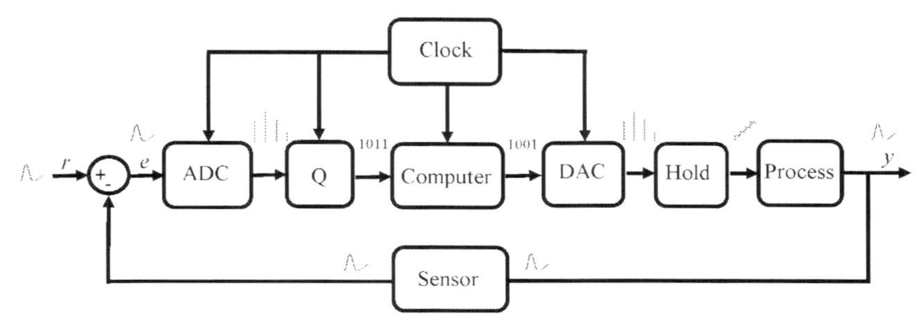

Fig. 6.7 Block diagram of a closed-loop digital control system.

number of bits useless in practical applications. The DAC resolution is usually chosen equal to the ADC resolution, or slightly higher, to avoid introducing another source of error called quantization error. Once the ADC and DAC resolution has been selected, the resolution of the reference signal representation must be the same as that of the ADC and DAC. With the conversion time provided by standard modern ADCs, this is not a significant issue in most applications. The choice of the ADC and DAC word length is therefore mainly determined by the quantization effects. Typically, commercial ADCs and DACs are available in the range of 8–32 bits.

6.4.2 Sampling frequency

A sample is a piece of data taken from the whole data which is continuous in the time domain. The term sampled data covers normal analog (continuous-time) systems with the distinctive characteristic that the input and the output are piecewise constant signals. This discretization of the analog signal is called "sampling" which is necessary for the processing of analog data using digital elements.

The sampling frequency is the reciprocal of the sampling period. This sampling frequency can be simply called the sampling rate. The sampling rate denotes the number of samples taken per second, or for a finite set of values. For an analog signal to be reconstructed from the digitized signal, the sampling rate should be highly considered. The rate of sampling should be such that the data in the message signal should neither be lost nor it should get overlapped. The sampling rate denotes the number of samples taken per second, or for a finite set of values. For effective reproduction of the original signal, the sampling rate should be twice the highest frequency. If the sampling frequency does not satisfy the sampling theorem (i.e., the sampled signal has frequency components greater than half the sampling frequency), then the sampling process creates new frequency components. This phenomenon is called aliasing and must be avoided in a digital control system. Hence, the continuous signal to be sampled must not include significant frequency components greater than the Nyquist frequency $f_s/2$. In digital control, the sampling frequency must be chosen so that samples provide a good representation of the analog physical variables.

6.4.3 Control algorithm

As in conventional types of controllers used in buildings, PID also plays a very important role in most control loops controlled by a digital controller. The PID algorithms used in actual controllers from different manufacturers differ greatly. The controller could be added in the forward path, in the feedback path, or in an inner loop. A prefilter could also be added before the control loop to allow more freedom in design. Several controllers could be used simultaneously, if necessary, to meet all the design specifications.

In a digital control system, the control algorithm is implemented in a digital computer. The designer of a digital control system must be mindful of the fact that the control algorithm is implemented as a software program that forms part of the control loop. In the context of control and communication, sampling is a process by which a

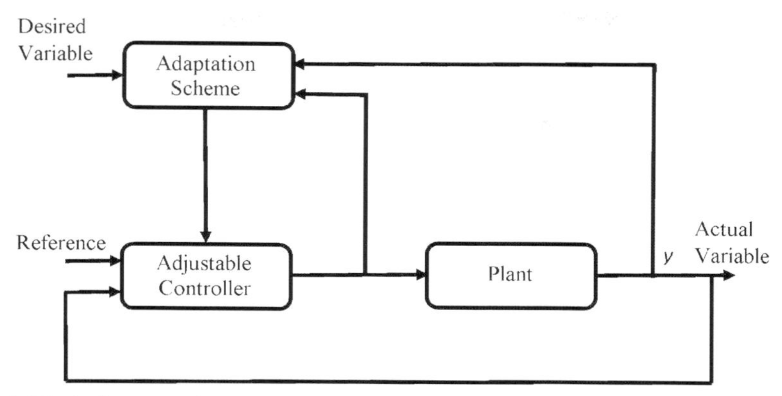

Fig. 6.8 Block diagram of an adaptive control system.

continuous-time signal is converted into a sequence of numbers at discrete time intervals. It is a fundamental property of digital control systems because of the discrete nature of the operation of digital computers. The implementation of a digital control system requires skills in software engineering and computer programming. There are well-known programming guidelines that help minimize execution time and control jitter for the control algorithm (Fadali and Visioli, 2013). Digital control systems have additional requirements such as data storage and user interface, and their proper operation depends not only on the correctness of their calculations but also on the time at which their results are available.

6.5 Adaptive control systems

Adaptive control covers a set of techniques that provide a systematic approach for automatic adjustment of controllers in real-time, to achieve or to maintain the desired level of control system performance when the parameters of the plant dynamic model are unknown and/or change in time.

In everyday language, "to adapt" means to change behavior to conform to new circumstances. Conceptually, an adaptive controller is thus a controller that can modify its behavior in response to changes in the dynamics of the process and the character of disturbances. An adaptive controller can be defined as a controller with adjustable parameters and a mechanism for adjusting the parameters (Wang, 2010). The reason adaptive control is needed is that the constant parameters of a controller might provide satisfactory performance in one condition but not in another due to significant changes like the system.

The dynamic model of the plant can be identified from input/output plant measurements obtained under an experimental protocol in an open or in a closed-loop. One can say that the design and tuning of the controller are done from data collected on the system. An adaptive control system can be viewed as an implementation of the above design and tuning procedure in real-time. The tuning of the controller will be done in real-time from data collected in real-time on the system. The corresponding adaptive control scheme is shown in Fig. 6.8.

An adaptive control system, which contains in addition to a feedback control with adjustable parameters a supplementary loop acting upon the adjustable parameters of the controller, will monitor the performance of the system in the presence of parameter disturbances. How information is processed in real-time to tune the controller for achieving the desired performances will characterize the various adaptation techniques.

6.6 Intelligent control systems

Intelligent control is transdisciplinary as it combines and extends theories and methods from areas such as control theory, mathematics, engineering, and computer science. It refers to approaches to control system modeling, design, and operation that use AI techniques, such as FL, neural networks, ML, evolutionary computation, and genetic algorithms.

6.6.1 Artificial neural networks (ANN)

ANN, as a data-driven technique and as part of model-based predictive control, is a powerful method that can deal with linear and nonlinear characteristics. This black-box model is a self-learning controller that can be used for building energy prediction and HVAC system control. It is one of the most prominent forecasting ML algorithms being applied and made by regular computer programming as if it is mutually associated with brain cells.

ANNs are commonly classified by their network topology, node characteristics, learning, or training algorithms. On the other hand, the potential benefits of neural networks extend beyond the high computation rates provided by the massive parallelism of the networks. They exploit massively parallel local processing and distributed representation properties that are believed to exist in the brain. The parameters of the ANN-based controller (Fig. 6.9) are designed based on both the identified model of the system and the linear controller (Jafari and Xu, 2018). Furthermore, the robustness of the proposed controller concerning variations in system parameters has been demonstrated through different simulations.

Motivated by the structure and working of the human brain, neural networks are composed of interconnected simple elements, so-called neurons, of the order of 10^{11} with about 10^{15} connections ANN is inspired by a human nervous system with its neurons, axons, dendrites, and synapses. A neural network is a parallel distributed processor that stores knowledge from experience and makes it available to use (Haykin, 1994). ANN resembles the human brain in two ways: the network acquires the knowledge through the learning process, and interneurons connection strengths (synaptic weights) are used to store the knowledge (Ahmad et al., 2016).

While ANN models have many advantages, they suffer from a few limitations. First, as with all data-driven models, ANNs do not perform well outside of their

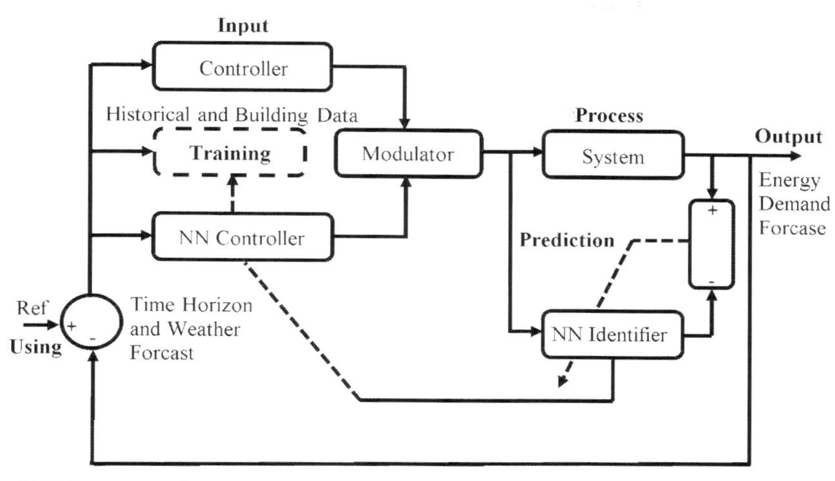

Fig. 6.9 The proposed neural network-based control architecture.

training range. For example, a model trained within a certain data set of winter, might not perform well outside that data set, e.g., in summer. Accordingly, models are limited to the range of values that occurred in training.

6.6.2 Fuzzy logic (FL)

FL is a mathematical logic, which attempts to solve problems with an open, imprecise spectrum of data that makes it possible to get an array of precise findings. FL is designed to be considered the best possible decision by considering all available information and looking at the input.

An FL controller (Fig. 6.10) utilizes FL to convert the linguistic control strategy based on expert knowledge into an automatic control strategy. To use FL for control purposes, it is needed to add a front-end "fuzzifier" and a rear-end "defuzzifier" to the usual input-output data set as shown in Fig. 6.10. The FL controller contains four components: rules, fuzzifier, inference engine, and defuzzifier. Once the rule has been established, it can be considered as a nonlinear mapping from the input to the output.

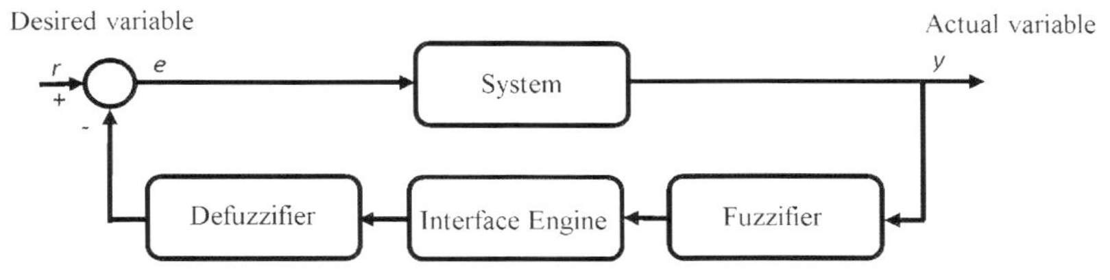

Fig. 6.10 Block diagram of a fuzzy logic controller.

6.6.3 Hybrid systems

Hybrid intelligent control, together with the rapid development of neural computing and learning, has recently gained broad attention and remarkable applications in various areas, such as autonomous underwater vehicles, unmanned aerial vehicles, self-driving cars, and smart manufactories. As a consequence, increasing demands for productivity, safety, system stability, and reliability, are posing challenging theoretical and technological problems in modeling, control, decision making. Neural computing and learning researchers have found new efficient solutions for improving the performance and safety of control systems. Novel intelligent control topologies turn out to be imperative to integrate neural networks, deep learning, evolutionary computation, and other artificial intelligence methods for control systems that have strong heterogeneity and nonlinearity.

Hybrid application methods have proven to be effective in designing intelligent control systems. As it was shown in recent years, FL, ANN, and evolutionary computations are complementary methodologies in the design and implementation of intelligent systems. Each approach has its merits and drawbacks. To take advantage of the merits and eliminate their drawbacks, several integrations of these methodologies have been proposed by researchers during the past few years. These techniques include the integration of ANNs and FL techniques as well as the combination of these two technologies with evolutionary methods.

6.7 Building automation and control systems (BACS)

BACS, or building automation system (BAS) or building management system (BMS), comprises hardware and software that allow for automated control and monitoring of various building systems by integrating and connecting various technologies through information flow. BACS technology and its connectivity extends across all types, sizes, and functions of facilities for not only automation but the free flow of information.

The BACS technical architecture is divided into three layers, namely management, automation, and field as shown in Fig. 6.11. The management layer contains the human interface, generally on the organization's enterprise network including operator and monitoring units, and other peripheral computer devices connected to data processing devices. This layer is generally an information system software package, located and alongside many other business systems within the corporation's ICT.

The automation (supervisory) layer provides the primary communication and control. It consists of hardware and software where the software part represents the messages or the signals carrying the information about the status of the device. Feedback loops and control logics are subject to this layer which is associated with controllers that serve plants like the air handling units, chillers, and boiler units, etc.

The field layer is associated with the physical input sensors and output activators, application-specific controllers, connected to plant and equipment to monitor and control the environment. This layer provides connectivity from the many field devices to

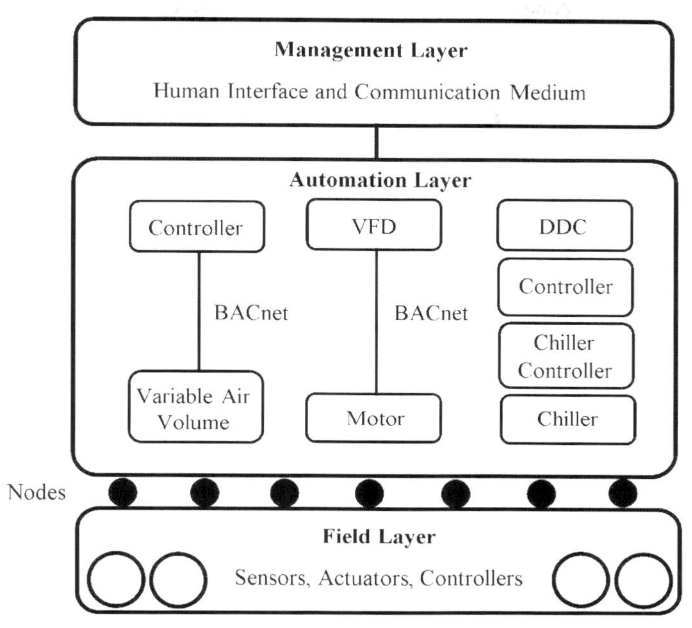

Fig. 6.11 The three major layers of BACS.

the automation layer controllers. The field devices are spread throughout all parts of the building and provide monitoring and control functions. They are installed throughout the BACS to continuously measure and monitor the physical parameters.

BACS is modular, formed from the integration of devices, equipment and communication platform networks with open communication protocols. Different levels of language can be used to communicate instructions to a computing system. Typical examples include machine language (expressed in machine code), assembler language (expressed in mnemonics), high-level languages (expressed in codes closely related to English words) (Wang, 2010). Fig. 6.12 shows a centralized architecture of a BACS. For BACS to function, there is a requirement for connectivity and common language communication. Most BACSs consist of a primary and secondary bus that connects high-level controllers, commonly PLC to control multiple loops, lower-level controllers, input/output devices, and a user interface. Connectivity is achieved via various communication networks that link and integrate the many discrete devices. Such a requirement has led to several building automation networks and communication protocols being established. These include ASHRAE's Building Automation and Control Networks (BACnet), a standard protocol that allows control units from different manufacturers to share information starting from what type of cable should be used to particular requests and commands; Local Operating NetWork (LonWorks); Dynalite network and protocol (Dynet); European Home System; ZigBee (an industry-standard protocol for wireless networking), etc.

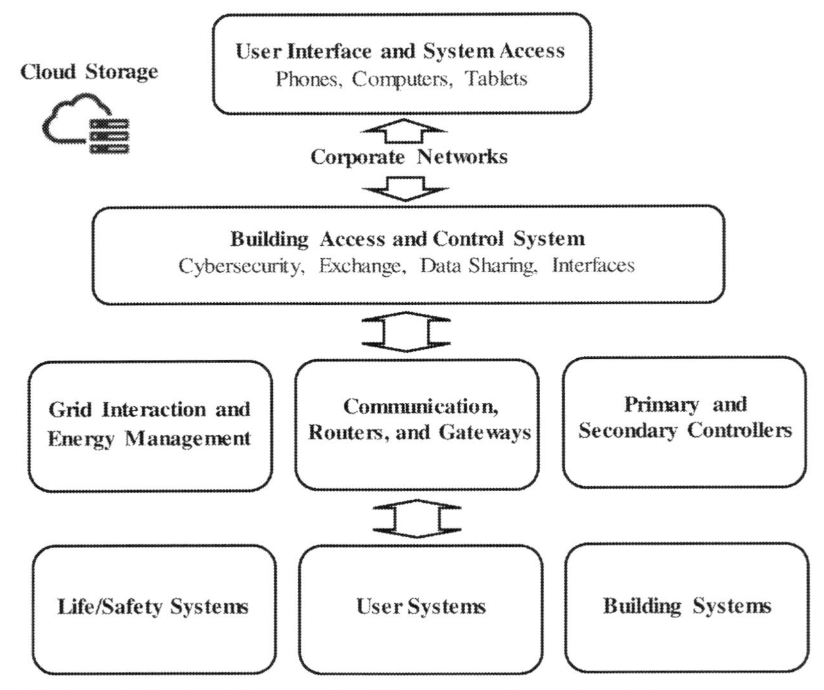

Fig. 6.12 A centralized architecture of building automation and control system.

The building is equipped with an on-premises building energy management system (BEMS) that handles the usual building automation tasks for various subsystems (e.g., HVAC and lighting). The BEMS software runs in a distributed computing environment, including the IoT gateway software components. On gateways, it executes rules on sensor data, and on a server connected to the Intel enterprise network, it analyzes filtered data from the gateway. The BEMS does not control the systems; instead, it monitors, measures, collects, analyzes, and optimizes data of energy consumption. Therefore, BEMS is normally considered as part of the BACS. As a core function, BEMS monitors the heating system, boilers, pumps, chillers, cooling systems, and the systems that spread air throughout the building such as fans or opening/closing dampers. It offers also monitoring, metering, as well as submetering, functions that help collect energy data, giving property managers and owners a comprehensive insight into the building's energy usage.

An effective, next-generation BACS is a platform for integrating building, business, device IoT data and segment-specific specialty systems such as air quality monitoring. Such systems are far broader in scope and capabilities than traditional systems originally contemplated by ASHRAE (Dovan, 2020). Further, EN ISO 16484-1 is an international standard to guide principles for project design and implementation and for the integration of other systems into the BACS.

6.8 HVAC control

The invention of the first modern air conditioning system by Willis Carrier in 1902 created the ability to have a fully controllable thermally stable environment in buildings. Mostly, HVAC systems are categorized based on the fluid media used in the thermal distribution system. The most common way of moving heat energy outside North America is through the use of water. Airflow is the most common means of distributing energy in North American buildings. Air-water also exists for a higher cooling load. All-air systems are often used in buildings that require individual control of multiple zones like schools, hotels, hospitals, and office buildings. The major advantage of these systems is the potential for use of outside air for ventilation and cooling. The major drawback of using air as a heat transfer fluid is its very low heat capacity larger than if water were used.

6.8.1 System components

An HVAC system is like the pillar of a building ecosystem that provides more than human health and comfort. The HVAC system produces heat, cool air, and ventilation, and helps control dust and moisture, which can lead to adverse health effects. It is a building's HVAC system that makes sure its occupants are comfortable in the space they are in. HVAC services involve heating, ventilating, zoning, air conditioning, addition or removal of humidity, filtering and cleaning, and pressure control as shown in Fig. 6.13. These are independent functions, although they are often combined within a specific system. Since each of these functions may generally be undertaken in several ways, the total number of system sequences is large.

The HVAC system is a complex nonlinear system that has different variables as the parameters of the systems. It consists of equipment, sensors, and controllers that control several variables of the system. While HVAC systems have certain features like nonlinearity, time-dependent, time-varying system dynamics, insufficient data, complex interactions between the components, and limited supervisory controls, thus modeling the HVAC systems is a very characteristic and challenging process (Afram and Janabi-Sharif, 2014). Therefore, while developing an HVAC system

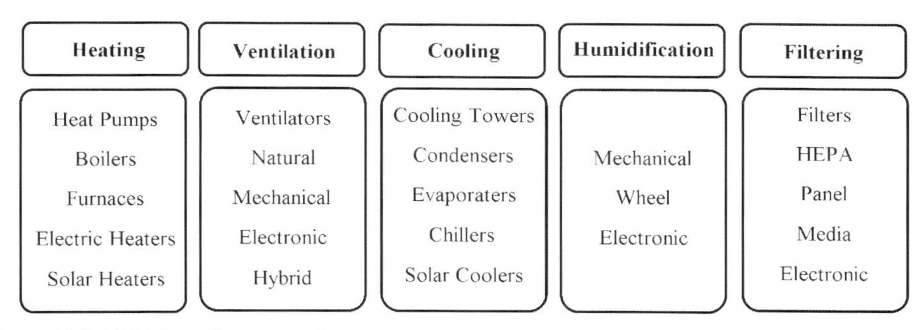

Fig. 6.13 HVAC equipment and components.

and component model, close attention should be given to deciding on the accurate model structure, model parameters, and constraints.

HVAC system large-scale nonlinearities include large thermal inertia, time variability, nonlinear constraints, uncertain disturbance factors, and coupled properties for both temperature and humidity. The variables to be controlled are temperature, air quality, air motion, and relative humidity. Several factors influence thermal comfort in buildings including air temperature, air velocity, relative humidity, radiant environment, clothing, and activity level. Air temperature is the most common measure of comfort and the one that is most widely understood (Wang, 2010). HVAC system provides the controlled environment in which the following parameters are maintained within desired ranges: temperature, humidity, air distribution, and IAQ to building up a comfortable and healthy indoor environment for people to work or live in.

6.8.2 Control system

Today, with SD as a goal, the top priority is to optimize the control of the HVAC system while creating a healthy and comfortable indoor environment, to achieve maximum comfort with the minimum energy consumption, reduce the use cost, reduce the energy consumption, and solve the problem (Shao, 2021).

The existing HVAC control systems, generally, can be divided into air temperature regulator type, which is the most widely used, and thermal comfort regulator type which mainly uses the predicted mean vote index to help to save energy of the HVAC system (Conceição et al., 2018). The traditional HVAC system uses air temperature regulator type which employs temperature and humidity as control parameters; therefore, it has a single control goal, low comfort, and high energy consumption. However, this type cannot ensure the user's best thermal comfort accordingly to the thermal comfort concept. The predicted mean vote index, which is widely used, is a nonlinear function of the indoor air temperature, mean radiant temperature, indoor air relative humidity, indoor air velocity, activity level, and clothing level.

Building HVAC controllers typically utilize multiple PI/PID controllers to regulate these indices by individual single input single output control loops (Anderson et al., 2005). Early investigations into HVAC control focused on distributed PI controllers. The dynamics of a multizone HVAC system are too complex for a PI/PID control to be a sufficient controller for an HVAC system. The software interprets information from various sensor points to optimize the HVAC system's operation while improving occupant comfort. Smart HVAC controls can limit energy consumption in unoccupied building zones, detect and diagnose faults, and reduce HVAC usage, particularly during times of peak energy demand.

Closed-loop control is accomplished by the control signal being sent to the controlled device with constant feedback from the sensor/status device providing input to the controller. The air conditioning system (Fig. 6.14), with its control loops for a temperature cascade control, is used for education and research on building energy management systems. In its structure, it is equivalent to common industrial

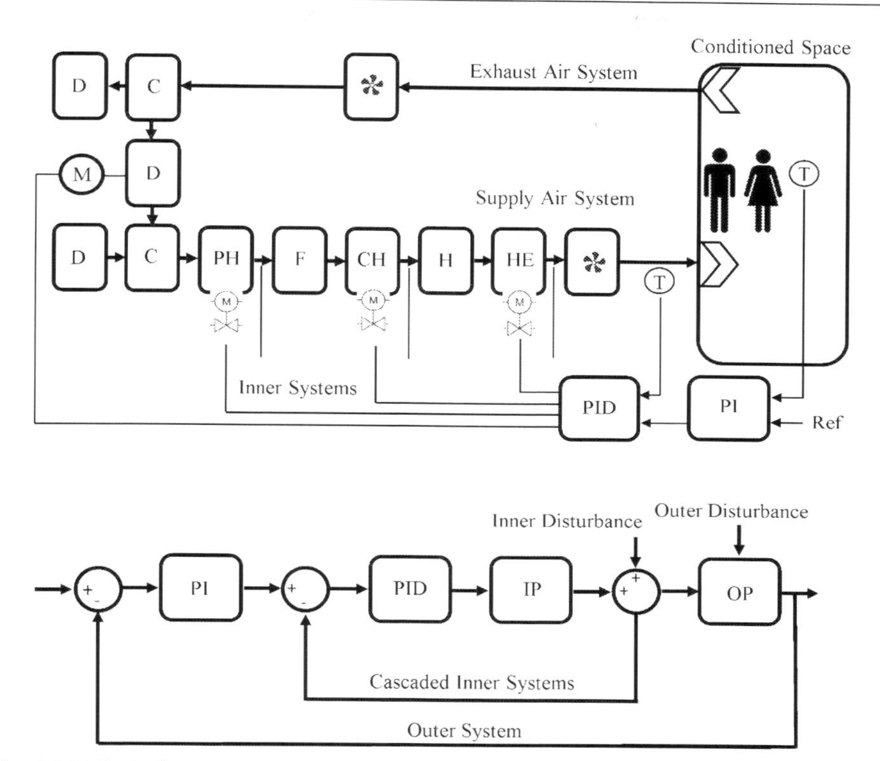

Fig. 6.14 Block diagram of the building air-conditioning system. *IP*, inner process; *OP*, outer process.

applications. The room temperature (T) is controlled by conditioned supply air. The supply air system consists of motorized dampers (D), a preheat exchanger (pH), a chillier (CH), and a second heat exchanger (HE) to condition the air temperature. The steam humidifier (HU) and filter (F) are not taken into account for the temperature control (Zilouchian and Jamshidi, 2001).

The HVAC system operation becomes difficult when it is needed to optimize simultaneously the building energy consumption, ensuring the occupants' comfort and preserving the indoor air quality (Conceição et al., 2018). The system itself relies on the principle of air movement to work. In this way, it can reduce the power consumption of air conditioning and increase the temperature. At this level, airflow is no longer the external environment of the system, but a parameter within the system (Lv et al., 2021). After the introduction of this index, the reference point of the automatic control system of air conditioning is not only the parameters of the indoor air environment but also the temperature experience of indoor personnel (Shao, 2021). This will undoubtedly make the control of air conditioning humanized and bring greater comfort to people.

6.8.3 Optimization techniques

Optimization is a process in mathematics that is used to maximize and minimize a specific function. Optimization techniques have been expansively considered and practiced on HVAC design problems. Regular simulation-based optimization, especially at the HVAC system configuration level is a unique concept. HVAC design optimization problems can be classified into two types. The first type is the optimization of stationery design parameters and the second type is the optimization of the dynamic input variables, which typically include control scheduling and set points. The stationary variables are generally system design parameters that are unchanging in each simulation (Zhang, 2005).

Optimization works related to HVAC systems have been done mainly to reduce the energy consumption of the system while maintaining a comfortable indoor environment. The optimization of HVAC system operations had moved from the basic based-on-system model optimization to control system optimization and then to building design optimization. HVAC's operational parameter optimizations are done based on the system model to find suitable parameter settings for low energy consumption operation. Meanwhile, optimizations based on the control system design of HVAC systems have the purpose of improving the system response to avoid unnecessary energy usage (Selamat et al., 2020). The building design optimizations aim to find suitable building parameters and designs that will enable the efficient use of energy for the HVAC system.

A wide range of optimization algorithms is used in various HVAC optimization problems. However, the decision about the selection of a suitable algorithm is vital for the optimization of HVAC configurations. One of the most popular optimization methods is the genetic algorithm, which is an optimization technique that is based on the theory of natural selection. Optimization problems can be solved using a genetic algorithm to reduce the error between the simulated and actual data for the testing period. Algorithms has been used to optimally tune the fuzzy control rules and membership functions in controlling the set point temperature to minimize energy consumption while providing predicted mean vote values that are within the comfortable range (Ahmad et al., 2016).

6.9 Lighting control

Lighting is a key factor influencing both occupants' comfort and energy consumption in buildings, the fact that requires effective control strategies to enhance energy efficiency. Such strategies decrease the operating time of light fixtures based on various factors like occupancy, time of day, and availability of daylight (Asif ul Haq et al., 2014). In addition to energy efficiency, lighting controls also help ensure that lighting is delivered at the right level for certain zones or places. Today, various approaches (Fig. 6.15) exist for lighting control in buildings. The control network typically is established through proprietary digital software and cabling. The existing lighting control systems incorporate load management and interconnect building spaces from a centralized control system.

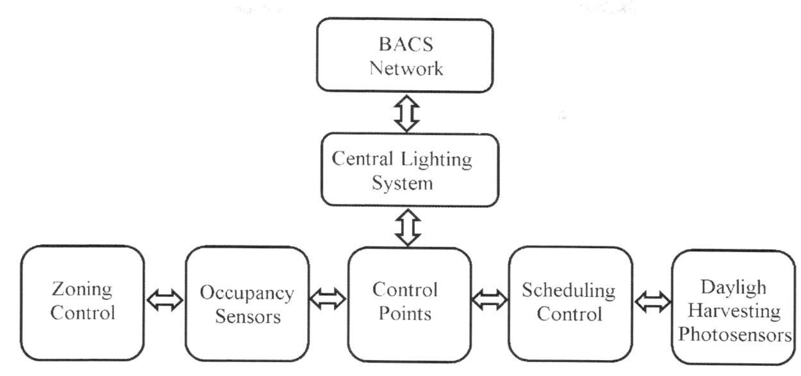

Fig. 6.15 Block diagram of a central lighting system.

6.9.1 Lighting control approaches

As far as the choice of automation technology is concerned, occupancy sensors are significantly ahead of scheduling and daylight sensors. Occupancy sensors employ some sort of motion-sensing technology to detect the presence of occupants in a given range of space, so the lights are switch on when it detects any occupant and switched off when there is no occupant within a prefixed delay period. Passive infrared occupancy sensing in areas that are occupied intermittently, occupancy sensors can be used to indicate whether or not anybody is present and switch the light on or off accordingly. A built-in sensor in an automatic lighting control will help energy efficiency, as it can be preset to natural light and detect when someone is entering or exiting. Standalone occupancy and light sensors may be mounted in or attached to luminaires for autonomous luminaire control.

Utilizing daylight offers a natural possibility for reducing energy consumption and enhancing visual comfort. For buildings with provision to receive daylight, the lighting control schemes that are linked to daylight availability can provide the maximum amount of savings, given that the factors related to daylight availability like orientation (Williams et al., 2018). Daylight harvesting controls can be divided into two mechanisms including switching and dimming. Switching can control the lights by switching between "on" and "off" states based on available daylight. The systems also depend on the algorithm of control such as open-loop and closed-loop. A closed-loop system continuously detects light levels of the control zone, which includes light from both daylight sources and light fixtures (Fig. 6.16). The change in the light levels of the light fixtures due to the availability of daylight is feedback to the control system continuously, and it can make necessary adjustments based on the feedback (Asif ul Haq et al., 2014). On the other hand, an open-loop system does not receive any feedback from the level of electrical lighting; it only detects available daylight levels.

Timing or light scheduling systems are based on time is another control method. It is useful in areas where the occupancy pattern is accurately predictable. Properly commissioned time-based control systems can provide substantial savings. These systems run on an easy principle based on fixing the operating time of the light fixtures.

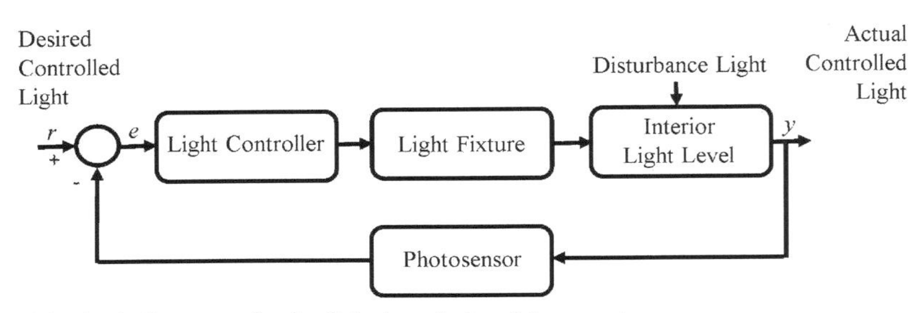

Fig. 6.16 Block diagram of a daylight-based closed-loop system.

The lights that are controlled by the control system are switched on and off based on a prefixed schedule. For scheduling purposes, BAS and time switches are both used commonly. But BAS is preferred for larger buildings, particularly for new construction projects, since the initial cost is much higher in this case, and also the commissioning process is more complicated (Asif ul Haq et al., 2014). Therefore, for smaller buildings and retrofit projects, time switches (mechanical, electromechanical, or even digital) are often used for scheduling due to comparatively lower costs and easier implementation.

Zoning control is an important aspect of lighting control system design as zoning is the process of dividing the floor plan into smaller zones that have separate light controls assigned to lighting loads. The smaller the control zones, the more adaptable the control system's response, but it may increase cost and complexity. It is used where lights are switched to the use and layout of the lit areas, to avoid lighting a large area if only a small part of it requires light. Another approach is time control to switch on and off automatically in each zone to a preset schedule for light use. For this reason, the majority of energy codes regulate control zoning by imposing area limits. Light level monitoring, which involves the automatic switching or dimming of lights in each zone to maintain a light level concerning the time of the day measured by a photosensor. This avoids overlighting an area that is already appropriately lit with natural light (Dugar, 2010). For example, the lights inside a corridor that is brightly lit with natural light do not come on until evening when it becomes darker.

6.9.2 Integrated and intelligent lighting control

Integrated lighting (Fig. 6.17) is an approach that delivers daylight and electric light separately using various control strategies that maximize the use of available daylight and minimize electricity consumption. This may be achieved by using custom-made daylight strategies with an intelligent electric lighting system and enhanced controls that seek the maximum benefit from any source of daylight. The control strategies can actively adapt to the changing outdoor or indoor environment of the building, which amounts to significant energy cost savings. Such a control strategy would integrate occupants' preferences, the state of their immediate indoor and outdoor environment, and the operation goals, in terms of energy efficiency.

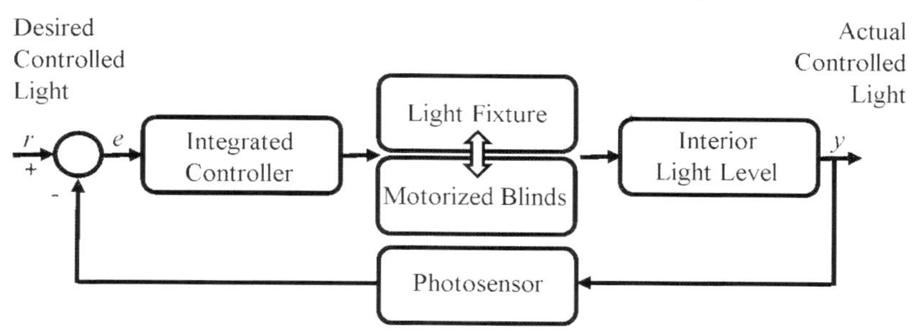

Fig. 6.17 Block diagram of a daylight-based closed-loop system.

Advanced sensing may create energy-friendly climate ceilings and walls with responsive intelligent systems that change colors and intensities throughout the day to replicate the effects of natural light. Such systems can support the human biological cycles in the same way as natural light by mimicking the rhythm of sunlight to support occupants to be active during the day and sleep well at night. The use of movement detection, daylight equalization, occupancy, and vacancy sensing, and control technology for scheduling power management, via dimming and software solutions, can effectively improve the efficiency of lighting systems. These systems integrate smart LED fixtures, being built-in occupancy and daylight harvesting sensors, that sense when natural daylight is available and automatically reduce the use of artificial lighting.

Intelligent control enables the creation of lighting networks that communicate locally as well as to remote locations to increase energy savings further and reduce maintenance costs via items such as fixed light output in lighting arrays, monitoring light output, and providing real-time operating issues. Additional sensors embedded in the luminaire may collect data such as occupancy and temperature.

The challenges to developing such an integrated control strategy include recognizing the preferences of occupants as preferences change over time, gathering information about the immediate indoor and outdoor environment of the occupants, and optimizing the trade-off between meeting occupants' preferences and energy efficiency. Some lighting control systems allow data collection from control points connected via digital networks. The system may directly measure or estimate energy consumption or monitor operating parameters. No matter which lighting control approach is used, the ability to effectively control the light source is essential. In essence, this is a lighting system with intelligent control strategies such as FL and FL-PID, which seek the maximum benefit from any source of light. In an intelligent lighting control system, the lighting controller may be the software itself. ANN is widely used in the control system in multiple areas due to its most appropriate and well-performance method.

These systems also seem to be a promising field as more and more researchers are mixing up various methods of control. Combinations between multiple control strategies are becoming easier to achieve with the growing eminence of building automation platforms. Combining various inputs and outputs results in several unique lighting control strategies available that can address visual needs, energy management needs, or both (Asif ul Haq et al., 2014). In turn, control strategies may be combined in the same space with other services like occupancy sensing, time scheduling, daylight response, institutional task tuning, color tuning, data generation, and demand response. These control systems can make sure an acceptable amount of daylight enters the room without causing glare as well as controlling the dimming level of the interior lights at the same time.

Integrating the building systems allows them to communicate with one another through the BACS, thereby facilitating the control of various systems concurrently, either in preprogrammed scenarios or operating modes. By such integration, a synergistically better result may be achieved. For example, lighting can be integrated with fire alarm control, security systems, and IT network through shared protocols. Visually, lighting control supports custom mood lighting designs which can vary brightness, color, and color temperature across large lighting arrays. Such a requirement has led to several networks and communication protocols being established. These include a digital addressable lighting interface as a standard lighting control protocol for large networked lighting systems and can provide separate controllability to each fixture driver within the system.

Further, in lighting system design and retrofit, the practice should comply with the standard, for instance, standards developed by Illuminating Engineering Society of North America (IESNA), and the European Standard EN12464-1.

6.10 Research trends

The increasing complexity of BAS requires essentially more intelligent control and scheduling strategy to manage the interaction with the smart grid and provide the information and communication among various components of the building. The main research trends in the field of advanced control evolve around several major approaches including model-based, model-predictive, data-based, agent-based, and exergy-based.

6.10.1 Forward model-based HVAC system

The majority of building thermal controls are based on a model-based approach where the controller is designed based on the mathematical model of the plant, assuming that the model represents the actual plant (Schmelas et al., 2015). Also, the most commonly used controller is still PID and on/off control. However, model uncertainties and modeling errors always exist in the above modeling process. Due to the complexity of the learning-based methods, although effective but turn out to be complicated and time-consuming in the parameterization process.

The conventional energy consumption model is one of the simplest ways to optimize the operational parameters of an HVAC system. It is based on the model that relates the operational parameters such as the air temperature, airflow rate, and air static pressure with the energy consumption of the system. The energy consumption of an HVAC system (*Esystem*) is conventionally defined as the total power usage from its components, such as the chiller (*Echiller*), fan (*Efan*), pump (*Epump*), and the cooling tower (*Ecooling tower*) as shown in Eq. (6.1). Many parameters affect the amount of energy used by these components and several are used as the optimization variables. For example, *Echiller* is affected by the coil temperature and position, while *Efan* is affected by the air supply and return speed.

$$Esystem = Echiller + Efan + Epump + Ecoolingtower \tag{6.1}$$

By minimizing the energy consumption of the HVAC system, the optimized operational parameter values can be obtained and can result in an energy-saving operation.

Forward models are based on engineering principles and usually required thorough physical information. In common, HVAC modeling approaches can be classified into three classes, components, control, and systems.

In the process of modeling, the energy conservation law that the energy storage change rate in the constant temperature chamber is equal to the energy entering the constant temperature chamber per unit time minus the energy flowing out from the constant temperature chamber per unit time (Krishnaraj et al., 2020).

$$a\frac{dT}{dt} = (Q_1 + Q_2) - (Q_3 + Q_4) \tag{6.2}$$

Q_1 is the heat entering the room, Q_2 is the heat generated by the indoor equipment and human body, Q_3 is the heat dissipation of the room (mainly outside the air), and Q_4 is the heat dissipation of the room due to maintenance construction.

6.10.2 Data-driven HVAC system

The most common data-driven computerized decision support system is built using a data warehouse product and a report and query product (Fig. 6.18). Data-driven models do not need any information on the system, and they can simply be used for such real-time applications. As real-time performance data are available in most modern building automation systems BASs, data-enabled model-based techniques may be the most effective way to achieve optimally secured and demand flexible building energy system operations (Afram and Janabi-Sharifi, 2015). Since knowledge-based models are uncertainty models that are often applied for intelligent decision support and control, they are suitable for modeling the increased complexity in IBs and therefore are being extensively applied in such fields. Common techniques may include ANNs, FL, and ML.

Typically, a PID controller is the widest controller for HVAC systems but it cannot deal with the nonlinearity present in HVAC systems. Several predictive modeling techniques such as ANN and support vector machines, and aggregated bootstrapping are widely used. The ANN technique makes predictions without having any knowledge of the system and like other data-driven methods; it heavily relies on the quality of the training data. An ANN is carried out using several steps including extracting the

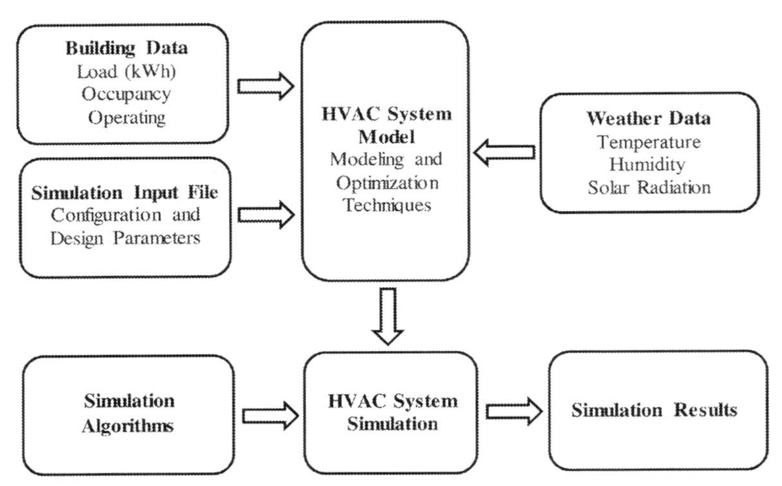

Fig. 6.18 Modeling and simulations of HVAC optimization and system configuration.

results or data, training the network using experimentally or theoretically predicted values, testing the network with the data that are not used for training, and identifying the best network structure.

For control and fault detection and diagnosis purposes, FL-based controllers and detectors are widely used. FL systems imitate the human thinking and execution ability that helps to deal with uncertainty and fuzziness. When a situation is unclear, the computer may not be able to produce a result that is true or false, similar to Boolean Logic where the value 1 refers to true and 0 means false. In an FLC system, multiple objectives can be easily included in the design. FL is used to control building indoor temperatures, CO_2 concentrations in air handling units, and fan speeds. FL control works through the manipulation of dampers, fans, and valves to adjust the flow rates of water and air. Another aspect of the FL application is obtaining optimum performance of the elevator system in buildings such as minimizing waiting and riding times as well as leading to several passengers in the elevator, comfortability, and safety issues.

ML is a highly effective technique that can be embedded directly into the BAS for immediate system control. This algorithm flows throughout the BAS to realize when occupants arrive in different parts of the building and put different strains on its systems. Training and testing are the primary two steps in creating any data-driven model. Training is a phase in creating the ML models. A set of examples is used to fit the parameters of the model. The models are trained to entail an input and a corresponding output or target. The created model will be run with the training dataset and produces a result, which is then compared with the output or target for each selected input. Based on the results, the model parameters will be adjusted. Testing is the closing phase after training. A dataset is used to provide an unbiased evaluation of the final selected model from the training set. The dataset used for testing the models is different from the one used for training.

Smart HVAC controls feature ML to optimize building temperature automatically, without human intervention. In this context, ML is such a great tool that has transformed how owners perceive BAS systems. For instance, ML-powered HVAC units leverage an online automated monitor that evaluates the internal/external conditions of a building, its control actions, and the results of those actions 24/7 to ensure maximum reliability. An ML-based HVAC system can characterize data visualization and clustering, demand response, demand-side management, energy forecasting, peak shaving, outage protection, anomaly detection, predictive control, fault localization and optimization, process control and optimization, and renewable energy integration.

6.10.3 Model predictive control (MPC)

MPC is seen as one of the key future enablers for intelligent energy management in buildings to meet inhabitants' comfort needs in a more efficient way (Afram and Janabi-Sharifi, 2015). MPC is a control algorithm that follows the principles of the classical controls to optimize a sequence of manipulated variable adjustments over a prediction horizon by utilizing a process model to optimize forecasts of process behavior based on a linear or quadratic objective. A major challenge in applying MPC to building automation and control is the development of a simplified mathematical model of the building for real-time control with fast response times. However, building models are highly complex due to nonlinearities in heat and mass transfer processes of the building itself and the accompanying air-conditioning and mechanical ventilation systems (Yang et al., 2018). In this scenario, MPC can be applied to the efficient management of energy distribution in buildings. However, the approach requires significant computational requirements, which grow exponentially with the number of building zones and subsystems.

6.10.4 Agent-based control

The agent-based control technique is currently, getting more popular among researchers due to its particular assigned task capabilities (Zheng et al., 2016). Control agents are autonomous in that they do not require human supervision or monitoring to function. They are responsive to information provided by operators as needed and react to operator demands and objectives. In general, an agent-based system is formed by a set of two or more software or hardware agents embedded in an environment, and able to perceive it and autonomously react to changes in the environment. Software agents are software processes that can analyze and act upon their environments.

6.10.5 Exergy-based control and optimization

Complex systems like campuses consisting of components like power plants, energy storage devices, and even autonomous vehicles, have multidomain energy flows. Reducing energy consumption and waste to increase efficiency while maintaining the effectiveness of these networks is a challenge as they become more integrated and complex (James et al., 2020). The body of research investigation where exergy is used in conjunction with efforts to perform optimization and control in energy systems is quite small. However, exergy-based optimization and hybrid controller combining model predictive and agent-based control is usually used to improve energy efficiency in buildings.

References

Abhinandana, B., Beddiar, K., Benamour, A., Amirat, Y., 2018. Intelligent systems for building energy and occupant comfort optimization: a state of the art review and recommendations. Energies 11, 2604. https://doi.org/10.3390/en11102604.

Afram, A., Janabi-Sharif, F., 2014. Theory and applications of HVAC control systems—a review of model predictive control (MPC). Build. Environ. 72 (November), 343–355.

Afram, A., Janabi-Sharifi, F., 2015. Black-box modeling of residential HVAC system and comparison of gray-box and black-box modeling methods. Energ. Buildings 94, 121–149.

Ahmad, M.W., Mourshed, M., Yuce, B., Rezgui, Y., 2016. Computational intelligence techniques for HVAC systems: a review. Build. Simul. 9, 359–398. https://doi.org/10.1007/s12273-016-0285-4.

Albin, S., 1997. Building a system dynamics model: part 1: conceptualization. In: MIT System Dynamics in Education Project. https://ocw.mit.edu/courses/sloan-school-of-management/15-988-system-dynamics-self-study-fall-1998-spring-1999/readings/building.pdf.

Amara, F., Agbossou, K., Cardenas, A., Dubé, Y., Kelouwani, S., 2015. Comparison and simulation of building thermal models for effective energy management. Smart Grid Renew. Energy 6, 95–112.

Anderson, M., Buehner, M., Young, P., Hittle, D., Anderson, C., Tu, J., Hodgson, D., 2005. MIMO robust control for HVAC systems. IEEE Trans. Control Syst. Technol. 6 (3), 477–483.

Asif ul Haq, M., Hassann, M.Y., Abdullah, H., Abdul Rahman, M., Abdullah Md, P., Hussin, F., Said, D.M., 2014. A review on lighting control technologies in commercial buildings, their performance and affecting factors. Renew. Sustain. Energy Rev. 33, 268–279.

Conceição, E.Z.E., Gomes, J.M.M., Ruano, A.E., 2018. Application of HVAC systems with control based on PMV index in university buildings with complex topology. IFAC-PapersOnLine 51 (10), 20–25.

Dounis, A.I., Caraiscos, C., 2009. Advanced control systems engineering for energy and comfort management in a building environment—a review. Renew. Sust. Energ. Rev. 13, 1246–1261.

Dovan, P., 2020. Three essential elements of next generation building management systems (BMS). White paper https://download.schneider-electric.com/files?p_enDocType=White+Paper&p_File_Name=WP500V1.pdf&p_Doc_Ref=Buildings_WP500_EN.

Dugar, A.M., 2010. Intelligent lighting for intelligent savings. Lighting India 5 (6), 48–53.

Fadali, M.S., Visioli, A., 2013. Digital Control Engineering Analysis and Design, second ed. Elsevier, Oxford.

Haykin, S., 1994. Neural Networks: A Comprehensive Foundation. Macmillan, New York.

Jafari, M., Xu, H., 2018. Intelligent control for unmanned aerial systems with system uncertainties and disturbances using artificial neural network. Drones 2 (3), 30.

James, C., Kim, T.Y., Jane, R., 2020. A review of exergy based optimization and control. Processes 8 (3), 364. https://doi.org/10.3390/pr8030364.

Kerschen, G., Worden, K., Vakakis, A.F., Golinval, J.C., 2006. Past, present and future of nonlinear system identification in structural dynamics. Mech. Syst. Signal Process. 20 (3), 505–592.

Krishnaraj, N., Elhoseny, M., Lydia, E.L., Shankar, K., ALDabbas, O., 2020. An efficient radix trie-based semantic visual indexing model for large-scale image retrieval in cloud environment. Softw. Pract. Exp., 489–502. https://doi.org/10.1002/spe.2834.

Lu, X., Clements-Croome, D.J., Viljanen, D., 2011. Past, present and future mathematical models for buildings: focus on intelligent buildings (part 1). Intell. Build. Int. 1 (1), 23–38.

Lv, Z., Han, Y., Singh, A.K., Manogaran, G., Lv, H., 2021. Trustworthiness in industrial IoT systems based on artificial intelligence. IEEE Trans. Ind. Inf. 17 (2), 1496–1504.

Macas, M., Moretti, F., Fonti, A., Giantomassi, A., Comodi, G., Annunziato, M., Pizzuti, S., Capra, A., 2016. The role of data sample size and dimensionality in neural network based forecasting of building heating related variables. Energy Build. 111, 299–310. https://doi.org/10.1016/j.enbuild.2015.11.056.

Marvuglia, A., Messineo, A., Nicolosi, G., 2014. Coupling a neural network temperature predictor and a fuzzy logic controller to perform thermal comfort regulation in an office building. Build. Environ. 72, 287–299. https://doi.org/10.1016/j.buildenv.2013.10.020.

Salakij, S., Yu, N., Paolucci, S., Antsaklis, P., 2016. Model-based predictive control for building energy management. I: Energy modeling and optimal control. Energy Build. 133, 345–358.

Schmelas, M., Feldmann, T., Bollin, E., 2015. Adaptive predictive control of thermo-active building systems (TABS) based on a multiple regression algorithm. Energy Build. 103, 14–28.

Selamat, H., Haniff, M.F., Sharif, Z.M., Attaran, S.M., Sakri, F.M., Abdul Razak, M.A., 2020. Review on HVAC system optimization towards energy saving building operation. Int. Energy J. 20, 345–358.

Shao, T., 2021. Indoor environment intelligent control system of green building based on PMV index. Adv. Civil Eng. 2021, 1–11. https://doi.org/10.1155/2021/6619401.

Wang, S., 2010. Intelligent Buildings and Building Automation. Spon Press, London.

Williams, A., Pe, B.A., Garbesi, K., Pe, E.P., Rubinstein, F., 2018. Lighting controls in commercial buildings. J. Illum. Eng. Soc. N. Am., 161–180.

Yang, S., Wan, M.P., Ng, B.F., Zhang, T., Babu, S., Zhang, Z., Chen, W., Dubey, S., 2018. A state-space thermal model incorporating humidity and thermal comfort for model predictive control in buildings. Energy Build. 170, 25–39. https://doi.org/10.1016/j.enbuild.2018.03.082.

Zhang, Y., 2005. Synthesis of optimum HVAC system configurations by evolutionary algorithm. PhD dissertation, Loughborough University, United Kingdom.

Zheng, Z., Wang, L., Wong, N.H., 2016. Intelligent control system integration and optimization for zero energy buildings to mitigate urban heat island. Procedia Eng. 169, 100–107.

Zilouchian, A., Jamshidi, M., 2001. Intelligent Control Systems Using Soft Computing Methodologies. CRC Press, Boca Raton.

Building as a bioelectromagnetics ecosystem

7

Riadh Habash
School of Electrical Engineering and Computer Science, University of Ottawa, Ottawa, ON, Canada

If you want to find the secrets of the universe, think in terms of energy, frequency, and vibration.

Nikola Tesla

7.1 Nonionizing electromagnetic fields

Although the convenience of the vast proliferation of electrification and ICTs is offered by EM signals and fields, often little thought is given to their possible impact on human health, which should not be ignored. This is a shortfall that justifies considering the EM environment in building design to significantly minimize occupant exposure. The question is how the built environment responds to health concerns from the potential for the aggregation of different EM signals? This requires a detailed investigation into how to biologically optimize EM exposures in the built environment at a relatively modest cost at the planning phase.

In this regard, a distinction should be made between nonionizing vs higher frequency ionizing radiation that has enough energy to displace electrons and "ionize" atoms and molecules. Fig. 7.1 shows a graphical representation of the EM spectrum in ascending frequency (decreasing wavelength). The general nature of the effects is noted for different ranges. Ionizing radiation includes UV light, X-ray, and gamma-ray. It means that electrons are removed from the atoms/molecules in the material by the radiation. It can cause DNA damage and mutations; therefore, exposure to such sources as radioactive materials should be avoided. Below these frequencies, nonionizing radiation includes visible, infrared light, microwaves, and THz, RF, VLF, and ELF fields including power frequency 50/60 Hz. However, given modern technology, nonionizing radiation from power lines and home appliances, ICTs including wireless devices and cellular towers, is practically unavoidable.

In classical physics, the flow of energy at the speed of light through free space or a material medium in the form of the electric and magnetic fields makes up EM fields. In such a case, time-varying electric and magnetic fields are mutually linked with each other at right angles and perpendicular to the direction of motion. An EM field is characterized by its intensity and frequency. In terms of modern quantum theory, the EM field is the flow of photons (also called light quanta) through space. EM fields in the nonionizing region comprise photons that do not have sufficient energy to break chemical bonds or ionize biological molecules. The energy of a photon of an EM wave

Sustainability and Health in Intelligent Buildings. https://doi.org/10.1016/B978-0-323-98826-1.00007-7

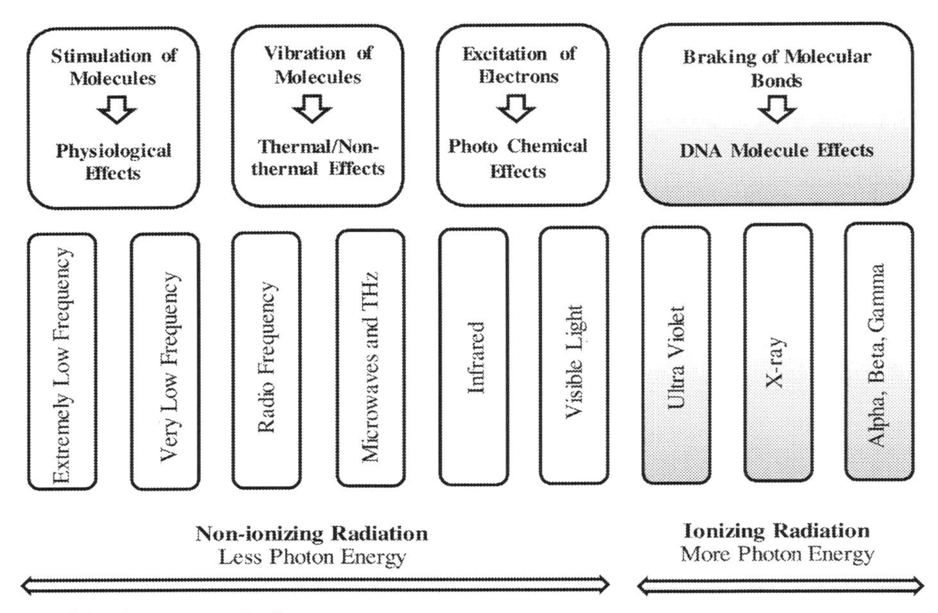

Fig. 7.1 Electromagnetic frequency spectrum.

is given by $E = hf$, where h is Planck's constant and f is the frequency of the EM fields (Stuchly, 1979). Therefore, the energy of a photon in the radiofrequency energy varies from approximately 4.1×10^{-6} eV at 1 GHz to 1.2×10^{-3} eV at 300 GHz.

ELF fields are created by the alternation of polarity (positive to negative to positive) of AC electricity on power lines and circuits at 50/60 Hz. In North America, power systems operate at 120 V and have a frequency of 60 Hz. However, utilities in Europe, Asia, and other places in the world supply users with 220 V and 50 Hz of electrical power. This means that North American systems are associated with higher currents and accordingly higher magnetic fields. ELF fields come into buildings from neighborhood transmission lines, household wiring and grounding systems, and electrical devices, particularly ones that have motors, transformers, and coils. Two types of ELF fields usually emanate from AC wiring and power lines at 50/60 Hz. First, AC electric fields are produced by differences in electric voltage and measured in volts per meter (V/m). Second, magnetic fields, which occur only when current is flowing and measured in units of amperes per meters (A/m) and milligauss (mG) (Habash, 2020). While the only safe level is zero mG, the more realistic levels should be several mGs for prolonged exposure.

RF fields occupy a wide range of frequencies extending from 300 MHz to 300 GHz and may be sometimes called microwaves. Existing ICTs operate up to several GHz; however, emerging 5G wireless networks will gradually utilize the MMW region of the EM spectrum, extending up to 300 GHz in terms of frequency. This leads to renewed interest in MMW wireless research scenarios regarding biological and health effects. Increased exposure to RF fields may result not only from the use of much higher frequencies in 5G but also from the potential for the aggregation of different signals, their dynamic nature, and the complex interference effects that may result, especially in dense urban areas.

As more products and services are developed and used in everyday applications, the potential for human exposure to EM fields will increase. When technologies change every 10 years but buildings may last anywhere from 30 to 50 years (or even more), designing infrastructure to adapt to evolving requirements is critical to ensuring the building will be able to meet the needs of its occupants both today and for decades to come. One of the key changes with the evolution to 5G is that the new standards rely heavily on optical fiber infrastructure to achieve the required data transmission speeds, rather than traditional copper infrastructure (Rodriguez, 2019). This means that new construction should plan to install spare capacity to accommodate future generations of technology.

7.2 Interaction mechanisms

Living organisms, including humans, are complex systems that evolved over billions of years in a world of a limited number of EM emitters. These living organisms interacted with and adapted to this EM environment to regulate various critical cellular systems. One example of this adaptation is the visual system developing filtering systems in the eye and the skin to protect itself from the impact of EM fields in the bands of visible light and UV radiation (Habash, 2020). Therefore, it is not surprising that the massive introduction of EM fields in an enormous range of new frequencies, modulation, and intensities in recent years will be impacting living organisms.

The term "interaction" is critical, meaning that the result depends not only on the action of the field but is influenced by the reaction of the living system which has a great capacity for compensating for the effects induced by external influences, including EM sources (Frey, 1988). The interaction of EM fields with living systems can be considered at the molecular, subcellular, cellular, organ, or system level, as well as the entire body. While the established mechanisms have been proposed over the years that attempt to explain how weak EM fields might affect biological systems, they typically have remained controversial because of failures to properly account for clear effects.

7.2.1 ELF fields

Electric fields can apply forces on charged and uncharged molecules or cellular structures within living systems. These forces can cause movement of charged particles, orient or distort cellular structures, orient dipolar molecules, or induce voltages across cell membranes (membrane sensitivity syndrome). Magnetic fields can also apply forces to cellular structures and may induce electric fields in the living system; however, since biological materials are largely nonmagnetic, these forces are usually very weak (Habash, 2020). Subjecting the human body to ELF fields can lead to induced electric fields and circulating currents in conductive tissues. These induced fields are an established mechanism that forms the basis for most established exposure limits.

In the human body, cells that make the tissues of various organs may be divided into those that are excitable and those that are nonexcitable. Excitable tissues consist of nerve cells (neurons) and muscle cells can conduct electrical impulses as part of their natural performance. These impulses, which consist of momentary reversals of the normal 0.1 V or so which exists across the cell membrane, are the mode by which messages are passed from one nerve cell to the next and by which muscle cells coordinate their contraction (Wood, 2017). To assess whether an electric or magnetic field can cause electrostimulation, it is necessary to determine the amount of electrical current that the external fields induce within the tissues of the body to stimulate the nerves or nerve endings that innervate the muscle beneath the skin. At low frequencies, induced currents can produce the effect of stimulation of nerve and muscle cells. Above these frequencies, muscle and nerve cells become increasingly less receptive to electrical stimulation. Usually, electric fields induced in biological systems by exposure to low frequencies stimulate single myelinated nerves in a biophysical manner when the internal field strength exceeds a few V/m. Myelinated nerves are thicker and more sensitive to ELF fields than unmyelinated nerves, as in the gray matter of the central nervous system (Hocking and Gobbo, 2011). The excitable organism subject to efficient stimulus produces a special bioelectrical response, followed by contraction and secretion. Any efficient stimulus must have stimulus intensity, duration, and intensity-time rate of change.

Free radicals which are atoms or molecules with unpaired electrons are linked to pathological modifications that may trigger cellular malfunction or mutation as well as protein degradation. They also play a large role in causing damage to all cells of the body but in particular the immune system and neurodegenerative diseases. The radical pair mechanism is a way in which ELF fields can affect specific types of chemical reactions, generally increasing concentrations of reactive free radicals in low fields and decreasing them in high fields (Habash, 2020). Since the lifetime of these free radicals is so short compared with the cycle time of the ELF fields in general and power frequency (50/60 Hz) fields in particular, the applied fields act like static fields during the time scale over which these reactions occur.

7.2.2 RF fields

A distinction that is often used concerning exposure to higher frequency EM fields is that between thermal and nonthermal effects (Fig. 7.2). However, this division is imprecise since interaction with the EM field always includes energy transfer and therewith usually a local temperature rise. Thermal mechanisms have been known since investigations into therapeutic applications of electricity were carried out based on studies in electromagnetism by Faraday, Ampere, Gauss, and Maxwell, and the development of AC sources by d'Arsonval and Tesla. Heating is the primary interaction of EM fields at high frequencies. Excessive heating may lead to biological effects

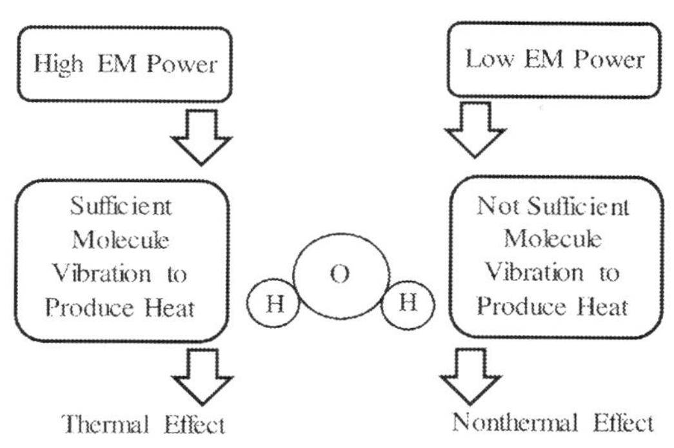

Fig. 7.2 Two separate paths for thermal and nonthermal effects.

that may cause health implications or may be used in therapy. Temperature is a macroscopic, average parameter of a system in mutual interaction and can be related to the average kinetic energy of the particles. Thermal effects result from the fact that EM fields with high frequencies may be partly absorbed by materials containing water, such as biological tissues, and be converted into heat. Heat resulted from the absorption of EM energy depends on the electrical conductivity which is partly due to the translational motion of charged particles.

Nonthermal effects have been increasingly becoming the norm of the current EM research. Occasionally, complaints are made that these nonthermal effects due to low-level RF fields are not being considered in the analysis of the scientific information, because they are not used as a basis for establishing EM exposure limits. The second meaning is that intermediate RF fields may cause biological effects, without the involvement of heat. This is sometimes referred to as an "athermal effect." In this case, the thermoregulatory system maintains the irradiated body at its normal temperature (Habash, 2020). Meanwhile, the macroscopic behavior of the body emerges out of quantum dynamics, producing the physics of living matter to a point where biochemistry has to be considered.

By increasing the frequency toward MMWs as is the case for 5G technologies, most of the energy will be absorbed within the human skin and by the shell of the cornea. Since skin contains blood vessels and nerve ends in its dermis layer, effects may be transferred through molecular mechanisms or the nerves.

7.2.3 Cell membrane and the chemical link

Many life scientists, through a series of findings, believe the cell membrane plays a principal role in the EM interaction mechanism with biological systems. Indications point to cell membrane receptors as the probable site of initial tissue interactions with EM fields for many neurotransmitters, growth-regulating enzyme expressions, and cancer-promoting chemicals (Habash, 2020). Scientists theorizing this mechanism conclude that biological cells are bioelectrochemical structures, which interact with their environment in various ways, including physically, chemically, biochemically,

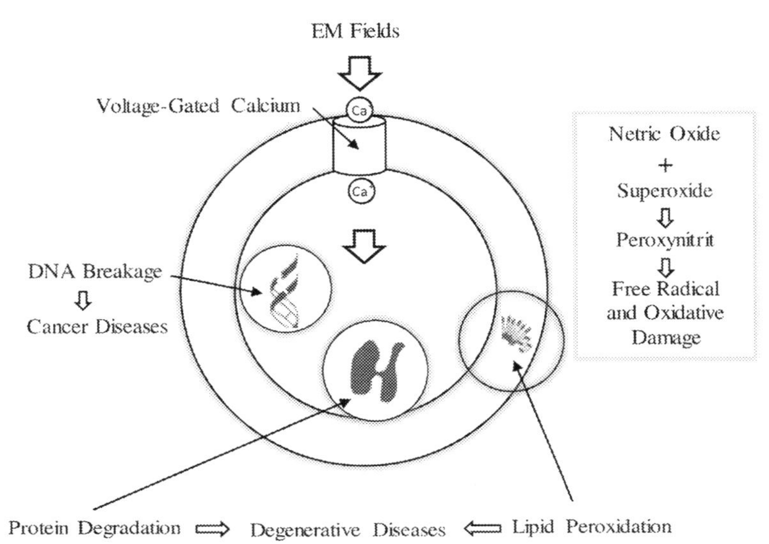

Fig. 7.3 Effects of electromagnetic fields at the cellular level.

and electrically. The role of increased intracellular calcium (Ca^{2+}) following EM exposure is well documented in the literature. Voltage-gated calcium channels are key transducers of cell surface membrane that allow intracellular Ca^{2+} in and out of cells and regulate calcium levels. They are present in the highest concentrations on neurons, neuroendocrine cells, and muscle cells, including smooth muscle and cardiac muscle. The channels use electrical signaling and processes for allowing certain amounts of calcium to pass through the membrane. They use this electricity to change membrane potentials and therefore cause ions to move through the membrane.

Calcium performs as a signaling molecule in cells, activating enzymes and molecular pathways that lead to an increase in nitric oxide, which then combines with superoxide to form peroxynitrite, which acts as a free radical and also leads to the formation of other free radicals which increase oxidative stress in the cell. Oxidative stress directly damages the structural molecules of the cell, such as the cell membrane and proteins, and even causes DNA strand breakage. Oxidative stress and the accumulation of damaged proteins are known to be at least one factor in cell aging and the development of age-related disease (Greenfield, 2021). Oxidative damage is implicated in the development of atherosclerosis, neurodegenerative diseases, and cancer (Fig. 7.3).

7.3 Safety standards and protection guidelines

Several decades of research in the area of bioelectromagnetics have led to a consensus on the safety of EM fields. Expert committees reflect this consensus when developing exposure standards including regulations, recommendations, and guidelines, which are an approved way to identify measurable EM field values that limit human exposure to levels below those considered hazardous to human health. Historically, many institutions and organizations throughout the world have recommended safety exposure limits for EM fields. These include the Institute of Electrical and Electronics Engineers (IEEE) (IEEE, 2019), the International Commission on Nonionizing Radiation Protection (ICNIRP) (ICNIRP, 2002), Safety Code 6 of Health Canada, the Canadian national public health authority, and the Australian Radiation Protection and Nuclear Safety Agency.

7.3.1 IEEE standard

The IEEE is the world's largest professional organization dedicated to advancing technological innovations for the advantage of humankind. The first formal standard project was initiated in 1960 when the American Standards Association (now the American National Standards Institute, or ANSI) approved the radiation hazards standards project. Considering ELF fields, the IEEE C95.6-2002 covers frequencies from 0 to 3 kHz and therefore includes power frequencies 50/60 Hz. This standard is designed to keep the induced current in the human body at least a factor of 10 below the lowest reported stimulation thresholds for electrically excitable cells. The recommendations are intended to apply to exposures of the general public, as well as to individuals in occupational environments.

The newly released IEEE Standard C95.1-2019 specifies exposure criteria and limits to protect against established adverse health effects in humans associated with exposure to electric, magnetic, and EM fields in the frequency range of 0 Hz–300 GHz. The limits, incorporating safety margins, are defined in terms of dosimetric reference limit replacing basic restriction and exposure reference level. The exposure limits to electric and magnetic fields are defined to protect against painful electrostimulation in the frequency range of 0 Hz–5 MHz. In the transition region of 100 kHz–5 MHz, protection against both electrostimulation and thermal effects is provided through both sets of limits. Below 100 kHz, only the electrostimulation limits apply, while above 5 MHz, only the thermal limits apply, and both sets of limits apply in the transition region (100 kHz–5 MHz) (IEEE, 2019).

7.3.2 ICNIRP guidelines

The ICNIRP was established in 1992 as an independent organization with a mandate of providing advice to international and national establishments as well as the public, on exposure to nonionizing radiation and related biological and health effects. The ICNIRP's mission is to coordinate knowledge of protection against various EM exposures in the development of internationally accepted recommendations. In particular, the ICNIRP develops guidelines for the safe exposure of workers and the general public. These guidelines are based on established scientific literature and are developed following well-explained criteria.

In 1998, the ICNIRP issued guidelines on exposures that covered the frequency range up to 300 GHz. In 2009, ICNIRP issued a draft revision for consultation, and in 2010 a new guideline for the frequency range 1 Hz–100 kHz was released. In general, the ICNIRP guideline specifies "basic restrictions" and "reference levels." Basic restrictions on exposure to EM fields are based on established adverse health effects. ICNIRP has issued new guidelines for ELF fields with frequencies between 1 Hz and 100 kHz in 2010, and for frequencies between 0 and 1 Hz in 2014, but these have not yet led to changes in the EU recommendation. The limits for static and ELF fields in the EU directive are derived from the 2009 and 2010 ICNIRP guidelines. The European Parliament and Council of the EU have also issued a directive on the marketing of low voltage electrical equipment (2014/35/EU), which requires that such equipment does not endanger the health or safety of persons.

7.3.3 Building biology institute (IBN) evaluation guidelines

According to Winfried Schneider, Director of the IBN, building biology is the study of the holistic relationships between humans and their built environment. The aim is to create a healthy, natural, sustainable, and beautifully designed living and working environment. In building biology, buildings and rooms are referred to as "third skin," which reflects how closely humans are connected to the built environment.

The IBN evaluation guidelines are based on the precautionary principle. They are specifically designed for sleeping areas associated with long-term risks and a most sensitive window of opportunity for regeneration. They are based on the experience and knowledge of the building biology community and focus on achievability. Also, scientific studies and other recommendations are consulted (Habash, 2020). With its professional approach, building biology testing methods help identify, minimize, and avoid environmental risk factors within an individual's framework of possibility. It is the Standard's goal to identify, locate and assess potential sources of risk by holistically considering all subcategories and implementing the best possible diagnostic tools available with analytic expertise to create indoor living environments that are as exposure-free and natural as practicable (IBN, 2015). Figs. 7.1 shows IBN evaluation guidelines for AC electric and magnetic fields (50/60 Hz) in sleeping areas. No anomaly reflects the unexposed natural conditions or the common and nearly inevitable background level of our modern living environment. Slight anomaly calls a precaution especially for sensitive and ill people. Severe anomaly means not acceptable from a building biology point of view and calls for action. Extreme anomaly calls for immediate and rigorous action.

In the case of RF fields, the quantity to be measured is power density in the units of $\mu W/m^2$. Power densities less than 1 indicate no concern, between 1 and 10 indicate a slight concern, 10–1000 indicate a severe concern and values greater than 1000 indicate extreme concern (Maes, 2008). According to the standard, the values mentioned above refer to peak measurements and are applicable to single RF sources (Table 7.1).

Table 7.1 IBN evaluation guidelines for AC electric and magnetic fields in sleeping areas.

	No anomaly	**Slight anomaly**	**Severe anomaly**	**Extreme anomaly**
AC electric fields				
Field strength with ground potential in V/m	<1	1–5	5–50	>50
Body voltage with ground potential in mV	<10	10–100	100–1000	>1000
Field strength potential-free in V/m	<0.3	0.3–1.5	1.5–10	>10
AC magnetic fields				
Flux density in nT	<20	20–100	100–500	>500
Flux density in mG	<0.2	0.1–1	1–5	>5

7.4 Electromagnetic living environment

Human exposure to EM fields comes from many different sources and occurs in various situations in everyday life. Current technologies have become a source of pervasive pollution from generated EM fields.

7.4.1 Potential ELF sources

ELF fields in buildings can originate either inside or outside the building. Main sources of ELF fields from outside buildings include electrical transmission and distribution lines, substations and transformers, high-current powered railway systems, and electrical cabling in buildings. Internal sources of magnetic fields are generated in buildings from grounding practices, wiring practices, and electric devices (appliances, heating elements, office equipment, etc.).

ELF fields from home appliances are produced by the electric current which is usually the low source of fields because any current within the appliance is balanced by a return current a relatively short distance away. Appliances that operate via an electric motor including vacuum cleaners, washing machines, hair driers, electric shavers, bench grinders, and overhead projectors are associated with fairly strong magnetic fields, but these fall off with the inverse cube of distance ($1/R^3$) (Wood, 2017). Typical electric field values from appliances at homes include electric oven (8 V/m), refrigerator (120 V/m), vacuum cleaner (50 V/m), color TV (60 V/m), and stereo receiver (180 V/m) (Habash, 2020).

Electric fields, not like magnetic fields, are quickly disturbed by materials with a high relative permittivity (dielectrics) and even more significantly by conducting materials. A conducting enclosure eliminates the electric field within it. A conducting object also perturbs the field outside it, increasing it in line with the field and reducing it to the sides. For ELF fields, buildings are sufficiently conducting to reduce the electric field within them from an external source. Electric fields are particularly affected by earthed conducting objects including not just the ground, but also trees, hedges, fences, buildings, and human beings.

The stray magnetic field from those appliances depends strongly on the designs, which aim to keep stray fields as low as possible. ELF field exposures within residences vary from over 150 μT and 200 V/m a few centimeters from certain appliances to less than 0.02 μT and 2 V/m in the center of many rooms. Appliances that have the highest magnetic fields are those with high currents or high-speed electric motors (for example, vacuum cleaners, microwave ovens, electric washing machines, dishwashers, blenders, can openers, electric shavers) (Eichwald and Walleczek, 1998).

Background magnetic fields are in general between 0.1 and 0.3 μT. In residential areas, maintaining a distance from visible sources may be relatively easy, but avoiding fields generated by concealed wiring is much more difficult. Furthermore, because building materials like concrete cannot block magnetic fields and because each floor serves as the ceiling of another, fields from the wiring system in a floor or apartment may reach the occupants of adjacent floors and apartments (Habash, 2020). Apartment residents may therefore be at higher risk of exposure to ELF fields than residents of detached houses.

7.4.2 Potential RF sources

The sun is the most important source of RF fields including ultraviolet, visible light, and infrared radiation. Sunlight is the driving force of the weather and all life on Earth. Many scientists are concerned about the consequences of further temperature increase on Earth. The question is whether this is due to an increased intensity of solar radiation or increased release of GHG that reduces the reflection of solar radiation back to space (Brune et al., 2001).

The use of RF fields in buildings has increased rapidly during the last decades, mainly due to the widespread of environmental RF sources. Broadcast stations are usually located near densely populated areas so that large audiences can receive the signals. They transmit at various RF frequencies, depending on the channel, ranging from about 550 kHz to about 800 MHz for some TV stations. Operating powers can be in the range of a few kW for certain radio and TV stations. The radiation patterns from broadcast antennas are not as highly collimated as those from other RF sources such as dish antennas used for satellite earth stations. Therefore, exposure to main-beam radiation intensities near the broadcast antenna is possible, especially if individuals are at eye level with the antenna bays (e.g., residents of high-rise buildings).

The rapid growth of wireless communication devices and infrastructure resulted in the installation of a large number of cellular base stations which are mounted on free-standing towers, rooftops, or the sides of buildings. A base station refers to the antennas and their associated electronic equipment. It may contain more than one transmitter, with the output of each transmitter fed to the antenna on top of the tower. Today, 5G is the newest wireless network that is replacing the current 4G technology by providing many improvements in speed, coverage, and reliability. The current 4G networks use frequencies below 6 GHz, while 5G uses extremely high frequencies in the 6–300 GHz range. This means that 5G antennas can be much smaller than existing antennas while still delivering specific directional control (Habash, 2020). These mini-5G stations may be placed on top of street lights or the sides of buildings in cities.

The number of sources has increased indoors. The installation of access points and short-range base stations, WiFi hotspots, Bluetooth, security purposes, as antitheft and personal access control has given rise to exposure at very close distances (within 1 m), whereas farther away the emitted EM fields do not exceed the common background levels. Several frequency bands within the ISM bands are used for WiFi and within these, many channels have been designated. The main frequency bands used for carrying WiFi are 2.4 GHz, and 5 or 5.8 GHz bands.

Today's electricity meters commonly called smart meters, record the consumption of electricity, water, and natural gas. They transmit customer information including electricity consumption wirelessly to the utility company for billing and other purposes. Many different wireless technologies can be used, including cellular and WiFi. Most smart meters transmit in the 900 and 2.4 GHz frequency range and communicate with a utility access point that can be located on transmission line poles that are high above the ground. Smart electricity meters emitting RF fields generate a low level of exposure with a power of 1 W or less and an average power density of less than 5×10^{-5} W/m^2. However, they only emit for short periods during the day, ranging from 1 to 6 min.

Residential microwave ovens, which operate at a frequency of 2.45 GHz, may leak some RF energy. The Food and Drug Administration has regulated the manufacture of microwave ovens since 1971. Typical average microwave oven leakages are in the range of 0.1–0.5 mW/cm^2. There are several types of biomedical equipment that are of interest, including physiotherapeutic use of diathermy, hyperthermia, ablation treatments, electrosurgery, cosmetic medicine, transcranial magnetic stimulation, magnetic resonance imaging, and switched gradient field scanners.

MMW and THz applications are expected to be available soon in various industrial environments, such as for imaging systems used for nondestructive quality control, as well as for short-range broadband telecommunications. These applications will operate with low power and, due to the small penetration depth of the radiation, expose only superficial tissues. The emitted EM fields from the above devices and technologies when combined will result in a marginally higher exposure compared to single sources.

7.5 Biological and health effects

Health is a state of total physical, mental, and social well-being, and not just the absence of disease or sickness. A biological effect occurs when exposure to EM fields causes some noticeable or detectable physiological change in the living system. Such an effect may sometimes, but not always, lead to an adverse health effect, which means a physiological change that exceeds the normal range for a brief time. However, increased biological action does not necessarily cause noticeable health effects, since adaptive mechanisms are operating at cellular-tissue-organism levels in response to ever occurring changes (Panagopoulos et al., 2015). However, these mechanisms may not always be effective, especially when the organism is under extra stress or has increased metabolic needs.

7.5.1 Neuropsychiatric symptoms

Some members of the public have recognized a spread of symptoms to low levels of exposure to EM fields. This may include headaches, anxiety, sleep disruption, suicide and depression, nausea, fatigue, difficulty concentrating, poor memory, and loss of libido. To date, the relevance of these effects for the health of people may be attributed to the abnormal release of neurotransmitters and hormones. However, some of these health problems may be caused by other factors in the environment, or by concerns related to the presence of new technologies.

7.5.2 Effects on pregnancy outcomes

Different sources and exposure scenarios to EM fields in the living and working environment, including computer screens, water beds and electric blankets, RF welding machines, diathermy equipment, and radar, have been evaluated by the WHO and other organizations. The overall weight of evidence shows that exposure to EM fields at typical environmental levels does not increase the risk of any adverse outcome such as spontaneous abortions, malformations, low birth weight, and congenital diseases. However, there have been occasional reports of associations between health problems and presumed exposure to EM fields, such as reports of prematurity and low birth weight in children of workers in the electronics industry, but these have not been regarded by the scientific community as being necessarily caused by the EM field exposures (Habash, 2020).

Given the ubiquity of cell phone use, the potential role of this environmental exposure needs to be clarified. A specific concern is the possible effects of exposure to RF fields on fertility, in general, however, the majority of the literature on reproductive function describes the possible effects of RF fields on male sperm. There have been numerous studies on the effects of RF fields especially from cell phones and WiFi on fertility parameters. The measures are mainly motility, viability, and concentration, which are the parameters most frequently used in clinical settings to assess fertility.

7.5.3 Electromagnetic hypersensitivity (EHS)

EHS, also known as EM sensitivity and electrohypersensitivity, occurs when the amount of EM fields exceeds the body's ability to deal with it. As a result of rapid growth rates and the greater vulnerability of developing nervous systems, the long-term risks to children from RF exposure Several official bodies and researchers have expressed concern and caution about the potential vulnerability of children to RF fields from wireless technologies and the accompanying ELF resulting from modulation.

EHS is a recent phenomenon unparalleled in human history. It is a disorder in which people are physically sensitive to the effects of EM fields. EHS has been known since 1932 and has been identified by various names including microwave sickness, radio wave syndrome, and EMF intolerance syndrome. Some of the most common symptoms of EHS include headaches, sleep disorders, heart arrhythmia, ringing in the ears, fatigue, anxiety, concentration problems, skin rashes, nausea, cardiac, and others. The primary treatment for EHS is the avoidance of EM fields.

7.5.4 Electromagnetic fields and cancer

In spite of the general belief that most cancer is genetic, several studies ins have convincingly shown that cases of cancers result from some environmental exposure and are not mainly genetic (Lichtenstein et al., 2000). Despite many studies, the evidence for any effect remains highly controversial. However, it is clear that if EM fields do cause cancer, then any increase in risk will be extremely small. The results to date contain many inconsistencies, but no large increases in risk have been found for any cancer in children or adults. Several epidemiological studies suggest small increases in the risk of childhood leukemia with exposure to ELF fields in the home. Leukemia is the most common cancer to affect children, accounting for approximately a third of all childhood cancers. As with most other cancers, the mechanism by which cancer arises is likely to involve gene–environment interactions. DNA strand breakage, for example, might be a result of oxidative stress and cell aging and is known to be a potential cause of cancer, as the genes that regulate the cell's growth and cycle are damaged. In 2011, the International Agency for Research on Cancer categorized EM fields as a possible (level 2B) human carcinogen.

7.6 Electromagnetic safety in the built environment

EM field management in buildings includes engineering measures to reduce, avoid, or eliminate certain fields or field characteristics. These measures in the design of electrical installations can achieve low levels and can significantly reduce the potential for interference as well as health impact if any. This process requires mitigation and shielding techniques with a tremendous energetic extent and probably modifying the physical arrangement of the source. There are several mitigation measures to be considered, few are relatively straightforward, even obvious.

7.6.1 Reducing current

Wherever electricity is in operation, there are also electric and magnetic fields, unseen lines of force created by the electric charges. Electric fields are easily shielded, they may be distorted or blocked by conducting bodies such as earth, vegetation, and buildings, but magnetic fields are not as easily blocked. Magnetic fields may easily penetrate building materials. They are not shielded by most common materials and pass easily through them. In general, magnetic fields are strongest close to the source and diminish with distance. Usually, people are not able to sense the presence of magnetic fields.

High magnetic fields can exist in various buildings. In many cases, the problems are attributed to the inappropriate design and installation of electrical equipment or conductor systems. By using proper strategies in the design and installation of wiring systems, the magnetic fields in a building may be significantly reduced, and the magnetic interference to sensitive equipment avoided. One approach to reducing the level of magnetic fields is by reducing current in conductors. The magnetic field at any point is directly proportional to the currents in the conductors. Reducing the current may be feasible in practice if system voltage can be increased (e.g., 11 kV in building distribution) or more efficient load equipment installed (e.g., energy-efficient loads).

7.6.2 Increasing distance

The distance from the source of EM fields has a great influence on magnetic field reduction. For example, the magnetic field can reduce to one-fourth for a line source or one-eighth for a point source if the distance is doubled. The strategy is to distance sensitive equipment from the field-producing equipment. However, relocating may not always be practical. Relocating the source of EM fields is the right practice for consideration.

In large buildings, single-core cables are often employed for the secondary connections between transformers and switchboards. They are normally installed in a duct, which is very close to the floor underneath. If magnetic interference is of concern on that floor, an increase of separation from the source significantly reduces the magnetic field on the affected floor.

7.6.3 Reducing spacing

The impact of conductor spacing is straightforward. However, engineers in design and installation practices may not appreciate this. A magnetic field is created whenever an electric current flows in a wire. The magnetic field created by the current in the hot wire will be equal in strength but opposite in direction to the magnetic field created by its paired neutral. Therefore, it is always recommended to run the two conductors for each circuit as close together as possible right up to the breaker and neutral bar. Any physical separation between these two paired wires will effectively increase the magnetic fields. This is because their magnetic fields are equal but opposite. However, if there is any physical separation between the hot and neutral, there will be less "cancelation" between the two wires. This will result in a higher magnetic field from the panel. To reduce magnetic fields in respect of conductor spacing, installing phase and neutral conductors along the same route, and minimizing the spacing are recommended.

7.6.4 Phasing conductors

The magnetic field is both a spatial vector and a temporal phasor. By properly arranging the phases of multiple parallel conductors, the magnetic field can be reduced significantly. The field reduction is mainly achieved by the cancelation effect of the magnetic fields from the parallel conductor currents. This method is only recommended if there are multiple conductors per phase and the phase currents be out of balance by less than 10%. Normally, no extra cost is required to cover the implementation of this strategy.

7.6.5 Shielding techniques

Shielding is used in buildings to protect people and equipment from the effects of other nearby EM field sources. It is most often used in specific-service buildings like research laboratories and medical facilities due to the specialty equipment and compounds used in those types of facilities. Architects and engineers working in the healthcare and institutional sectors need to understand the various kinds of shielding available.

Shielding can be broken into two main categories, passive and active. Passive shielding is a result of one or both of the two shielding mechanisms, induced currents due to electrical conductivity or flux shunting due to ferromagnetic materials. Passive shielding occurs in response to an applied field. Active shielding involves the control of additional field sources (e.g., currents) that attempt to cancel the imposed magnetic field in some optimum fashion. Passive shielding is made of shields from materials that are both conducting and ferromagnetic. However, hybrid systems are combinations of active and passive shielding methods that make use of the strong points of each technique to form an extremely effective shield system.

All shielding materials work by diverting the EM fields to them, so although the field from a magnet will be highly reduced by shielding material, the shield material will itself be attracted to the magnet. Steel which is composed mainly of iron

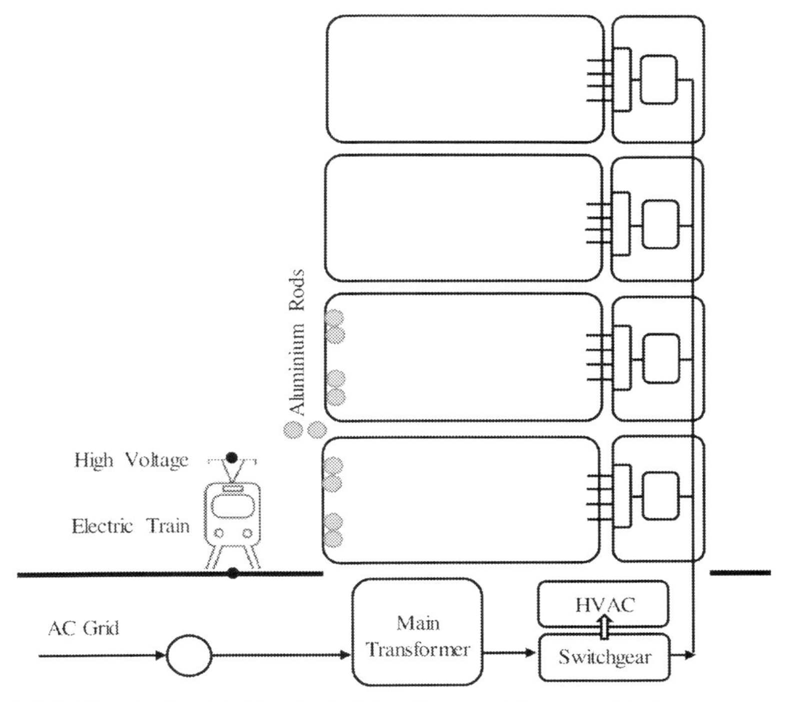

Fig. 7.4 Shielding the frontal side of a building from an AC-powered train.

(ferromagnetic) or aluminum (pure conductive) is the most commonly used material for shielding. Closed shapes are most effective for magnetic shieldings, such as cylinders with caps and boxes with covers (Salinas et al., 2007). Because of the large dimensions of buildings, steel would be excessively heavy, thus aluminum seems to be the obvious choice. Fig. 7.4 shows a train passing laterally by a frontal-side aluminum-shielded building. Trains sometimes are powered by AC overhead 15 kV voltage with a frequency of 16.67 Hz and an average current of about 200 A. The current's return path follows the rail track.

In general, conductive materials like metals, due to their high reflectivity, are widely used to isolate spaces from surrounding EM fields. This reflection shielding is based on the principle of the "Faraday cage," in which inside the cage, space is completely impermeable to external fields. On the other hand, absorption shielding is related to magnetic materials. Accordingly, metallic conductors suffer a lack of flexibility, heaviness, and high costs. Meanwhile, ferromagnetic materials have an intrinsic cut-off frequency, usually below the low GHz range, that hinders their use in EMI shielding over a broad GHz range. From this, there is a need to explore broadband shielding materials, which do not only work in the MHz range but also neutralize RF waves in the GHz range to include emerging 5G and IoT signals.

7.7 Design for minimum exposure

Assessment of EM field exposure in buildings has been most often conducted based on surrogate models. These include a wire code system that classifies exposure based on the calculation of the distance between households and energy transmission equipment, historical calculation through information provided by electric power companies, and personal measurements by dosimeters held by subjects during specific periods. EM fields from particular sources may vary greatly, depending on the way they are designed and manufactured.

7.7.1 Minimizing power frequency fields

Importantly for the user, there appear to be general procedures and suggestions to reduce the levels of electric and magnetic fields in homes and workplaces. The following are a few suggestions to minimize the level of ELF fields, as a procedure before resorting to various mitigation and shielding techniques, except when shielding is the most effective and least expensive alternative (Habash, 2020):

- Before sitting a building on a site, visually inspect potential sources of power-frequency fields including nearby power transmission lines, substations and locate buildings as far away as feasible. A simple Gauss meter could be used to determine the strength levels and locations of fields.
- If potential sources are found, sits the building to provide maximum distance from the sources or consider proper shielding techniques.
- Bring power into the building in a way that minimizes fields and ensures that any internal electrical transformers are not placed close to spaces where people will be using them daily.

- Cabling to and from the transformer can be a significant producer of fields so their location in the building requires attention or can be housed in metal conduits.
- In designing new buildings or renovating existing buildings, any transformer or distribution panel should be given special treatment in the initial phase of design. Areas immediately adjacent to the transformer should be assigned as storage areas and not living or working spaces.
- Keep the service panels and subpanels away from normally occupied rooms. To reduce field levels even further, special shielding using materials with high magnetic permeability, such as Mumetal, may be installed around the panel.
- Keep major high-load wiring from the main panel to a subpanel or high current equipment, such as elevators and rooftop air conditioners, away from frequently used spaces.
- Avoid separating hot and neutral wires, and ensure there is always a supply and return current in all wiring runs.
- Place high-load appliances such as electric dryers and electric water heaters away from bedrooms, kitchens, etc. Preferably, socket outlets should not be placed in close proximity to beds.
- In-floor electrical heating where electrical wiring is embedded in a concrete floor is of concern, especially in areas where occupants spend many hours a day. Measurement of magnetic fields is required to make sure that levels of fields are not of concern. Also is advisable to have it so that such spaces can be turned on to heat then turned off when occupied.
- Avoid placing heavily used spaces like offices and bedrooms near spots of elevated fields like placing electrical appliances such as refrigerators, stoves, TV, and entertainment equipment on the other side of the wall.
- Install energy-efficient equipment where any procedure to reduce currents will reduce levels of magnetic fields.
- The entire electrical installation should be protected using a ground fault circuit interrupter.
- As a last solution, measure levels of fields and modify exposures as much as possible by considering various forms of shielding.

7.7.2 RF survey and link planning

RF Survey is the gathering of data from the site to install a new building. This is important for site analysis, determining the coverage region of RF and cellular communication, identifying the potential RF obstacles, deciding the type of link connectivity with cellular towers, and determining locations of wired network access at the building. To achieve very low RF field levels, new buildings may be located in a low-RF environment, for example at a distance from cellular towers, radio and TV broadcast towers, and radar sites. The proposed location should be evaluated with professional-grade RF equipment to determine ambient RF field levels and potential sources (Clegg et al., 2020). Sometimes, it is a requirement by regulatory bodies for all new and existing buildings to have adequate RF signal strength for fire departments and public safety. This usually starts with performing a study of the building RF link planning.

RF signals that interact with a building will produce losses that depend on the electrical properties of the building materials and material structure. In general, the attenuation in buildings is higher than in free space, requiring more cells and higher power to obtain adequate coverage. Building entry loss can be measured as the difference, expressed in dB, between the spatial median of the signal level outside the illuminated

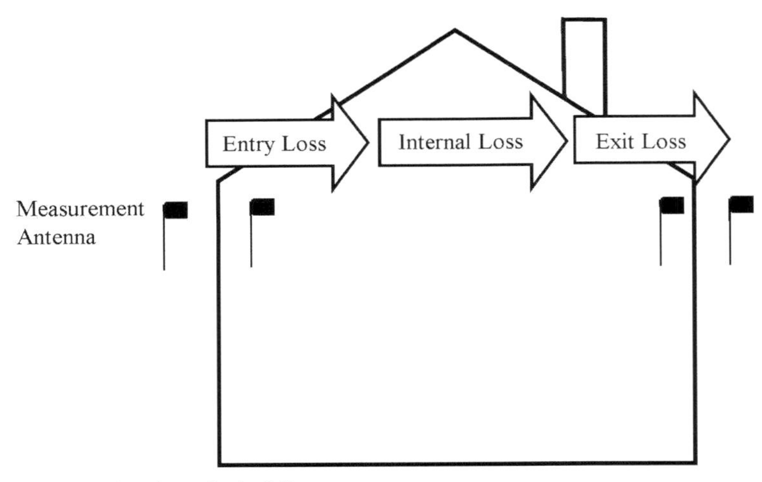

Fig. 7.5 RF link planning of a building.

face of a building and the spatial median of the signal level inside the building at the same height above ground as shown in Fig. 7.5. The outdoor field should be measured as close as possible to the building while ensuring that near-field effects are avoided and antenna characteristics are unaffected (ITU, 2015).

7.7.3 Metamaterials for shielding RF fields

Architectural shielding protects buildings from RF fields using a range of suitable materials. Commonly used materials include heavy-duty copper and aluminum foils and steel sheets as well as artificial materials. In particular, dynamic metasurfaces can artificially and dynamically control RF waves under external control signals.

To improve the design of buildings and create new constructions, two particular novel structures may be used: frequency selective surfaces building components such as walls, windows, and facades; and intelligent walls. The surfaces are simple printed metallic patterns that can be applied to insulation windows or as standalone panels and act as filters to wireless EM signals allowing some frequency bands to pass while rejecting others, for example passing cellular signals and rejecting WiFi signals. This technique allows the use of metal-backed insulation which does not interfere with some of the EM signals while rejecting others. Intelligent walls provide designers with the flexibility to reconfigure a building for different users to facilitate enhanced wireless connectivity in designated sections of the building.

The architectural design may be used to optimize the presence and distribution of EM fields in buildings. These structures could be designed to reflect EM radiation as Faraday cages, but also to actively change states between blocking and permitting EM waves to pass through them. Architects can account for the use of materials and

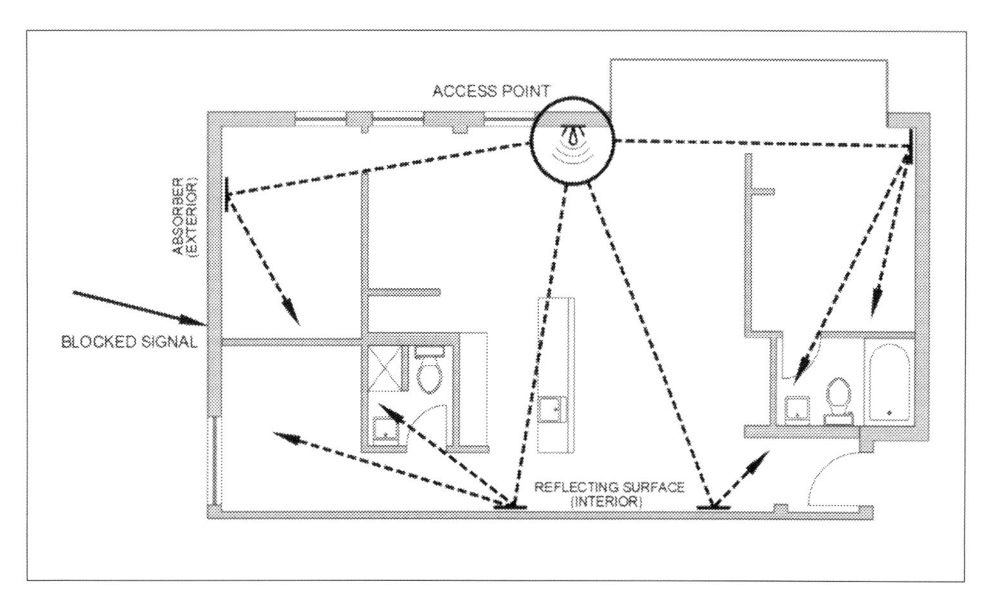

Fig. 7.6 The electromagnetic environment within and outside one unit of a building.

disposition of routers in a more active manner, resulting in better EM propagation. This requires detailed studies of network propagation.

Metasurfaces are 2D counterparts of metamaterials, in the sense of having small but not negligible depth. A reflecting surface can treat the EM wave propagation similar to the routing process in classic networking. Connecting two wireless devices becomes a problem of finding a route over reflecting surfaces while blocking signals is achieved by absorbing or deflecting its EM emissions. An example is given in Fig. 7.6 which studies a possible EM routing configuration to serve the objectives of the scenario (Habash, 2020). Using the above configuration in MMW and THz communication systems, which are severely limited in terms of very short distances and line-of-sight scenarios is of particular importance. The configuration can mitigate the acute path loss by enforcing the lens effect and any custom reflection angle per surface, avoiding the ambient dispersal of energy and non–line-of-sight effects, extending the effective communication range (Lee et al., 2012). In a simulated scenario, the functions can be applied to maximize the minimum EM received power over several receivers.

The propagation of low-frequency wireless is far superior to that at higher frequencies which is an important fact for WiFi and future 5G systems which operate at frequencies above 6 GHz. The key enabler for building programmable wireless networks is the concept of metamaterials and metasurfaces. EM metamaterials have emerged as innovations with the potential to change architecture toward providing management of indoor wireless networking (Savic, 2017). They provide invisible characteristics to control waves, such as negative refraction, perfect absorption, and transmission, through the periodical arrangement of metaatoms made by the size smaller than incident EM wavelength (Yannopapas et al., 2009). A basic design consists of an array of electrically small EM scatterers, called inclusions, embedded into a dielectric host material. From the form of their metaatom design (building blocks at the nano-, micro-, or macroscale), they are arrays of structured composite materials that can be engineered to have desired EM properties and functions including reflection, refraction, absorption, and polarization. When treated macroscopically, metamaterials display custom permittivity and permeability beyond those found in natural materials. As a result, metamaterials enable different interactions with impacting EM waves, being able to fully reengineer received waves.

7.8 Electromagnetic risk management

The power of technology and the wisdom of humans for how to manage the technology are needed to investigate ways in which the new EM technologies will impact the quality of life, positively by providing better life services or negatively through environment and health. Risk management is fundamentally a societal decision. It includes not only the outcome of risk assessment expressed in characterized risk, but also numerous other parameters, such as cost–benefit and risk–benefit analyses, views of stakeholders, and other nonscientific judgments.

7.8.1 Anticipatory ethics of EM fields

As society continues to explore the benefits and risks associated with emerging EM technologies, attention to transparency and the ethics of these systems' use needs special consideration. The mechanisms that enforce the ethical EM agenda need to be relevant to an ecosystem that includes humans and the evolving physical environment (Habash, 2020). With more ethical challenges, the industry must take a hard look at what it may have to do to promote positive applications of the technology, discourage negative ones, maintain transparency with end-users, and build health safety into systems from the beginning, and answer some challenging questions; for example, how to develop a moral code into energy technologies?

In a risk management model, the design is seen as an "ethical factor" because it is built to empower and entitle users to frame their ethical choices. The concept of ethical design in the EM domain shares some similarities with the concept of "anticipatory ethics" which involves early consequence anticipation of future applications, probably at the research and development stage of technology development. This concept should become a major component in the design of emerging technologies and lead to better ethical outcomes and health-responsible technologies (Habash, 2020).

The QoE of a user is an important metric to quantify how well a technology, in particular a wireless one, performs; and hence, its overall acceptability and usability can be assessed. QoE is defined as, the degree of delight or annoyance of the users of an application or service. Ethically, QoE evaluates as 5G and corresponding IoT technologies are impacting human health and comprises work–life quality, positively or negatively.

7.8.2 Precautionary approaches

There has been an increasing movement to consider precautionary approaches within a structured methodology for the management of risk in the face of scientific uncertainty. These approaches include both prudent avoidance and the precautionary principle.

Prudent avoidance became an attractive option because it serves to minimize exposure to the perceived problem with minimal costs. For instance, no radical changes to power lines or wireless base stations should be implemented until science shows clear evidence that there is a health risk. By acting prudently, management can embrace a wide range of sensible actions that take into account the research results and community concerns. In the context of EM fields, some national and local governments have adopted "prudent avoidance," a variant of the precautionary principle, as a policy option. It was initially adopted for ELF fields and is defined by using simple, low to modest (prudent) cost measures to reduce individual or public EM exposure, even in the lack of certainty that the measures would decrease risk.

The precautionary principle is another process that emerged in the 1970s in response to concerns about the extent to which complex and uncertain risks could be addressed within existing science and policy structures. It is usually applied when there is a high degree of scientific uncertainty and there is a need to take action for a serious risk without anticipating the outcomes of more scientific research. Under this principle, any claim that action might pose a risk to the environment or people's health, however unjustified, seeks the initiator to prove that the action will not harm being allowed to act.

The precautionary principle alters environmental policy-making by markedly changing the balance in the contest between opposing views on where to set the balance point when science is uncertain and the environment and human health are at stake (Balzano and Sheppard, 2003). This principle is an extremely conservative decision that leads to prudent actions in the face of uncertainty. It reflects the need to take action at a reasonable expense and with reasonable consequences for a potentially serious risk without awaiting the results of scientific research. The precautionary principle has been incorporated into numerous international treaties and declarations throughout Europe and several other countries. Many countries have adopted it to help set precautionary limits for EM exposure.

Several precautionary approaches for RF exposure have been adopted in Canada. For example, the city of Toronto has implemented a policy that it calls "prudent avoidance" in the siting of cellular base stations. This policy encourages the adoption of individual or societal actions to avoid unnecessary exposures to RF fields that entail little or no cost. Also, the "common sense" approach was adopted by the National Collaborating Centre for Environmental Health in Canada in dealing with RF energy. The Centre generated a "tool kit" for environmental health professionals which outlines a range of actions, including engineering controls, substitution, administrative controls, and protective equipment that can be suggested to individuals who wish to reduce their current RF exposure (Demers et al., 2014).

The American Academy of Environmental Medicine published recommendations regarding EM exposure in 2012. The Academy called for physicians to consider EM exposure in diagnosis and treatment and to recognize that EM exposure may be an underlying cause of the patient's disease process (Dean and Rea, 2012).

In 2015, an International Scientific Declaration on Electromagnetic Hypersensitivity and Multiple Chemical Sensitivity were published by the Scientific Committee following the 5th Paris Appeal Congress, which took place in 2015 at the Royal Academy of Medicine, Brussels, Belgium. It calls upon national and international agencies and organizations to recognize EHS and multiple chemical sensitivity as a disease and urges particularly the WHO to include EHS in the International Classification of Diseases. It also asks national and international agencies and organizations to adopt a precautionary approach to inform the public, and to appoint truly independent expert groups to evaluate these health risks based on scientific objectivity (RAM, 2015).

References

Balzano, Q., Sheppard, A.R., 2003. The influence of the precautionary principle on science-based decision-making: questionable applications to risks of radiofrequency fields. J. Risk Res. 5, 351–569.

Brune, D., Hellborg, R., Persson, B.R.R., Pääkkönen, R. (Eds.), 2001. Radiation at Home, Outdoors and in the Workplace. Scandinavian Science Publisher, Oslo.

Clegg, F.M., Sears, M., Friesen, M., Scarato, T., Metzinger, R., Russell, C., Stadtner, A., Miller, A.B., 2020. Building science and radiofrequency radiation: what makes smart and healthy buildings. Build. Environ. 176, 1–15. https://doi.org/10.1016/j.buildenv.2019.106324.

Dean, A.L., Rea, W.J., July 2012. American Academy of Environmental Medicine Recommendations Regarding Electromagnetic and Radiofrequency Exposure. Executive Committee of the American Academy of Environmental Medicine, Wichita, KS. https://www.aaemonline.org/pdf/AAEMEMFmedicalconditions.pdf.

Demers, P., Findlay, R., Foster, K.R., Kolb, B., Moulder, J., Nicol, A.-M., Prato, F., Stam, R., 2014. Expert Panel Report on a Review of Safety Code 6 (2013): Health Canada's Safety Limits for Exposure to Radiofrequency Fields. Royal Society of Canada, Ottawa, ON.

Eichwald, C., Walleczek, J., 1998. Magnetic field perturbations as a tool for controlling enzyme-regulated and oscillatory biochemical reactions. Biophys. Chem. 74, 209–224.

Frey, A.H., 1988. Evolution and results of biological research with low-intensity nonionizing radiation. In: Marini, A.A. (Ed.), Modern Bioelectricity. Marcel Dekker, New York, pp. 788–837.

Greenfield, J., 2021. EMF's Effects at the Cellular Level. https://www.hightechhealth.com/dangers-of-emfs/.

Habash, R., 2020. Bioelectromagnetics: Health Effects and Biomedical Applications. CRC Taylor and Francis, Boca Raton.

Hocking, B., Gobbo, F., 2011. Medical aspects of overexposures to electromagnetic fields. J. Health Saf. Environ. 27 (3), 185–195.

IBN, 2015. Building biology evaluation guidelines for sleeping areas, supplement to the standard of building biology testing methods SBM. https://buildingbiology.com/site/wp-content/uploads/richtwerte-2015-englisch.pdf.

ICNIRP, 2002. General approach to protection against non-ionizing radiation. Health Phys. 82, 540–548.

IEEE, 2019. IEEE Approved Draft Standard for Safety Levels With Respect to Human Exposure to Electric, Magnetic and Electromagnetic Fields, 0 Hz to 300 GHz, IEEE Standard C95.1–2019.

ITU, 2015. Effects of Building Materials and Structures on Radiowave Propagation above about 100 MHz. https://www.itu.int/dms_pubrec/itu-r/rec/p/R-REC-P.2040-1-201507-I!!PDF-E.pdf.

Lee, S.H., Choi, M., Kim, T.-T., Lee, S., Liu, M., Yin, X., Choi, H.K., Lee, S.S., Choi, C.-G., Choi, S.-Y., Zhang, X., Min, B., 2012. Switching terahertz waves with gate-controlled active graphene metamaterials. Nat. Mater. 11 (11), 936–941.

Lichtenstein, P., Holm, N.V., Verkasalo, P.K., Iliadou, A., Kaprio, J., Koskenvuo, M., Pukkala, E., Skytthe, A., Hemminki, K., 2000. Environmental and heritable factors in the causation of cancer, analyses of cohorts of twins from Sweden, Denmark and Finland. N. Engl. J. Med. 343, 78–85.

Maes, B., 2008. Standard of Building Biology Testing Methods, Institute of Building Biology + Sustainability IBN (Rosenheim, Germany, Technical Report SBM-2008).

Panagopoulos, D.J., Johansson, O., Carlo, G.L., 2015. Polarization: a key difference between man-made and natural electromagnetic fields, in regard to biological activity. Sci. Rep. 5, 14914.

RAM, 2015. International Scientific Declaration on Electromagnetic Hypersensitivity and Multiple Chemical Sensitivity. Royal Academy of Medicine, Brussels, Belgium. http://appel-de-paris.com/wp-content/uploads/2015/09/Statement-EN.pdf.

Rodriguez, K.O., 2019. 5G and the Future of Building Design. https://hhangus.com/5g-and-the-future-of-building-design/.

Salinas, E., Atalaya, J., Hamnerius, Y., Solano, C.J., Gonzales, D., Contreras, C., Leon, C., Sumari, M.A., Dimitriou, S., Rezinkina, M., 2007. A new technique for reducing extremely low frequency magnetic field emissions affecting large building structures. Environmentalist 27, 571–576.

Savic, S., 2017. Designing for connectivity: rethinking the interaction with the built environment and wireless communication infrastructure. Interact. Des. Archit. 32, 48–67.

Stuchly, M.A., 1979. Interaction of radiofrequency and microwave radiation with living systems: a review of mechanisms. Radiat. Environ. Biophys. 16 (1), 1–16.

Wood, A.W., 2017. Chapter 17: Electric and magnetic fields and induced current hazard. In: Wood, A.W., Karipidis, K. (Eds.), Non-ionizing Radiation Protection: Summary of Research and Policy Options. John Wiley and Sons, Hoboken.

Yannopapas, V., Paspalakis, E., Vitanov, N.V., 2009. Electromagnetically induced transparency and slow light in an array of metallic nanoparticles. Phys. Rev. B80, 035104.

Building as a hygiene system

Riadh Habash
School of Electrical Engineering and Computer Science, University of Ottawa, Ottawa, ON, Canada

An ounce of prevention is worth a pound of cure.

Popular saying

8.1 Historically thinking

Disease outbreaks, epidemics (regional), and pandemics (global) occurred throughout history and continue to happen, all the time, and all around the world with significant impacts on global economies and public health. Although diseases have long been a tragic scourge on cities and communities, they have also forced architecture and urbanization planning to evolve. Previous pandemics have all helped public health systems in preparing for major future outbreaks.

The history of disease outbreaks can be traced back to Ancient Rome and Greece through plagues in the middle ages. They have had devastating effects on populations. The more civilized humans develop in construction and services, the more likely disease outbreaks develop. At around 430 BCE, the earliest known pandemic occurred during the Peloponnesian War. The disease passed through Libya, Egypt, and Ethiopia. Then, it crossed the Athenian walls as the Spartans laid siege. The disease is suspected to have been typhoid fever, and almost, 60% of the population died.

In 165 CE, the Antonine plague was possibly an early appearance of smallpox that began with the Huns who then infected the Germans and passed it to the Romans, and then returning troops spread it throughout the Roman Empire. In 250 CE, the Cyprian Plague, named after the first known victim, the Christian bishop of Carthage, entailed diarrhea, vomiting, throat ulcers, fever, and gangrenous hands and feet. There were recurring outbreaks over the next three centuries in Britain. In 541 CE, Justinian Plague, first appearing in Egypt, then spread through Palestine and the Byzantine Empire, and then throughout the Mediterranean (History, 2021).

Eleventh-century leprosy, 14th-century bubonic plague, and the 17th-century great plague of London, which wiped out large numbers of Europe's population, helped to inspire the radical urban improvements of the Renaissance. Cities cleared squalid and cramped living quarters, expanded their borders, developed early quarantine facilities, opened larger and less cluttered public spaces, and deployed professionals with specialized expertise, from surveyors to architects (Lubell, 2020). The word "quarantine," which means restricting the movement of people or goods, is rooted in the Latin word "Quaranta" or "40 days," a reference to preventative measures taken in Venice during the middle ages to stop the spread of the bubonic plague. Ships arriving from

Sustainability and Health in Intelligent Buildings. https://doi.org/10.1016/B978-0-323-98826-1.00008-9

areas affected by the "Black Plague" were required to anchor for 40 days in a quarantine island before the crew could disembark (Budds, 2020).

Girolamo Fracastoro (1478–1553) proposed in his book titled "Contagion De Contagione et Contagiosis Morbis; 1546" that tiny particles can cause epidemic diseases through direct or indirect contact or even without contact over a distance. Likewise, 18th-century yellow fever and 19th-century cholera and smallpox outbreaks helped to catalyze innovations like broad boulevards, citywide sewer systems, indoor plumbing, disease mapping, and the early suburbs.

In the 20th century, influenza, tuberculosis, typhoid, polio, and Spanish flu breakouts prompted urban planning, slum clearance, tenement reform, waste management, and, on a larger level. Then came threatening viruses from various sources: Asian flu (H2N2), Hong Kong flu (H3N2) in 1968, severe acute respiratory syndrome (SARS) in 2002–03, the swine flu (H1N1) influenza in 2009, Middle East respiratory syndrome-coronavirus (MERS-COV) epidemic in 2012, and Ebola epidemic in 2013. But the outcome of each disease depends mainly on the way it was caught, how contagious and fatal it was, the hygienic level of people and facilities, and how quickly a vaccine or cure became available.

Previously, health systems managed to prevent coronavirus infections from becoming more than localized epidemics. COVID-19, on the other hand, has reached every continent. The main reasons for the increasing pandemic threat in the 21st century are the rapidly growing world population, global travel, urbanization trends, and industrialized food production in global value chains (Gössling et al., 2020). Urban planning may be another factor that can have a big influence on the spread of infectious diseases.

Learning from history, each outbreak led to great changes in the built environment system. For example, as a preventative response to London's Cholera outbreaks in the early 19th century, advances in water-based sewage and sanitation systems were created to fight the pathogens that cause cholera and typhoid. This led to the sanitary reform movement, which created drinking water and sewage infrastructure worldwide. The outbreak of plague in Paris in the 18th and 19th centuries led to the practice of the numbering of houses and the full mapping of every building in urban neighborhoods. From the 19th century, hospitals began to be laid out on more coherent lines, with wings so that wards could be naturally ventilated. Tuberculosis in New York in the early 20th century led to improvements in building regulations and public transit. SARS in Asia encouraged the development of medical technology to monitor and map diseases. Looking forward, the entire world is learning from the COVID-19 pandemic.

An eye-opener, the above series of historical health outbreaks and COVID-19 pandemic will show the world what the future of building design will be. The future will not happen in a vacuum, it takes awareness, work, and caring to create health- and comfort-focused ideas for change. The built environment in itself will not solve the pandemic problem, but architects working with engineers, physicians, health professionals, and politicians could start to address this complex health challenge. This team of professionals who knows the built environment, whether it is for living, working, or for healthcare, needs to renew and reshape itself for healthy living and to take ownership of the future.

8.2 Infectious disease outbreaks

An infectious disease is a viral, bacterial, parasitic, or fungal disorder that can be transmitted between people. The typical epidemiologic triad model holds that infectious disease outbreaks result from the interaction of agent, host, and environment. More specifically, transmission occurs when the agent (e.g., viral or bacterial) leaves its reservoir (habitat) or host through a portal of exit, is conveyed by some mode of transmission and enters through an appropriate portal of entry to infect a susceptible host. This sequence is sometimes called the "chain of infection" (CDC, 2021), which is a dynamic process determined by multiple factors originating from disease pathogens and/or parasites, vector species, and human populations. These factors interact with each other and demonstrate the intrinsic mechanisms of the disease transmission temporally, spatially, and socially (Xia et al., 2017). Disease Outbreaks induce geopolitical instability, social and economic risks, public health, and medical resources and may include reactions such as infection control, contact tracing, quarantine, isolation, and social distancing.

Bacteria and viruses can travel through the air, causing and worsening diseases. They get into the air easily. When someone sneezes or coughs, tiny water or mucous droplets filled with viruses or bacteria scatter in the air or end up in the hands where they spread on surfaces like doorknobs. Crowded conditions with poor air circulation may promote this spread. Some bacteria and viruses thrive and circulate through poorly maintained building ventilation systems. Humid air may increase the survival rate of viruses indoors (CARB, 2005). Viruses have been identified as the most common cause of infectious diseases acquired within indoor environments, particularly those causing respiratory and gastrointestinal infections. The most common viruses causing respiratory infections include influenza viruses, rhinoviruses, coronaviruses, respiratory syncytial viruses, and parainfluenza viruses; whereas viruses responsible for gastrointestinal infections include rotavirus, astrovirus, Norwalk-like viruses. Some of these infections like the common cold are very widely spread but are not severe, while infections like influenza are relatively more serious (Kubba, 2017). Today, it is no surprise that respiratory droplets and direct contact are both conceived to contribute to infectious disease outbreaks (Ai and Melikov, 2018). However, for airborne transmission, there was a long-lasting dispute over the years. In general, three major infectious diseases are a threat to urban built environments. First, respiratory viruses, such as pandemic influenza, severe acute respiratory syndrome (SARS), and recently COVID-19. Second, Aedes-borne diseases, such as dengue, Zika virus disease, chikungunya, and yellow fever. Third, local diseases, such as plague and malaria (Fahey, 2020).

Transmission of infectious diseases has been a subject of interest over recent years with growing evidence that IAQ has a significant impact on the transmission of infection. Immunologists Erin Bromage at the University of Massachusetts Dartmouth proposes a model for typical risk for the spread of viral infection:

$$\text{Successful Infection} = \text{Exposure to Virus} \times \text{Time} \tag{8.1}$$

Eq. (8.1) is the basis of contact tracing. If a person is infected, the droplets in a single cough or sneeze may contain as many as 200,000,000 virus particles which can all be dispersed into the environment around them. A single breath releases 50–5000 droplets. Most of these droplets are low velocity and fall to the ground quickly. There are even fewer droplets released through nose-breathing while the respiratory droplets released from breathing only contain low levels of virus (about 33 particles per minute). Speaking increases the release of respiratory droplets about 10-fold (Bromage, 2020). This denotes a great concern about spaces that are occupied for extended periods, including hospitals, tower apartments, offices, and schools. This justifies using fresh air in a true displacement ventilation approach, which allows for the minimum amount of cross-contamination in occupied spaces and reduces concerns around recycling virus-laden air (Kuwabara et al., 2020).

The Wells–Riley equation, derived in 1978, was initially used to model measles outbreaks in schools. This approach is well suited for buildings in the COVID-19 pandemic crises.

$$P = 1 - e^{\frac{Ipqt}{Q}} \tag{8.2}$$

where P is the probability of infection, I represents the number of infectious people, p is the average breathing rate of individuals, q represents a unit of infection termed a quantum, introduced by Wells (1995) to express the response of susceptible individuals to inhaling infectious droplet nuclei (14–48/h), t is the exposure time, and Q represents the airflow rate from the HVAC system. By using this equation, it is possible to estimate the probability of infection for individuals within a defined space. It may be applied to whole buildings or rooms. The equation's significant shortcoming comes from only considering airborne particles as a pathway to infection, not accounting for fomite transmission, or even nonaerosolized respiratory droplets, which are both very important with COVID-19.

8.3 Routes of transmission

Building occupants are exposed to contaminants via mechanisms like inhalation, dermal absorption (through the skin), and ingestion. Professor Yuguo Li, of mechanical engineering at the University of Hong Kong, provided an overview of the mechanisms and implications of human exposure to microbes in urban buildings. He introduces three major routes of respiratory infection transmission: through the air, people breathe (airborne route), through the surfaces they touch (fomite route), and through other people, they meet (close-contact route). WHO recognizes all exposure mechanisms of COVID-19, it is therefore important to understand all possible mechanisms of transmission to realize proper engineering control approaches. Fig. 8.1 provides an idea about major routes of infection in the indoor environment. Transmission can take place through short-range or long-range routes, where short routes involve close contact of fewer than five feet with large droplets (>100 μm) transmitted through the air

Building as a hygiene system

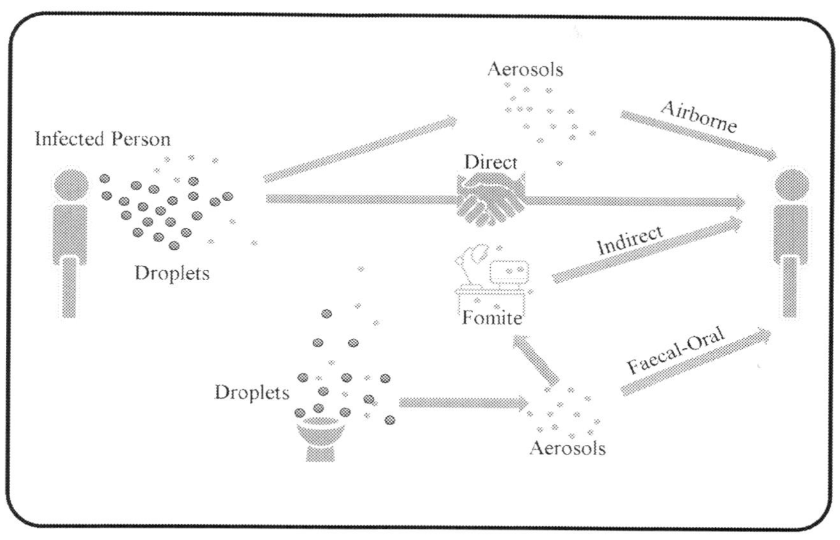

Fig. 8.1 Mechanisms of human exposure to microbes including coronavirus.

or via direct touching. Long-range routes, more than 5 ft, involve small droplets, droplet nuclei, or aerosols (<5 μm) transmitted through the air or in contact with fomites (Wei and Li, 2016). Viruses primarily spread when infected people breathe, talk, sneeze, or cough as well as plumbing including toilet flushing and splashing in sinks.

8.3.1 Respiratory droplets

Respiratory viruses are transmitted between individuals when the virus is released from the respiratory tract of an infected person and is transferred through the environment, leading to infection of the respiratory tract of an exposed and susceptible person (Leung, 2021). People can contract COVID-19 if they breathe in droplets from an infected person who coughs or exhales droplets. Those droplets can also land on objects and surfaces, and people can then catch the virus from touching those surfaces and then touching their eyes, nose, or mouth (GHN, 2020). Most droplet transmission (propulsion) likely happens at close range because of dilution and inactivation of viruses over distances. In respiratory exhalation flows, the large droplets (>100 μm) in size are expected to completely evaporate before traveling about 0.5 ft (talking) or 1.5 ft (coughing). Coughs and sneezes consist of a turbulent cloud of buoyant gas with suspended droplets. The largest droplets follow a ballistic trajectory relatively unaffected by the flow in the gas phase, while the smaller droplets are suspended to varying degrees within the turbulent gas cloud, thereby having their range extended (Bourouiba et al., 2014). These actions expel respiratory droplets containing particles of the virus in combination with mucus, saliva, or fecal matter. If these droplets land on or are inhaled by others nearby, they could transmit the virus.

8.3.2 Aerosol and airborne

Airborne transmission implies the spread of germs over a distance of more than several feet between the source and the victim. Questions remain, however, about whether tiny virus particles, of about 0.1 μm in size, can become airborne and travel greater distances. Although heavy droplets, of about 5–10 μ, usually travel less than 1 m before settling, smaller droplets may evaporate, leaving virus particles, referred to as aerosols (droplets evaporated to become droplets nuclei), suspended in the air (Balgeman et al., 2019). These very small particles may remain suspended in the air for hours and be disseminated by air currents in space or through a facility and do not easily fall to the forces of gravity (Tierno, 2017). The distinction between droplets and airborne is due to particle size, if droplets are small enough, they can become airborne and even remain so for several hours under the right conditions.

In a COVID-19 outbreak, a Chinese study associated with air conditioning in a restaurant in Guangzhou, China, Lu et al. (2020) concluded that droplet transmission was prompted by air-conditioned ventilation (split unit). The key factor for infection was the direction of the airflow. It was found that nine people from three families who had eaten in the same air-conditioned restaurant were infected by the virus (Fig. 8.2). The index patient and his family were sitting in between two other tables for lunch on January 24, 2020. The families at the table in front and behind the index patient were infected by the virus. The research pointed out that droplet transmission of the virus from the index patient, who was asymptomatic at the time, was due to the direction of the airflow. It was concluded that in this outbreak, droplet transmission was prompted by air-conditioned ventilation.

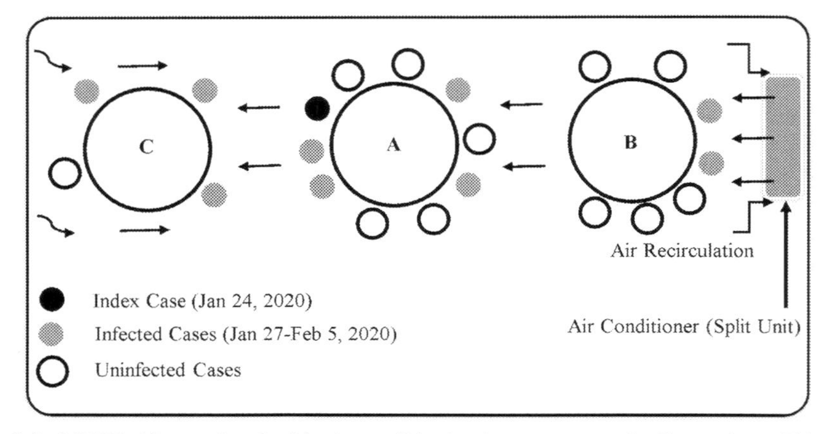

Fig. 8.2 COVID-19 associated with air conditioning in a restaurant in Guangzhou, China.

Several studies have demonstrated that viruses are characterized by fecal-oral transmission (Drexler et al., 2014; Yeo et al., 2020; Xiao et al., 2020). Therefore, blocking the path of fecal-oral transmission can reduce the probability of cross-infection in surrounding areas, thus suppressing the global spread of emerging and reemerging viruses. Li et al. (2020) indicated that fecal-oral transmission is a common transmission route for many viruses, including COVID-19. Suggestions concerning safer toilet use and recommendations for better toilet design are also provided.

A group consisting of 239 scientists from over 30 countries claims that the virus can float inside of air droplets and spread that way. Their finding would support the idea of COVID-19 being airborne. This is evidence that the virus can spread indoors through aerosols that remain in the air and can be infectious even in smaller quantities (Morawska et al., 2020). The group gives practical advice to prevent airborne spread such as by providing efficient ventilation (clean outdoor air and reduced air recirculation) particularly in community facilities (buildings, workplaces, schools, hospitals, and nursing homes); supplementing general ventilation with airborne infection controls such as local exhaust, high-efficiency air filtration, and germicidal UV lights; and avoid people congestion, mainly in public buildings and transportation systems. The study demonstrates that viruses are released during exhalation, talk, and cough in aerosols to remain in the air and pose a risk of exposure at distances beyond 6 ft from an infected individual. Recognizing the above fact helps target the control more precisely.

8.3.3 Direct and fomite

Contact is one of the common modes of transmission of infection which may include direct contact, indirect contact, and contact with droplets. Direct contact refers to the person-to-person spread of microbes through physical contact between the infectious agent including the contaminated hands or gloves of the agent with the skin or mucous membranes of the recipient. Indirect contact occurs when a susceptible person comes in contact with a contaminated object (Van Khai, 2016).

Surface-to-person, also known as fomite transmission occurs when a person touches a surface where large droplets of the virus have landed and then touches their mouth, nose, or eyes. Fomite is a type of large droplet transmission in that when infected individual talks, sneezes, coughs, or vomits nearby surfaces can become contaminated by large droplets that fall from the air. It can also potentially occur via contact with an airborne virus that settles after disturbance of a contaminated fomite, such as shaking a contaminated blanket (Boone and Gerba, 2007). Fomites are inactive objects that can carry infectious agents and serve as a mechanism for transfer between hosts. They may include a wide variety of objects such as walls, doors, doorknobs, keyboards, tables, chairs, pieces of clothing, elevator buttons, handrails, phones, keyboards, etc. Surfaces of materials of architectural elements such as walls, doors, partitions must be solid and smooth enough to be able to prevent the suspension of droplets.

8.4 Infection control strategies

Knowing how, when, and in which settings infected people transmit the virus is key for designing and implementing control measures to interrupt chains of transmission. Current control strategies mainly target the interruption of transmission through body contact and exposure to a short-distance, large-droplet spray. These strategies do not address the need to immobilize tiny, aerosolized droplets, which can spread infectious microorganisms across significant distances and for extended periods through the air (TVR, 2019).

After vaccines and the physical removal of viruses, the most effective means to prevent exposures to COVID-19 is through source and pathway control strategies. For workplaces deemed necessary, such as hospitals and supermarkets, this means workers are not attending to their duties when they are sick or have potentially been exposed to others who are sick. For the general public, this means eliminating needless trips to the places like stores or banks. It also means remaining at home if the person is sick or around someone who is sick. Social isolation is significant in the context of pandemics because many people are asymptomatic and can, unintentionally, spread the virus while outside their homes.

Pathway engineering control is the next level in separating people from the hazard. This is accomplished in places with a high ventilation rate that removes the air from rooms where viruses may be found. Receptor control is the last level which includes handwashing, physical distancing, usage of personal protective equipment, isolation, and quarantine. Fig. 8.3 shows the hierarchy of infection control pyramid that may be

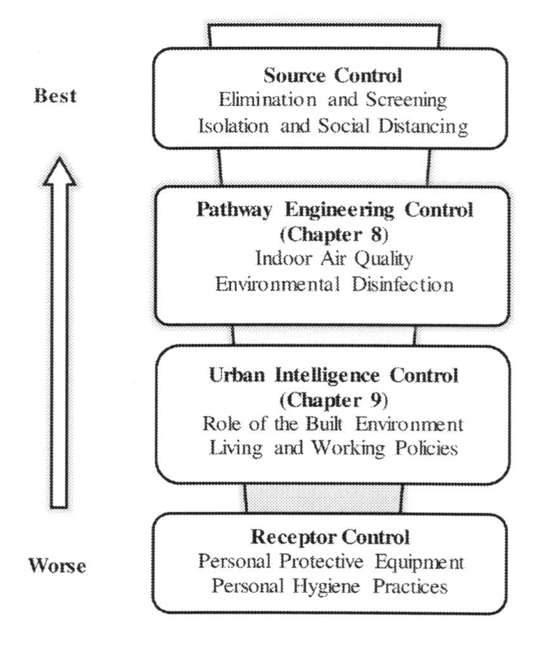

Fig. 8.3 Hierarchy of infection control strategies.

used to determine effective strategies needed to effectively reduce exposures. The top levels are considered more effective at minimizing pathogen exposure than the lower levels. Integration of control strategies is required to efficiently decrease exposures. However, the second level is unique (urban intelligence) at enhancing the role of the built environment and the way people live and work. A historical understanding of the interlinkages in the way cities and communities are designed, built and impacted human health should be realized. Many cities and communities have already started exploring public health opportunities and implementing various aspects of intelligence for sustainable and healthy living.

All possible engineering control strategies against airborne transmission in indoor scenarios should be taken. These include increased ventilation rate, using natural ventilation, avoiding air recirculation, avoiding staying in another person's direct airflow, enhancing humidity levels, and minimizing the number of people sharing the same environment (Alter, 2020b). Even in nonpandemic scenarios, it may be a smart idea to spend some time focusing on healthy indoor environments for occupants. Morawska et al. (2020) argue that existing evidence is strong to warrant engineering controls targeting airborne transmission as part of an overall strategy to limit infection risk indoors. Appropriate building engineering controls include sufficient and effective ventilation, possibly enhanced by particle filtration and air disinfection, avoiding air recirculation, and avoiding overcrowding. Often, such measures can be easily implemented and without much cost, but if only they are recognized as significant in contributing to infection control goals.

The use of engineering controls in buildings, along with the effective application of other control strategies (including isolation and quarantine, social distancing, and hand hygiene), would be an additional important measure globally to reduce the likelihood of transmission and thereby protect the general public. Identifying scenarios where infectious doses could potentially occur in various building types should be a priority of any control strategy. On a risk management platform, common approaches including naturally ventilated buildings and isolation should be considered for applying controls to emerging pathogens, based on the prospect and duration of exposure and the infection of the pathogen.

8.5 Design considerations for indoor air quality

Understanding and controlling building ventilation can improve the IAQ and reduce the risk of indoor health concerns including preventing viral diseases from spreading indoors. To achieve this goal, a balance between energy efficiency and IAQ is needed.

8.5.1 Hygiene ventilation

Indoor ventilation is part of a comprehensive package of prevention and control measures that can limit the spread of certain respiratory viral diseases. However, ventilation alone, even when properly applied, is inadequate to provide a satisfactory level of protection (WHO, 2021). A key technique to slow the spread of viruses in buildings is

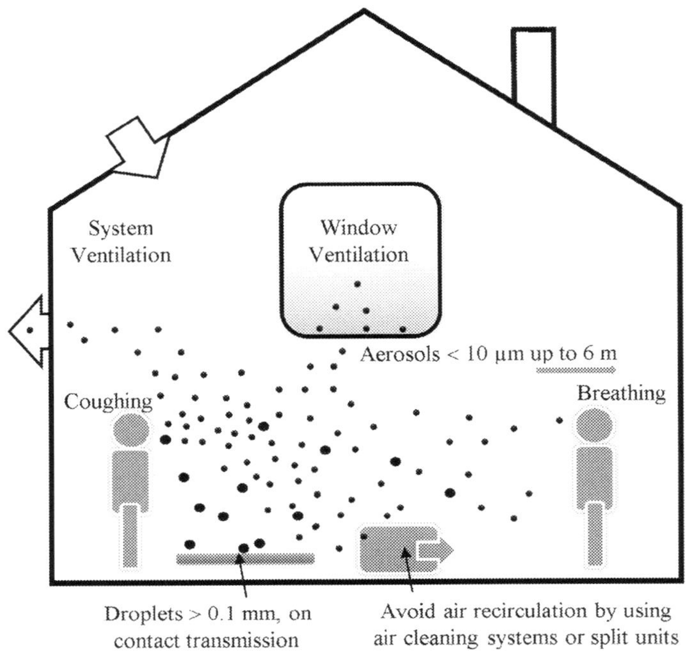

Fig. 8.4 Impact of ventilation on transmission via respiratory droplets and aerosols.

to bring outdoor air in by using windows openings or running exhaust fans that may create a negative air pressure for a building and therefore helps outside air to be taken and spread inside. Bringing fresh air (with caution in polluted areas) from the outdoors into a building is significant because constant air circulation prevents air particles, such as viruses, from staying in one place (Smieszek et al., 2019). One model (Fig. 8.4) shows how ventilation could prevent aerosol transmission, suggesting that for diseases transmitted by droplet nuclei, having good ventilation would be effective.

Hygiene ventilation is a term used to describe the process of providing adequate IAQ for all occupants while also remaining comfortable and energy-efficient. It is an approach that uses 100% fresh air (balanced ventilation) to ensure that recirculated air does not contribute to human health risks. This is partly because many HVAC systems, whether in residential and office buildings, lecture halls, shopping malls, campuses, airports, or airplanes, make use of recirculated air. Typically, air breathed indoors includes about 20% fresh air, meaning that at least 80% of that air is circulated.

HVAC systems can provide engineering controls via various techniques including economizers and purge ventilation. Dedicated outside air systems with energy recovery, which has considerable implications for how IAQ in buildings is treated, is one approach to improve ventilation, preferably "displacement ventilation" systems that introduce fresh air low and exhaust stale air high, so that people get the good air in the occupied zone and the rising air carries the contaminants up to the exhaust registers and out. High-rise multifamily buildings face challenges beyond ventilation namely maintaining building pressure (make-up and exhaust air) and the natural-stack effect. In residential houses and apartments, normal practices such as opening windows and doors, and using portable air-cleaning devices when practical to ensure healthy indoor air, should be implemented.

Based on how COVID-19 transmission, it seems logical that moving more air and providing more outdoor air would be beneficial. The Centers for Disease Control and Prevention (CDC)'s interim guidance for businesses and employers to plan and respond to COVID-19 specifically mentions "increase ventilation rates" and "increase percentage of outdoor air" (CDC, 2021). Also, REHVA (2020) recommends the use of more window airing. Even, in buildings with mechanical ventilation, window airing can be used to further boost ventilation. Therefore, it is time to take steps to make real changes to manage indoor airflows and optimizing ventilation and airflow indoors to limit viral spread. ASHRAE Standard 62.1 provides details in this regard. This standard is updated regularly. Also, many buildings designed in recent years according to LEED or WELL building standards already adhere to higher thresholds of outdoor airflow and other strategies to improve IAQ for occupants.

8.5.2 Air recirculation

In 1974, a young girl with measles went to school in upstate New York. Even though 97% of her fellow students had been vaccinated, 28 ended up contracting the disease. The infected students were spread out across 14 classrooms, but the young girl, the index patient, spent time only in her classroom. A ventilation system operating in a recirculating mode sucked in the viral particles from her classroom and spread them around the school (Allen, 2020). Recirculation of air using air cleaning systems or split units is fundamentally unhygienic and accordingly, a cause of concern. This was recommended also by the European and American engineering associations (Alter, 2020b). Virus particles in return ducts can reenter a building when centralized air handling units are equipped with recirculation sectors. In case this leads to problems with cooling or heating capacity, this has to be accepted because it is more important to prevent contamination and protect public health than to guarantee thermal comfort (REHVA, 2020).

During an epidemic, including the current COVID-19, air should not be recirculated as far as practically possible, to avoid the distribution of virus-laden particles throughout the indoor environment. As for split unit systems, which have no fresh air intakes at all, these should just be turned off or supplemented by lots of fresh air through open windows. Again, this will be very hard in summer in the hotter parts of the country (Alter, 2020b). If the recirculation method cannot be enhanced, maximizing ventilation by using natural ventilation through opening windows may be considered.

In Europe, REHVA (2020) is calling for engineers to stop recirculating air in buildings where there has been a COVID-19 outbreak as virus particles in return ducts can reenter a building if centralized air handling units have recirculation. In most buildings, the key function of the HVAC system is heating or cooling; therefore, there is a need for a lot of air to do this, so it mostly is recirculating since bringing in the fresh air and moderating will take too much energy. As buildings get more efficient, the hygiene ventilation or fresh air becomes more dominant, as less energy is needed to heat and cool while maintaining optimal human health environment.

8.5.3 Filtration

To assist the ventilation process, it is essential to consider using air cleaners with filters. Air filtration may facilitate the removal of airborne biological particles from the air as it enters a building or is circulating within it. Particulate filters and disinfection equipment in recirculated air streams can reduce this risk, but they need to be purposely designed to control the risk of airborne infection and need regular service to maintain their effectiveness. In addition to removing larger and small tiny-size particles associated with adverse health effects, the industry is developing a chemical coating for air filters that kills viruses and bacteria. This practice avoids costly upgrades for HVAC systems to handle denser filtration.

While changes to air filtration practices may take additional effort to implement, it is worth mentioning as this approach can help property managers, architects, and engineers plan for the future and prioritize efforts to maximize IAQ. Research suggests that filtration of recirculated air may be effective in reducing the transmission of airborne infectious diseases. When operating at their full potential high-efficiency particulate air (HEPA) filters can remove 99.97% of particles that are 0.3 μm or larger. These filters remove dust, vapors, bacteria, and fungi, and also effectively capture viral particles spread by droplet nuclei (Mangili and Gendreau, 2005). Systems that are compatible with HEPA filters, control humidity levels, and facilitate a high number of outdoor and recirculated air changes per hour can reduce some of the risks posed by bio-aerosols. A high-efficiency filter may have a high initial pressure drop and/or load with dust and particles very quickly, thus requiring frequent filter changes. A high-pressure drop filter can also cause more air to bypass the filter if it is not properly installed and well-sealed (NAFA, 2020).

ASHRAE Epidemic Task Force (ASHRAE, 2021) update noted that airborne transmission of COVID-19 is significant, including HVAC systems. It advises using minimum-efficiency reporting value (MERV)-13 or -14 filters in HVAC systems. Both LEED and WELL standards use MERV-13 (or better) and HEPA as the guidance for top-level filtration (efficiencies of the typical HEPA filter are based upon particles at 0.3 μm). The main disadvantage to utilizing HEPA or other filters in an existing HVAC system is the high-pressure drops of these closely weaved filters (BALA, 2020). To further enhance the effectiveness of filters, it is advisable to utilize filters with antimicrobial agents and/or chemical coatings to kill dangerous microbes on contact. This practice avoids costly upgrades of HVAC systems to handle denser filtration.

8.5.4 Authoritative positions

The AEC community is feeling massive pressure about COVID-19, especially since the CDC finally joined in the scientific consensus about airborne transmission. ASHRAE introduced the "ASHRAE Epidemic Task Force" to respond to the COVID-19 pandemic and provide guidelines on how to ensure the preparedness of buildings for future epidemics. This was established to help deploy ASHRAE's technical resources to address the challenges of the current pandemic and future epidemics as it relates to the effects of HVAC systems on disease transmission in healthcare facilities, the workplace, homes, public and recreational environments (NAFA, 2020). ASHRAE (2020) stressed the need for changes to building operations, including HVAC systems that can reduce airborne exposures.

In Europe, the Federation of European Heating, REHVA (2020) released a document that summarizes advice on the operation and use of building services in areas with a COVID-19 pandemic, to prevent the spread of the virus depending on HVAC or plumbing systems related factors.

8.6 Environmental electromagnetic disinfection

While EM fields in the built environment have increased significantly over the past century with lots of health concerns; however, little consideration is dedicated to the possible gains of creating EM microenvironments that may develop friendly hygiene zones aligned more closely to their biophysical principles. To bring normalcy in daily life, it becomes important to frequently disinfect places of high risk like parks, public transport, and open market, etc. Therefore, there is a need for effective solutions to disinfect the environment so that people feel more comfortable being indoors.

8.6.1 Vertical potential gradients (VPG)

A VPG, a property of air, exists in nature during good weather times, with the positively charged ionosphere acting as an anode and the earth as a cathode. During such conditions, VPGs of 80–150 V/m can occur in low-lying areas (Dolezalek, 1985), and VPGs of $\leq$5000 V/m on high elevations such as on mountains. At the surface, VPG is, conceptually, the voltage difference between a flat surface and a point 1 m vertically above it. In good weather conditions, the electric potential increases progressively with height. In general, VPG at 1 m above the surface is about 150 V/m. During poor-weather conditions, triboelectric inversion can occur as a result of air below positively charged areas of clouds becoming negatively charged to the ground directly beneath it causing the direction of current flow to reverse. Some buildings, for instance, those predominantly of timber construction, can allow the passage of VPGs into their interiors to a high degree. In comparison to this, most modern buildings, particularly those of steel and concrete construction, tend to act as "Faraday cages" substantially shielding/masking individuals from such fields (Jamieson et al., 2010).

The creation of VPGs indoors, to replicate those found in nature, has shown to help reduce concentrations of airborne contaminants and surface deposition in individual microenvironments, and might be used as a technique in EM hygiene initiatives to help substantially reduce individuals' exposures to contaminants while increasing general well-being (Jamieson et al., 2010), especially as a disinfection tool against COVID-19 pandemic. Stersky et al. (1970) demonstrated that VPGs could significantly reduce airborne concentrations of microbes. Even in instances where natural or artificial VPGs exist indoors, their presence may be masked in some microenvironments within rooms by highly localized incidences of EM-pollution as a result of fields created by equipment, finishes, and/or materials.

8.6.2 Bipolar ionization (BPI)

BPI is an emerging disinfection technology that can be used in HVAC systems to generate positively and negatively charged particles. It is created when an AC voltage is applied to a special tube with two electrodes where an ionization field is produced around the tube. Such ions, as natural disinfectants, occur naturally especially on mountain tops and waterfalls, where the production of both positive and negative ions purifies the air. When a molecule is decomposed into atoms, free electrons are generated, releasing positive and negative ions into the airstream as shown in Fig. 8.5. As air passes, the ions are formed. These bipolar ions disperse into the occupied space through the duct system, proactively attacking airborne contaminants. They are unstable and will seek out atoms and molecules in the air with which to trade electrons. The ions are attracted to airborne particles like dust, smoke, allergens, and other air pollutants. The result is purified air since ions can neutralize any contaminants they come into contact with.

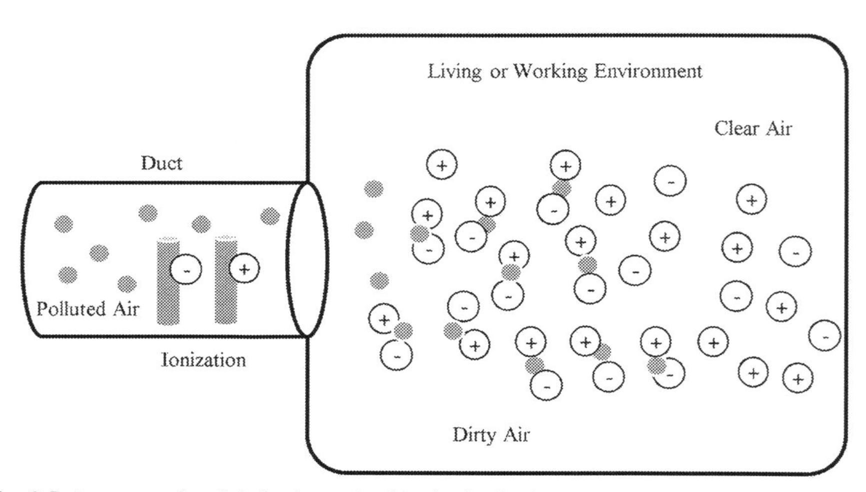

Fig. 8.5 A concept for disinfection using bipolar ionization.

Such a system has significant applications including buildings. The airflow distributes the energized ions into all spaces served by the duct system (Tierno, 2017). Regarding particles, oppositely charged ions cause particles to attract to other particles and become bigger and heavier. These bigger and heavier particles can now be better trapped by HVAC system filters so the filters operate more efficiently. Also, many small particles that are generated within a space by people and their activities may never get to system filters and ordinarily stay suspended in the air for long periods and can be breathed in, increasing the chance of illness and respiratory distress. The BPI process will drop these to the floor quickly taking them away from where people breathe.

BPI has proven to be effective against several viruses: N1N1 influenza, H5N1 avian influenza (bird flu), and corona. The ions destroy the virus's surface structure (Clark, 2020). As a result, the BPI is effective in making viruses inactive. Therefore, ionization is a critical tool when blended with other IAQ practices in combating COVID-19. An advantage of the BPI technology is that it requires no reengineering of the existing HVAC system.

8.6.3 Electrostatic disinfection

Electrostatics is a phenomena and property of stationary or slow-moving electric charges. It involves the buildup of charge on the surface of objects due to contact with other surfaces. Using Coulomb's law, these systems place a positive or negative charge on the chemical disinfectant as it leaves the spray nozzle (Tang and Smith, 2001). Because most surface areas are neutral or negative, a positively charged electrostatic spray application system optimizes adhesion and attraction (electromagnetic theory).

Electrostatic misting for surface disinfection is a method of applying disinfectants to a target surface area by utilizing the electrostatic force of attraction. When a disinfectant is sprayed onto a surface, each positively charged disinfectant particle tries to pair with a negatively charged surface particle. This pair causes positively charged spray particles to be drawn to the nearest available negatively charged particle at high pressure. Misting systems that use high-pressure misting pumps can produce very small droplets of chemical liquid. This electrostatic system places an electrical charge on the droplets and disperses them across a target surface area, providing widespread and smooth coverage. Compared to typical spray-and-wipe, and UV lighting, electrostatic disinfection application systems present a complementary and cost-effective approach to environmental surface disinfection techniques.

8.6.4 Ultraviolet sterilization

Within the light range of the EM frequency spectrum, which is optically invisible to the human eye, the UV region covers the wavelength range 100–400 nm and consists of three subregions. UVA (long-wave; 400–315 nm) is the most abundant in sunlight, responsible for skin tanning and wrinkles. UVB (medium-wave; 315–280 nm) is primarily responsible for skin reddening and skin cancer. UVC (short-wave;

280–200 nm) is the most effective wavelength for germicidal control. Wavelengths in the UVC range are destructive to cells because they are absorbed by nucleic acids. The penetration depth of UV radiation into the human body is between 0.1 and 1 mm, so the tissues at risk are the skin and the eyes.

While UV radiation is available in the form of UVA and UVB from the sun, UVC, the UV light of choice for purifying the air and for inactivating microbes, can be generated only via artificial processes. Germicidal or UVC lamps emit predominantly 253.7 nm radiation and are effective in destroying bacteria, viruses, and molds by destroying the molecular bonds that hold their DNA together. They are used extensively in hospitals, cold-storage facilities, and food handling and processing rooms.

A properly designed and maintained UV system, often in concert with filtration, humidity control, and airflow management, has been shown to reduce infections from other viruses. However, the details of the system are very important such as the design of fixtures, lamp type, lamp placement, airflow amount and mixing, etc. (Dixon, 2020). Within the HVAC system, the application of UVC lamps should be at strategic locations to disinfect lighting coils or within the ductwork for airstream sterilization.

Research has begun to show the efficacy of UV light when it comes to mitigating the spread of airborne aerosolized viruses, as well as cleaning surfaces. A study found that a very low dose of UV light (2 mJ/cm of 222-nm light) successfully inactivated >95% of aerosolized H1N1 influenza viruses. It suggests that far-UVC light is a promising method for limiting aerosol transmission (Welch et al., 2018). Studies have also begun to show how UV light can safely reduce pathogens on surfaces, with one article finding that low disease of UVC light (10–300 mW/cm^2 at 260 nm) effectively inactivated foodborne viruses, including murine norovirus-1 and hepatitis A virus (Park et al., 2015), where the later was very resistant to UVC radiation, indicating that viruses react to UV light differently.

A study by researchers at Columbia University Medical Center (Welch et al., 2018), showed that "far-UVC efficiently inactivates airborne aerosolized viruses, with a very low dose of 2 mJ/cm^2 of 222-nm light inactivating >95% of aerosolized H1N1 influenza virus" and concluded that "continuous very low dose-rate far-UVC light in indoor public locations is a promising, safe and inexpensive tool to reduce the spread of airborne-mediated microbial diseases."

HVAC system upgrades are perhaps the most widely discussed means of enhancing IAQ (BALA, 2020. Equipping these systems with UV lighting or bipolar ionization tubes that deactivate airborne viruses can provide an additional line of defense (JLL, 2020), as can incorporating ductless mini-splits that prevent air from circulating across suites in multitenant buildings (Cohen, 2020).

8.6.5 Microwave sterilization

Microwaves are EM fields within a frequency band of 300 MHz to 300 GHz. Microwave sterilization is a heating process due to polarization effects at a selected frequency band in nonconducting materials. The commonly used frequency bands are 915 and 2450 MHz with penetration depths ranging from 8 to 22 cm at 915 MHz and 3–8 cm at 2450 MHz depending on the moisture content. Microwave energy is supplied to objects through the molecular interaction with the EM fields through molecular friction resulting from dipole rotation of polar solvents and the conductive movement of dispersed ions. The heating effect can have a major sterilization advantage over conventional techniques because of the relatively short heating time. Dielectric properties of bio-germs play a critical role in determining the interaction between the electric field and the objects, and they are dependent on composition, moisture content, frequency, thermal, and dielectric properties of materials.

Microwave sterilization has been widely implemented in the food industry. However, its application in buildings is new. In a study by Kang and Kao (2014), microwave sterilization was implemented in a humidifier of a central air conditioning system. It was observed that microwave sterilization has a slight influence on the air supply parameters and was found to be effective between 50°C and 59°C. In addition, microwave radiation power of 900 W and more can quickly reach the effective sterilization temperature in 5 min, and the sterilization rate can reach 100%.

8.7 Pandemic intelligent solutions

The spread of COVID-19 appears to have finally provided a sound reason for the public health system to fully embrace an intelligent digital ecosystem that can rapidly contribute to the management of the crisis. Intelligent technologies play the main role in controlling the pandemic situation and beating infectious diseases. These include AI and ML applications, sensing and mobile technologies, drones, and robots services as shown in Fig. 8.6.

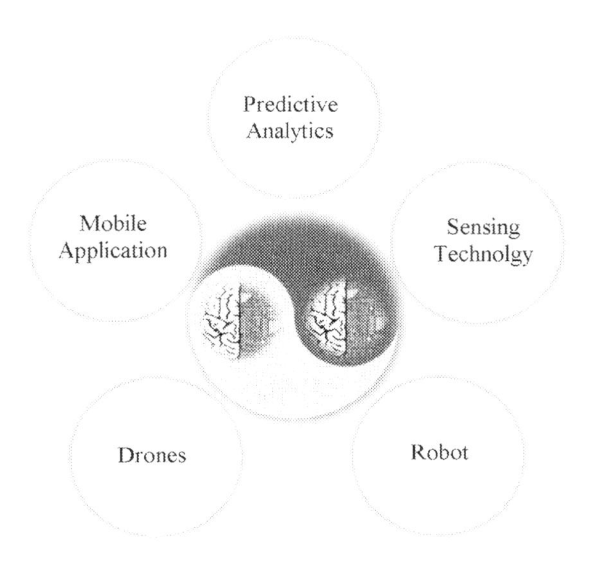

Fig. 8.6 Intelligent technologies that can assist in controlling the pandemic situation.

8.7.1 *Predictive analytics*

AI and other forms of data collection are highly useful and may be integral to the fight against any pandemic crises. They can be used to track the source of infectious diseases as they may emerge and their spread in terms of speed and location. AI can also provide advance notice for preparedness measures by relevant institutions, in particular, hospitals and other healthcare agencies (Obeidat, 2020). In the field of diagnosis, digital solutions combined with conventional methods, such as imaging and clinical data, appear to be encouraging.

Predictive analytics depict the use of statistics and modeling to verify future performance based on current and historical data. This type of AI looks at existing data to find patterns and then uses those patterns to make predictions. It is possible to use AI to build a huge firewall to safeguard against new infections? This requires the ability to predict and track new infections, identify symptoms in patients, detect new high-risk areas, and in fast-tracking the development of the new medication. AI can facilitate early detection of, and preemptive action against, the onset and spread of viruses and other infectious diseases generally and within any geographical location (Chik, 2020). The first detection of the COVID-19 outbreak came from the AI vendor BlueDot in 2019. This Canadian startup used an AI-based algorithm to warn people against traveling abroad due to the possibility of a flu-like outbreak. The startup, which uses data analytics to predict the spreading of infectious diseases, rang the warning bells before WHO and CDC gave their warnings. BlueDot's software can scan foreign news articles, animal and plant disease data, and government proclamations to warn their clients of potentially dangerous areas (Faggella, 2020).

ML, harnessed with the right computing power, can be used to understand the structure of the virus and present the results to researchers to develop vaccines or treatments. MIT researchers used ML to identify an antibiotic that can be used to fight against disease-causing bacteria. While the development of new drugs takes a long, predictive analytics can be applied to go through the list of existing drugs and propose which might be useful to manage the pandemic (Garbade, 2020).

8.7.2 *Mobile applications*

The development and use of mobile applications in the health care field is a reality and has been widespread. These applications have been leveraged in several ways to control the spread of diseases. They have been implemented for information distribution, risk assessment, self-management of symptoms, and home monitoring. Mobile applications are accessible, acceptable, and easily adopted, and can support social distancing efforts. As such, they have been widely developed and implemented in an attempt to flatten the curve of the increasing number of pandemic cases, providing knowledge and information to civilians while attempting to relieve the pressure from health care systems (Kondylakis et al., 2020). COVID-19 self-assessment applications for mobile applications help people to evaluate their health and receive instructions and recommendations on diseases. The aim is to offer a degree of understanding to people, who seem to have infection symptoms and favor ailment through sets of questions (Khan et al., 2021). In general, mobile applications act as a resource to educate people about diseases and provide them with preventive approaches. It may assist in resolving several pandemics–related challenges by reaching consistent information to both public and health professionals, decreasing confusion, tracing symptoms, discovering new predictors, and reducing the concern of hospitalizations.

8.7.3 Sensing technologies

Smart biosensors are at the heart of numerous rapid and essential diagnostic tools for providing accurate and timely guidance for the identification, prevention, and treatment of infectious diseases. Advances in IoT and nanotechnology are stimulating the proliferation of such smart sensor technology. Emerging technologies hold the promise to develop tests that are quick, highly sensitive, and effective.

Worldwide organizations are conducting a large number of tests to track the spread of diseases and isolate infected persons. In response to tackle the COVID-19 pandemic, scientist and the healthcare industry are developing new point-of-care biosensors to detect coronavirus as soon as it is present in the human. The intelligent sensors and IoT technology provide real-time monitoring of vital signs of symptoms which include respiratory rate, heart rate variability and pulse rate, etc. The technology helps to receive real-time patient data from the sensors and wirelessly transmits these data for continuous monitoring.

Wearable sensors such as smart watches, oxygen monitors, or thermometers are suitable for monitoring the populations at risk and those in quarantine, both for evaluating the health status of caregivers and management personnel and for facilitating triage processes for admission to hospitals (Ding et al., 2021). Also, sensing systems are used for detecting the disease and for monitoring patients with relatively mild symptoms whose clinical situation could abruptly deteriorate in unprepared hospitals.

8.7.4 Drones

The use of drones has a crucial role in responding to pandemic crises. They assist authorities and people in different ways to prevent the spread of infection and stop unwanted movements of people during the lockdown (Khan et al., 2021). Drones have become one of the main tools in such battle due to their versatility and the fact that they are a fairly extended technology that is easy to implement and already in the hands of many entities around the world. They have been extensively adopted for the faster and safe transfer of essential products. The COVID-19 crisis has served to propel the use of drones to deliver essential materials. They are also being used in broadcasting messages and imparting awareness among the public through loudspeakers and digital displays.

Disinfecting large areas that are difficult to access is also one of the important tasks during pandemic crises. Dispensing with human resources in some extreme places, or reducing human involvement to the least are some of the advantages of using drones for these duties. Drone technology has also assisted in the symptom-based bulk screening of individuals. These drones are installed with the temperature sensor to screen the people from their homes, cameras to take pictures of people, and GPS to track their location.

8.7.5 Robots

The application of robots in healthcare systems has shown encouraging solutions for avoiding the escalation of COVID-19. With the concurrent integration of AI and ML algorithms, robots have turned out to be an effective tools. They have proven their capability and potential in many applications like cleaning, disinfecting, surveillance, screening of patients, and logistic tasks. One of the most noticeable applications for robots in response to the pandemic is to disinfect hospital rooms and other spaces with ultraviolet radiation or chemical sprays. For surveillance, robots are being used when areas need to be monitored to guarantee that social distancing or lockdown guides are being respected. In spite of the success, however, problems related to privacy, equity, fairness, and usability still remain. Robots can also interact directly with symptomatic patients carrying out logistics tasks and handling resources in hospitals.

References

Ai, Z.T., Melikov, A.K., 2018. Airborne spread of expiratory droplet nuclei between the occupants of indoor environments: a review. Indoor Air 28 (4), 500–524.

Allen, J.G., 2020. Your Building Can Make you Sick or Keeping you Well. https://www.nytimes.com/2020/03/04/opinion/coronavirus-buildings.html.

Alter, L., 2020b. New Research Raises More Concerns about Airborne Transmission of COVID-19. https://www.treehugger.com/new-research-raises-more-concerns-about-airborne-transmission-covid-4848040.

ASHRAE, 2020. ASHRAE Position Document on Airborne Infectious Diseases. https://www.ashrae.org/File%20Library/About/Position%20Documents/Airborne-Infectious-Diseases.pdf.

ASHRAE, 2021. ASHRAE Epidemic Task Force Releases Updated Airborne Transmission Guidance. https://www.ashrae.org/about/news/2021/ashrae-epidemic-task-force-releases-updated-airborne-transmission-guidance.

BALA, 2020. COVID-19 and the Impacts to the Workplace. https://www.bala.com/sites/default/files/uploads/COVID-19%20and%20Impacts%20to%20the%20Workplace-Bala5-12.pdf.

Balgeman, S., Meigs, B., Mohr, S., Niemöller, A., Spranzi, P., 2019. Can HVAC Systems Help Prevent Transmission of COVID-19? https://www.mckinsey.com/industries/advanced-electronics/our-insights/can-hvac-systems-help-prevent-transmission-of-covid-19.

Boone, S.A., Gerba, C.P., 2007. Significance of fomites in the spread of respiratory and enteric viral disease. Appl. Environ. Microbiol. 73 (6), 1687–1696.

Bourouiba, L., Dehandschoewercker, E., Bush, J.W.M., 2014. Violent expiratory events: on coughing and sneezing. J. Fluid Mech. 745, 537–563.

Bromage, E., 2020. The Risks - Know them - Avoid them. https://www.erinbromage.com/post/the-risks-know-them-avoid-them.

Budds, D., 2020. Design in the Age of Pandemics. https://www.curbed.com/2020/3/17/21178962/design-pandemics-coronavirus-quarantine.

CARB, 2005. Report to the California Legislature: Indoor Air Pollution in California. http://www.arb.ca.gov/research/indoor/ab1173/finalreport.htm.

CDC, 2021. Interim Infection Prevention and Control Recommendations for Healthcare Personnel During the Coronavirus Disease 2019 (COVID-19) Pandemic. https://www.cdc.gov/coronavirus/2019-ncov/hcp/infection-control-recommendations.html.

Chik, W., 2020. Coronavirus: Pandemics, Artificial Intelligence and Personal Data. https://lawgazette.com.sg/feature/coronavirus-pandemics-artificial-intelligence-and-personal-data/.

Clark, L., 2020. UVI, BPI among Industry Responses to Kill, Corona Virus. https://www.hpac.com/covid-19/article/21128399/sustainability-health-and-safety-ultraviolet-light-bipolar-ionization.

Cohen, D., 2020. Vogel Realty Installing New HVAC Systems to Historic D.C. Office Building Over COVID-19 Concerns. https://www.connect.media/vogel-realty-installing-new-hvac-systems-to-historic-dc-office-building-over-covid-19-concerns/.

Ding, X., Clifton, D., Ji, N., Lovell, N.H., Bonato, P., Chen, W., Yu, X., Xue, Z., Xiang, T., Long, X., Xu, K., Jiang, X., Wang, Q., Yin, B., Feng, G., Zhang, Y.-T., 2021. Wearable sensing and telehealth technology with potential applications in the coronavirus pandemic. IEEE Rev. Biomed. Eng. 14, 48–70. https://doi.org/10.1109/RBME.2020.2992838.

Dixon, K., 2020. Air Filtration and COVID-19: Indoor Air Quality Expert Explains how to Keep you and your Building Safe. https://news.engineering.utoronto.ca/air-filtration-and-covid-19-indoor-air-quality-expert-explains-how-to-keep-you-and-your-building-safe/.

Dolezalek, H., 1985. Remarks on the physics of atmospheric ions (natural and artificial). Int. J. Biometeorol. 29, 211–221.

Drexler, J.F., Corman, V.M., Drosten, C., 2014. Ecology, evolution and classification of bat coronaviruses in the aftermath of SARS. Antivir. Res. 101, 45–56.

Faggella, D., 2020. Artificial Intelligence for Pandemic Response and Outbreak Tracking. https://emerj.com/ai-sector-overviews/ai-for-pandemic-response/.

Fahey, R., 2020. Coronavirus Could Be Spread by Air-Conditioning and May Be more Contagious than Previously Thought, Scientists Believe after Finding Traces of the Virus in Hospital Air-Duct. https://www.dailymail.co.uk/news/article-8086457/Coronavirus-spread-air-conditioning-contagious-previously-thought.html.

Garbade, M.J., 2020. Coronavirus: How Ready Is AI to Detect and Prevent Pandemics? https://www.entrepreneur.com/article/348804.

GHN, 2020. Coronaviruses: What You Need to Know. https://www.globalhealthnow.org/2020-04/coronaviruses-what-you-need-know.

Gössling, S., Scott, D., Michael, H.C., 2020. Pandemics, tourism and global change: a rapid assessment of COVID-19. J. Sustain. Tour. 29 (1), 1–20. https://doi.org/10.1080/09669582.2020.1758708.

Anon., 2021. History.com.

Jamieson, I.A., Holdstock, P., ApSimon, H.M., Bell, J.N.B., 2010. Building health: the need for electromagnetic hygiene? IOP Conf. Ser.: Earth Environ. Sci. 10. https://doi.org/10.1088/1755-1315/10/1/012007, 012007.

JLL, 2020. (Re) entry: A Guide for Working in the Next Normal. https://www.us.jll.com/content/dam/jll-com/documents/pdf/other/reentry-guide-for-working-in-the-next-normal.pdf.

Kang, Y., Kao, S., 2014. A study on the effectiveness of microwave heating for disinfection of humidifier elements. HVAC & Res 20 (1), 113–120. https://doi.org/10.1080/10789669.2013.834780.

Khan, K., Kushwah, K.K., Singh, S., Urkude, H., Maurya, M.R., Sadasivuni, K.K., 2021. Smart technologies driven approaches to tackle COVID-19 pandemic: a review. 3 Biotech 11 (2), 50.

Kondylakis, H., Katehakis, D.G., Kouroubali, A., Logothetidis, F., Triantafyllidis, A., Kalamaras, I., Votis, K., Tzovaras, D., 2020. COVID-19 Mobile apps: a systematic review of the literature. J. Med. Internet Res. 22 (12), e23170.

Kubba, S., 2017. Handbook of Green Building Design and Construction. Elsevier, Oxford.

Kuwabara, B., Turnbull, G., Opie, K., Hall, M., 2020. Pandemic effect: academic facilities. https://www.canadianarchitect.com/pandemic-effect-academic-facilities.

Leung, N.H.L., 2021. Transmissibility and transmission of respiratory viruses. Nat. Rev. Microbiol. https://doi.org/10.1038/s41579-021-00535-6.

Li, Y.-Y., Wang, J.-X., Chen, X., 2020. Can a toilet promote virus transmission? From a fluid dynamics perspective. Phys. Fluids 32. https://doi.org/10.1063/5.0013318, 065107.

Lu, J., Gu, J., Li, K., Xu, C., Su, W., Lai, Z., Yang, Z., Lai, Z., Zhou, D., You, C., Xu, B., Yang, Z., 2020. COVID-19 outbreak associated with air conditioning in restaurant, Guangzhou, China, 2020. Emerg. Infect. Dis. 26 (7), 1628–1631.

Lubell, S., 2020. Commentary: Past Pandemics Changed the Design of Cities. Six Ways COVID-19 Could do the Same. https://www.latimes.com/entertainment-arts/story/2020-04-22/coronavirus-pandemics-architecture-urban-design.

Mangili, A., Gendreau, M.A., 2005. Transmission of infectious diseases during commercial air travel. Lancet 365 (9463), 989–996.

Morawska, L., Tang, J.W., Bahnfleth, W., Bluyssen, P.M., Boerstra, A., Buonanno, G., Cao, J., Dancer, S., Floto, A., Franchimon, F., Haworth, C., Hogeling, J., Isaxon, C., Jimenez, J.L., Kurnitski, J., Li, Y., 2020. Loomans: how can airborne transmission of COVID-19 indoors be minimized? Environ. Int. 142 (September), 105832. https://doi.org/10.1016/j.envint.2020.105832.

NAFA, 2020. COVID-19 Special Feature: HVAC Filtration and COVID-19. https://facilityexecutive.com/2020/04/covid-19-special-feature-hvac-filtration-and-covid-19/.

Obeidat, S., 2020. How Artificial Intelligence Is Helping Fight the COVID-19 Pandemic? https://www.entrepreneur.com/article/348368.

Park, S.Y., Kim, A.-N., Lee, K.-H., Ha, S.-D., 2015. Ultraviolet-C efficacy against a norovirus surrogate and hepatitis a virus on a stainless steel surface. Int. J. Food Microbiol. 211, 73–78.

REHVA, 2020. REHVA COVID-19 Guidance Document. https://www.rehva.eu/fileadmin/user_upload/REHVA_COVID-19_guidance_document_ver2_20200403_1.pdf.

Smieszek, T., Lazzari, G., Salathé, M., 2019. Assessing the dynamics and control of droplet- and aerosol-transmitted influenza using an indoor positioning system. Sci. Rep. 9 (1), 2185.

Stersky, A., Heldman, D.R., Hedrick, T.I., 1970. The effect of a bipolar-oriented electrical field on microorganisms in air. Mich. Agric. Exp. Stn. J. 5120, 545–549.

Tang, K., Smith, R.D., 2001. Physical/chemical separations in the break-up of highly charged droplets from electrosprays. J. Am. Soc. Mass Spectrom. 12 (3), 343–347.

Tierno Jr., P.M., 2017. Cleaning Indoor Air Using Bi-Polar Ionization Technology. http://atmosair.com/wp-content/uploads/2020/03/Cleaning-Indoor-Air-Using-Bi-Polar-Ionization-Technology_Dr.-PhilTierno_NYU-SchoolMedicine_2017.pdf.

TVR, 2019. Higher Humidity in Buildings can Help Lower Infections. https://thevaccinereaction.org/2019/12/higher-humidity-in-buildings-can-help-lower-infections/#_edn2.

Van Khai, T., 2016. Adaptive architecture and the prevention of infections in hospitals. De Gruyter Civ. Eng. Ser. 16 (2), 165–172.

Wei, J., Li, Y., 2016. Airborne spread of infectious agents in the indoor environment. Am. J. Infect. Control 44 (9 Suppl), S102–S108.

Welch, D., Buonanno, M., Grilj, V., Shuryak, I., Crickmore, C., Bigelow, A.W., Randers-Pehrson, G., Johnson, G.W., Brenner, D.J., 2018. Far-UVC light: a new tool to control the spread of airborne-mediated microbial diseases. Sci. Rep. 8 (1), 2752. https://doi.org/10.1038/s41598-018-21058-w.

Wells, W.F., 1995. Airborne Contagion and Air Hygiene. Harvard University Press, Cambridge.

WHO, 2021. Roadmap to Improve and Ensure Good Indoor Ventilation in the Context of COVID-19. https://www.who.int/publications/i/item/9789240021280.

Xia, S., Zhou, X.-N., Liu, J., 2017. Systems thinking in combating infectious diseases. Infect. Dis. Poverty 6 (144), 1–7. https://doi.org/10.1186/s40249-017-0339-6.

Xiao, F., Sun, J., Xu, Y., Li, F., Huang, X., Li, H., Zhao, J., Huang, J., Zhao, J., 2020. Infectious SARS-CoV-2 in feces of patient with severe COVID-19. Emerg. Infect. Dis. 26 (8), 1920–1922. https://doi.org/10.3201/eid2608.200681.

Yeo, C., Kaushal, S., Yeo, D., 2020. Enteric involvement of coronaviruses: is faecal-oral transmission of SARS-CoV-2 possible? Lancet Gastroenterol. Hepatol. 5, 335–337.

Urbanization as an intelligent system

Riadh Habash
School of Electrical Engineering and Computer Science, University of Ottawa, Ottawa, ON, Canada

On Earth there is no heaven, but there are pieces of it.

Jules Renard

9.1 Urban history

History proves how the challenge of health and well-being inspires the built environment and the way spaces are designed and constructed. This leads to a change in conceptualizations of health from disease prevention to medication and health promotion in populations. This change reflects the increasingly environmental focus of the health promotion practice which should consider interventions across multiple scales, levels, and frameworks.

For thousands of years, humans have looked to physical space to treat and cure sickness. As early as 400 BCE, Hippocrates theorized that poor physical environments, like bad air and water, caused illness and disease, and believed that going to areas with fresh air and water were essential to health. The physical space can expose people to toxins or pollutants and influence lifestyles that contribute to infectious and chronic diseases. However, it can be a primary defense against infectious diseases.

During the 19th century, the connection between public health and the built environment became gradually evident as hundreds of thousands of workers crowded into unsanitary, industrial cities with a resulting increase in disease and epidemics and a decrease in life expectancy. As industrialization affected demographics, more people were in towns and cities without sufficient space for play and exercise. Society responded with the development of public parks and dramatic improvements in public health in industrialized nations were made possible by changes in the built environment. The construction of widespread sewer systems, improvements in building designs to ensure that residents had fresh air, and the movement of residential areas away from hazardous industrial facilities all created substantial improvements in health (Rosen, 1993). In many respects, sanitary engineers were the first urban planners in America. This shows that the history of urban planning in the past century highlights the effects that the built environment can have on both the prevention and containment of chronic and infectious diseases (Pinter-Wollman et al., 2018).

Sustainability and Health in Intelligent Buildings. https://doi.org/10.1016/B978-0-323-98826-1.00009-0

Frederick Law Olmsted, who was a sanitary officer during the Civil War, used public health to convince New York City to build Central Park, arguing that its open spaces would become "the lungs of the city." His belief in the medicinal qualities of green space also influenced his 1868 master plan for Riverside, Illinois, a "garden suburb" that was viewed as a healthier alternative to city life due to widespread access to recreational space (Fisher, 2010).

By the mid-20th century, the connection between public health and the built environment seemed to shrink. The infectious diseases had been brought under control, and as a result, the layout and planning of cities came to be viewed as a matter of esthetics or economics, but not health. Public health officials concentrated on human behaviors such as smoking and to the extent, they considered the built environment, the focus was on more discrete rather than larger-scale planning issues (Perdue et al., 2003).

In the preurban period, the light concentration of people let for movement away from pollution sources or allowed the dilution of pollution at its sources. Urbanization was relatively slow to start, but during the past hundred years, it accelerated quickly. Cities and towns have evolved into complex systems, since then and the movement of populations into urban settings placed individuals close to each other. Therefore, nearly all aspects of the built environment started to be shaped by law and governmental decisions. Urban planning started mostly about the design of systems, and historically the health of urban populations has been inherently bound to the practice of urban planning. Today, both buildings and locations are regulated by a complex set of local, state, and federal regulations and laws.

The pandemic has fast-tracked several transformative developments, such as remote working, automation, and digitization. COVID-19 delivered a strong lesson in the dangers of straight-line thinking where it is no longer valid to assume that what happened in the past will continue to happen in the future. The pandemic reminded humanity of its weakness and devastated life-as-usual thinking. The pandemic suddenly shook healthcare systems around the world. The question of the moment is, how can the world use this pandemic for learning to improve the health and comfort in the built environment?

Unlike the situation in the 19th and early 20th centuries, today's public health advocates have been largely missing from discussions about major planning or land-use decisions involving the built environment. Despite the existent of large planning departments or other bureaucracies that regulate land use and buildings. These frequently include urban planners, architects, engineers, lawyers, economists, environmental scientists, and demographers. They hardly include public health officials (Colmers and Fox, 2003). This may reflect a broader phenomenon of the increasing isolation of public health from the built environment. Nonetheless, public health can add an important voice to the decisions that shape the built environment in terms of healthy populations.

In recent years, calls for authorities to focus on sustainability and health in their planning have been growing. For a resilient built environment, urban plans need to be designed, evaluated, and approved using the lens of history, but probably in a different analogy.

9.2 Urban intelligence

The built environment may be described as the human-made or modified physical spaces of the environment where people live and perform their activities. These include physical spaces (buildings, open spaces, and infrastructures), human society spaces shaped and sustained by the culture and social interactions of urban residents, and cyberspace, which is comprised of computers, internet access, and the data flowing through these systems (Pan et al., 2016). These spaces provide a wide range of socio-economic benefits.

In the coming decades, urbanism will be a dominant concern of planners, governments, policymakers, investors, businesses, and communities worldwide. It is projected that by 2050, up to 70% of the global population will live in urban areas (Fudge and Fawkes, 2017). By concentrating people, investment, and resources, this infrastructure can reinforce the potential for economic development, habitat innovation, and social interaction. However, these urban infrastructures play a significant role in consuming energy, materials, and water, and they also generate a high output of CO_2, waste, and sewer water when it comes to their construction and exploitation (Habash, 2017).

Urban intelligence in the context of this discussion represents the capacity that analyzes city-level information using various intelligence (human, nature, and artificial) to explore its role in various aspects of living. Today, the world is in the middle of an urban intelligence transition with a paradigm that proposes an ecosystem of technologies to improve the urban environment, well-being, quality of life, and smart city systems. It fosters the definition of an urban digital twin, namely a CPS of the entire city or community. As the human has a brain and five senses, the built environment is adding a sensor layer to collect data, and insights from this data can help the environment to continuously learn and adapt. While urban planners have already been using 3D models and computer-aided design for many years, the integration of real-time data from IoT devices, weather, traffic, people movement, and other sources has been significant for urban planning, design, and operations (Castelli et al., 2019). The digital twin as a planning tool provides insight addressing energy management, resilience, sustainability, mobility, and economic growth by integrating spatial modeling of the urban environment, mathematical modeling of electromechanical systems, and real-time sensor data obtained from IoT devices. This process requires thorough planning that incorporates analysis of information, creating awareness of challenges and risks, identification of objectives, and quality which are mainly HI-based. The work of the AEC community alone is certainly not enough to accomplish planning quality goals without other efforts that fall directly under the control or influence of the city through its regulations, bylaws, and decisions about implementing projects. To speed up the process of planning, AI tools with new modes of interaction are required including predictive analytics, training, and learning, search, and planning. These tools like ML, FL, and urban big data are effective in mastering repetitive and automating tasks. Each kind of interaction is a response to the targets people want to achieve considering possible technology constraints. Fig. 9.1 presents the main

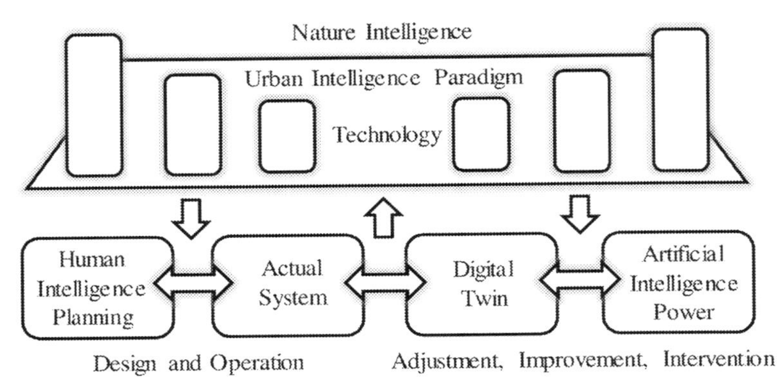

Fig. 9.1 Pathways to urban intelligence.

urban planning pillars for achieving sustainability and health encompassing their balanced accomplishment by articulating the wider influence across the environment, society, and economy. However, an intelligent component when exploring sustainability is data-driven urban planning connected to energy use, wind flow, and solar irradiance to build computer models to validate assumptions at all levels including city, community, and building.

The urban intelligence paradigm brings three trajectories including human intelligence (HI) through rising knowledge, ability to learn from experience, adapt to new situations, leadership, and emotional intelligence (EQ); nature intelligence (NI) through biodiversity and ecosystem; and the spread of AI predictive tools. As a middle position, to understand the context of urban intelligence, nature can be defined as living organisms and nonliving components of the ecosystem. This includes plants, water, animals, breezes, sounds, scents, as well as indirect evocations of nature. In the context of the modern built environment, part of nature is unsustainably designed. Urban intelligence is an innovative platform that uses ICT together with other tools to improve QoL and optimize the efficiency of urban operations, services, while addressing community and individual sustainability needs, for present and future generations. The above model helps design adaptation, functionality and ensures the best location and direction of facilities like parks, playgrounds, roof gardens, and PV panels. Technology is at the heart of the built environment where today more people than ever before are embracing smart technologies but more than that is the smart holistic-systems thinking. The possibility of hybrid decision-making, in which HI works with NI and AI algorithms, is universal. It is often provided as an intention to enter humans into the loop. This alliance between HI, NI, and AI is the essence of urban intelligence which should certainly adapt to humans and provide SD.

With AI, change management and process reengineering get reinvented. What was once a one-way street has become a two-way street: teaching technology to relate to

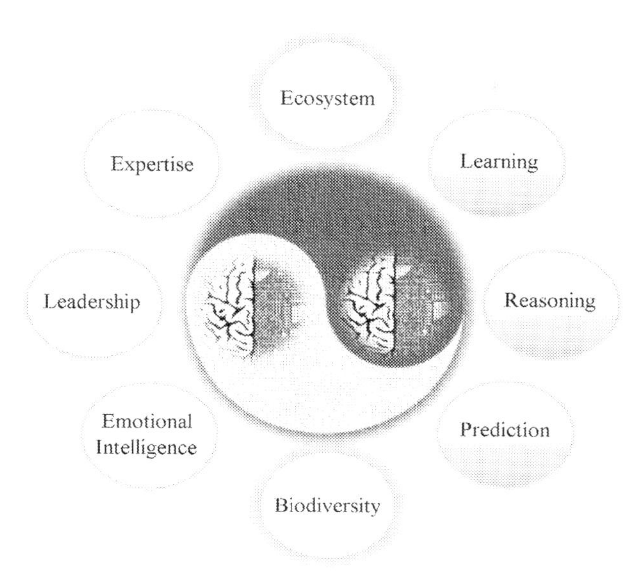

Fig. 9.2 A triadic relationship between human intelligence, nature intelligence, and artificial intelligence.

people, as much as training people to use the technology? Moving ahead, making this human-centric design right is the major factor in the success or failure of AI-driven transformations (Mantas, 2018). Many organizations are analyzing their situations through AI rather than HI and NI. HI has involved generations of the human race learning over thousands of years through discovery, investigation, and interaction. Though it is much debated whether working with AI in conjunction with HI and NI is better than working with HI and NI alone or not, we are slowly progressing toward an era where HI and NI are taking a back seat. Fig. 9.2 shows a sustainable-enhanced triadic relationship between HI, NI, and AI. Although there are some similarities between the three elements of intelligence, HI is far more complex than AI, since AI is so far just an extension of HI. Both are essential in urban planning where HI and NI contribute largely to the quality while AI speeds up the process.

9.3 Urban sustainability

Achieving the sustainability of cities can be conceived as entailing the integration of three pillars: social development, economic development, and environmental management (Habash, 2017). Social sustainability refers to the fairness, inclusiveness, and cultural adequacy of an intervention to promote equal rights that supports the livelihoods and lives of local communities. A sustainable economy is understood as the capacity and ability of a practice to be able to put resources to productive use for long-term benefit without increasing the ecological footprint. Yet, how a city can build sustainability will reflect its capacity to adapt, within the context of its particular history, by its urban governance and policy priorities.

The urban built environment emits more than 70% of global emissions and consumes around the same proportion of the world's primary energy. At the same time, they generate 80% of the global gross domestic product, which depends on healthy urban residents and functioning urban structures and services, many of which are increasingly vulnerable to the impacts of climate change such as heatwaves, storm surges, and rising sea levels (GCEC, 2016). While experiencing the impact of climate change, many of the cities have initiated policy processes around the issue and engaged in climate action planning in a practical sense.

Currently, there is increasing recognition that urban sustainability is tied directly to the quality of life of the population, which is improved by convenient, efficient, and accessible public transportation; recreation areas; convenient shopping; and suitable educational and health services. There is also growing interested in small communities within the urban landscape where all of these services are located close to one another, where residents can live and work without having to travel long distances (Habash, 2017). Sustainability issues for such cities tend to be focused on the delivery of these communal needs.

Sustainability principles must not be forgotten in the postpandemic world. Those urban spaces and cities that do thrive will need to be better prepared for a future pandemic and for climate change impacts, which requires not only an adaptable approach but mitigation measures also. There should be a renewed focus on sustainability features such as green façades, energy-efficient power systems, monitoring and evaluation systems for adaptation and efficiency.

An urban built environment is not a system that can be easily controlled or driven toward narrowly defined objectives but is a complex ecosystem of independent people, communities, and businesses, each pursuing its own goals using the resources available to them. Accordingly, the objective is not to make the environment smarter but to create one within which smart ideas are likely to flourish and succeed, wherever they occur (Robinson, 2014).

A modern focus of architecture is environmental design, in particular, the control of the quality of air and water in buildings and cities. In any urban design project, architects will need to carefully appraise environmental and health imperatives, where the advantages of sustainable mobility and technical systems in big networks are balanced with those of natural means of ventilation, heating, and cooling (Wood, 2020). The sustainability of the built environment concerns the capacity of an intervention to enhance the livability of buildings and urban infrastructures for city dwellers without damaging or disrupting the urban region environment. Finally, sustainability is concerned with the quality of governance system for guiding the relationship and actions of different actors among the previous four dimensions (Allen, 2009). Achieving sustainable buildings and communities should be a priority for urban planners, health professionals, and policymakers.

9.4 Urban health

Urbanization, as a social development motivator, acts as a two-edged sword as it leads to positive health outcomes on one side and raises the vulnerability to poor health on the other side because of several environmental and behavioral risk factors. Health is influenced by the built environment in different ways including green public open spaces, homes and schools, safe and stressless transport, proper jobs, and a hygiene-built food environment. This requires policymakers to make health a central line of government policy through an urban health agenda that incorporates health promotion.

Health promotion, coined in 1945 by Henry Sigerist, a Swiss medical historian, entails a pattern shift from an interpretation of health as the absence of disease (biomedical approach) to a socio-ecological understanding of health concentrating on resilience and assets to health. It enables people to increase control over their health and highlights the need to promote health and prevent disease through social and environmental interventions that tackle the causes of ill health. Health promotion is often used by architects designing healthcare facilities. Health promotion highlights the need to promote health and prevent disease through interventions that tackle the causes of ill health. It found reflections 40 years later in the Ottawa Charter for health promotion. In the year 1984, the WHO defined health promotion as the process of enabling people to increase control over, and improve their health. The health promotion movement, as reflected in the Ottawa Charter to achieve health for all by the year 2000 and beyond, is the process of enabling people to increase control over, and improve, their health (Asakura et al., 2015; Kumar and Preetha, 2012). The Charter was reconfirmed in the Jakarta Declaration on Leading Health Promotion into the 21st century in 1997. With roots in similar thinking, a city-based health promotion approach evolved in Western Europe and was extended to North America and later to other regions of the world. Also, the WHO's healthy city concept was promoted in several developing counties between 1995 and 1999.

In recent years, the WHO Regional Office for Europe has launched a "WHO European Healthy Cities Network" to promote health and well-being throughout Europe. Moreover, referring to the "Urban Health Rome Declaration" at European meeting "G7 Health," which defines the strategic aspects and actions to improve public health in the cities, and referring to the Agenda 2030, in particular SDG11 (sustainable cities and communities) (Gianfredi et al., 2021). This global movement works to put health high on the social, economic, and political agenda of city planners. In addition, many countries have developed health promotion policies and programs for schools, hospitals, workplaces, cities, and communities. In general, health promotion requires policy capacity to take actions on promoting health and prevention of diseases from a population-based approach to ensure healthy living and working in response to changing environments.

The built environment does offer opportunities to establish an important health-promoting context within resilient communities. During global pandemics, buildings play an even greater role in promoting public health, and public health can bring distinctive solutions that encourage healthy spaces, IAQ, and physical activity. Urban

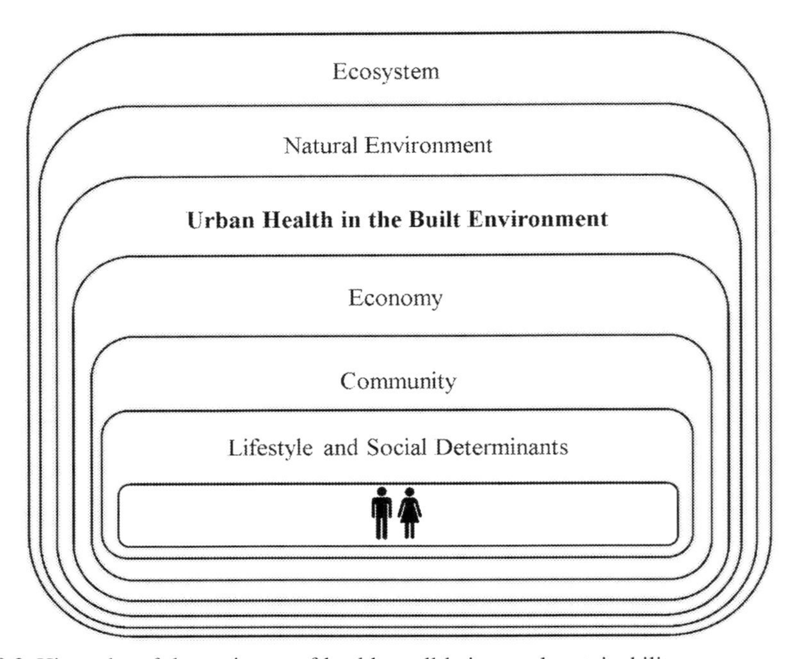

Fig. 9.3 Hierarchy of determinants of health, well-being, and sustainability.

planning to promote health and comfort, as well as community connection (not isolation) in the built environment, is one of the key challenges facing the international community during a pandemic (GCPH, 2013). The impact of health-supportive built environments on living and well-being varies extensively where cities and communities play a significant role in population health enhancement and have immense capacity to be health-generating sources through their various determinants of health.

Human health impacts in the built environment stem from diverse elements and aspects of the environment itself. The hierarchy of determinants of health, well-being, and sustainability are depicted in Fig. 9.3. The core is urban health as an intersectoral arena that links both the public health and the urban planning sectors, mainly captured by SDG3 (health and well-being) and SDG11 (inclusive, safe, resilient, and sustainable cities). Social determinants of health such as gender and socio-economic inequalities are captured by SDGs 1, 5, 10. In SDG11, health is present as a precondition and targets detrimental elements such as IAQ and positive environmental exposures, like green open spaces (Ramirez-Rubio et al., 2019). It has emphasized the interconnection between health and the built environment by outlining that health promotion efforts can contribute by making cities and human settlements inclusive, safe, resilient, and sustainable. All of these urban health-related SDGs are interconnected, where cities are an essential driving force to implement them that aim at inclusive, safe, resilient, and sustainable places.

Advancing population health involves thinking beyond public health and taking a contribution from sectors other than health to achieve progress toward healthy lives and well-being for all (SDG3). In this regard, the built environment must learn from the natural environment and ecosystem, which have been sustainable for millions of years. Some built environments affect the health of individuals and populations by influencing disease transmission dynamics because just as countries and large cities serve as hubs for migration and international travel, components of the urban built environment serve as hubs that drive the transmission of infectious disease pathogens (Alirol et al., 2011). However, the built environments can be modified to promote healthy behaviors and reduce the risk of contracting a disease.

While the AEC community has been working to design healthy buildings and spaces for their clients, the idea of an industry-wide push to use building design to advance health and wellness is still emerging. According to the Center for Active Design, this design practice is an evidence-based approach to development that identifies planning and architecture solutions to support healthy communities. Inspired by the legacy of urban activists who designed to solve infectious disease crises of the 19th century and the recent COVID-19, The Center's initiatives look at all aspects of design for health, physical, mental, and social aspects at home, work, and throughout a neighborhood, creating opportunity by optimizing a setting, no matter how incompatible it may seem (Spula, 2017). Based on emerging strategies from science, infectious disease transmission data, and epidemiological models, the American Institute of Architects, as an advocate for healthy design, has outlined five specific areas of opportunity to enhance physical and mental health. These five areas include safety and social equity, sensory environments, access to nature, physical activity, and environmental integrity (Householder, 2016). Buildings should be designed differently in the future to manage risk in the face of future pandemics. In shared buildings, architects might think about having more than one lift, and multiple communal staircases (Constable, 2020). One of the most inventive ways is through the polycentric model, in which self-sufficient districts are distributed across cities and function like urban villages. Such models have the potential to improve the quality of life, promote walking, and free up space for other uses, such as parks and gardens (Brumfield and Cubillos, 2020). Also, making different use of current spaces, implementing further sanitation, and transitioning toward more room for pedestrians are all going to be strategic features in a pandemic-resilient built environment of the future.

9.5 Urban adaptability

Design for temporary structures, adaptability, and disassembly is a straight design process that highlights the design of new buildings for future reuse and/or the reconditioning of building materials and components. This intelligent artistic approach may transform the built environment to more freely that allow the renovation of existing buildings to function new objectives. This includes the direct reuse, adaptation, or relocation of an existing building or its structure. Accomplishing adaptability needs a shift away from the current emphasis on form and function in response to immediate primacies, toward a context and time-based vision of design. In this regard, buildings should be resilient to climate change, and also to other future risks such as pandemics and other potential changes. Resiliency in this context describes the quality of being able to respond, recover, and renew no matter the pressure or the devastation. "Resilient," "responsive," and "adaptive" preparation are keys that should become new norms of living to stay. It means ensuring strong enough controls to respond and recover from a crisis and adaptability by being able to change and renew in response to new circumstances as well as learning from this experience to be better prepared in the future.

Climate change impacts on buildings vary according to the nature of the hazard, the type of building, the use made of them, the vulnerability of occupants, the dependency on infrastructure network services, the urban coping context, and their cultural and social value. Adapting buildings to climate change involves fostering a culture of risk and refining resilience, beyond guidelines. The impacts on buildings include the degradation of the structure, the deterioration of service networks, the downgrading uses of equipment, and adverse effects on users' comfort and health (GlobalABC, 2021). In buildings, design for adaptability avoids the significant impacts from demolition and landfilling of existing materials, and sourcing of new building materials. It also keeps in place all the natural resources that have been withdrawn to produce and install new materials as well as all the environmental releases to water, air, and land generated by the extraction, manufacturing, construction, and installation of those materials. The primary goal of design for adaptability is to lengthen a building's lifespan by making it possible to adapt to space with minimal disruption (Melton, 2020). In this regard, designers may consider designing products taking into consideration their end-of-use by following design for disassembly approaches. This is an effective methodology to avoid early demolition or producing waste at the end-of-service of a building. Combining design strategies in the initial life-stages of a building will ease the transformation of the building or the reuse of its elements at the end of its service (Fricke and Schulz, 2006). The end-of-service stage in the context of the circularity would mean that the buildings can be safely disassembled into several components; therefore, they can be repaired, remanufactured, adapted, or recycled.

The adaptive reuse of existing buildings for new purposes represents a highly sustainable approach to architectural design. Building adaptive reuse, in the context of a pandemic is typically seen as the reuse of an existing building in a manner that retains as much as possible of the original building while updating its performance to meet modern codes, standards, and new user requirements (Bullen, 2007). It is what happens when an existing building is already in place and ready to be altered to accommodate a new use. Adaptive reuse is a vital strategy because of the major impacts, especially the embodied carbon impacts of new construction. As awareness of climate change grows in the profession, it is increasingly important to cultivate the architectural skillset needed to support adaptive reuse (Melton, 2020). Adaptive reuse is an efficient and sustainable approach to creating new spaces, especially during urgent needs. The more adaptable the building is to different uses, the longer its useful life will be and that has various impacts over time. Along with modular construction, adaptive reuse is proven to be very effective in creating facilities that enable sustainably and cost-competitive designs. In addition to adaptability, the urban structure requires to become more efficient and environmentally friendly.

9.6 Urban therapy

The concept of therapeutic design or healing design is both modern and ancient. Today, therapeutic design may be defined as the evidence-based people-centered discipline of the built environment, which seeks to recognize and adopts approaches of integrating physiological and psychological spatial elements that interact with people into the design. Today, evidence-based design has become the hypothetical notion for what is called healing environments (Huisman et al., 2012). Consequently, this results in the creation of spaces considered to be healing environments. A therapeutic environment is a physical space that is supportive of each individual and recognizes that people with dementia are particularly vulnerable to chaotic environmental influences. It is individualized, flexible, and designed to support differing functional levels and approaches to care (Campernel and Brummett, 2010). Therapeutic design, which assimilates the five senses, creates a relieving and inspiring environment to stimulate healing.

In urban therapy, minds shift from preventing health problems to causing health enhancements. Combining spatial design with health parameters, architects can make decisions and take actions that protect the natural world and preserve the environment to support future life. Further integrating environmental sustainability with therapeutic technologies achieves healthier human environments (Youssef, 2021). This integration establishes a design index that supports and facilitates design' merger of the built and human–health environments. Fig. 9.4 illustrates the inception of how the idea of urban therapy can be realized. Components of both the built environment

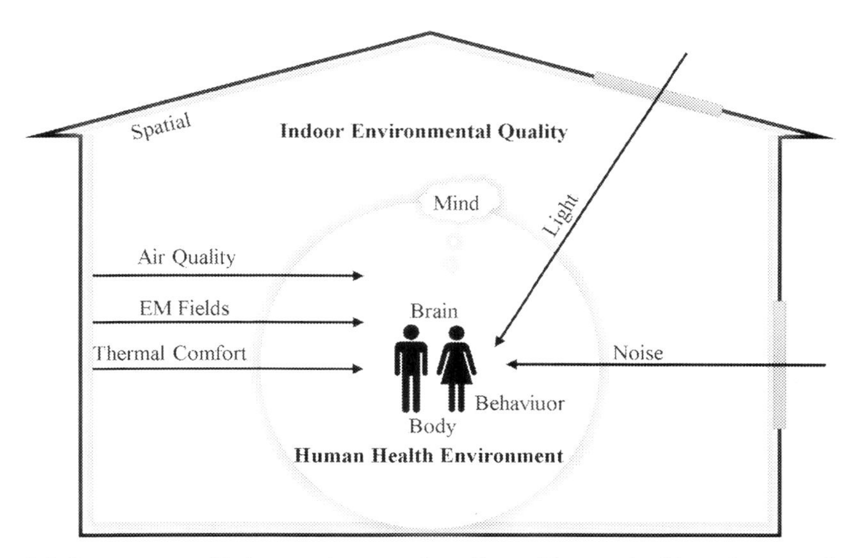

Fig. 9.4 Components of indoor environmental quality and human–health environment for urban therapy.

and the human–health environment illustrate spatial, light, thermal, acoustic, and EM fields design concerning the human brain, mind, body, and behavior.

With the ever-increasing population and their tolerance to using medicines to treat illness, it has become more important to rely on natural means, treatments, and remedies, which are being neglected due to lack of awareness among the masses (RTF, 2020). This means that architecture becomes therapeutic where the built environment is considered a way to therapeutically enhance the human body and mind as well as disease spread (Pinter-Wollman et al., 2018). In this regard, designers have been naturally exploring these capacities for their restorative effects, in particular, to manage stress and stress-related diseases by considering natural factors like sound, light, color, smell, and pleasant views certainly connect to human senses and prove to show more ability in the physical and psychological healing of patients (RTF, 2020).

Biophilia (meaning love of living systems) is a notion that centers around human's internal fascination with nature as a genetic connection. The term was first coined by social psychologist Eric Fromm in the 1960s and later popularized by American biologist Edward Wilson in the 1980s, who observed a link between increasing rates of urbanization and disconnectedness with nature (Cadena, 2020). Biophilic design is a novel architecture technique that conveys the relationships between nature, health sciences, and the built environment. It is designed for people as a biological organism and is linked to the stress-relieving benefits of nature and improving cognitive function, creativity, well-being, and expedite healing. Reducing stress is believed to affect mitigating pain. Essentially, the biophilic design emphasizes the requirement of maintaining, enhancing, and restoring the useful experience of nature in the built environment (Kellert, 2008). Extending the principles of bioclimatic design, architects are now starting to experiment with biophilic design principles. While both bioclimatic and biophilic architecture provides the opportunity to reach optimal human comfort and low levels of energy consumption (Miller and Burton, 2020), biophilic design specifically aims to move human activities from under architectural roofs toward the green and natural surrounds outside.

The built environment can also promote social interaction by providing recurring opportunities for individuals to have informal social contact with one another. Examples include providing gathering spaces on neutral territory, visual prospects, movable seating, and food or other features that generate activity (Brown et al., 2009). The legacy of modernist hygienist ideas for the prevention of epidemics can be seen in contemporary sanitary approaches to designing indoor environmental climates and in regulations regarding the health effects of various building materials. Environmental factors such as IAQ, lighting quality, thermal comfort, and acoustical quality are measured and controlled for their effects on the three systems of the human body, the nervous, immune, and endocrine systems, through which they influence physical and mental health (Bluyssen, 2009).

A considerable amount of evidence suggests that exposure to green space promotes healthy psychological development. Places that support or encourage physical activity can help to prevent and treat depression. Designers and planners can promote psychological health by creating places that are not noisy or crowded; that promote access to daylight; that encourage social interaction; and that invite people to walk, run, play, ride bicycles, and engage in other forms of physical activity (Sullivan and Chang, 2011). Green spaces, healing gardens, and nature are sometimes more powerful than medication, especially when considering the health benefits nature provides for human development and for sustaining resilient societies (Jones and Wilenius, 2020). Nature by itself is a distinct, collaborative, and intelligent system with therapeutic value.

9.7 Urban knowledge

Urbanization education is often associated with urban learning and pro-urbanization behaviors. Some approaches to urban education, however, need to acquire the knowledge, skills, and information to make sustainable and healthy choices. In this regard, young people need to have resources to make those choices and they need to be assured of a built environment in which people require further policies to further improve their living. Committed leadership at the city level and education field are critical to sustainable and healthy urban planning in building up learning resources in communities. Just as cities serve as hubs for sustainability and resilience innovations, urban education in cities has the potential to push knowledge toward innovative practices, including practices related to public health.

Technology is continuously changing the course of all disciplines and hence the potential for their interaction. Will any profession like architecture or engineering continue to exist by itself? In architecture, for example, the design is now understood to have a complex process of development across scales, disciplines, and context, the fact that depicts inspirations from all walks of life. This is reflected by the World Economic Forum (WEF, 2020), computing, mathematics, architecture, and engineering are among the most expected disciplines that will create new jobs due to advancement in technology. These jobs are distributed along a wide scale of sectors including climate change, natural resources, new energy supplies and technologies, big data, IoT, cloud technology, robotics, and AI. To address the substantial challenges facing society today, a holistic approach to create active linkages among disciplines should be established by education providers to ensure effective learning is provided to current students and future professionals.

An effective healthy built environment profession today rests on building a respectful relationship out of mutual understanding and fruitful, practical engagement across disciplines. Scholarship on how this is happening is emerging, and this acts as a forum for the interdisciplinary exchange of examples, ideas, and commentary to create built environments that can better promote human health and well-being (Kent and Thompson, 2012). Educators at various levels should boost the knowledge of their students about the interrelationships of humans with their built environments. They should develop critical thinking skills in response to health and environmental issues. And they should foster positive attitudes about environmental stewardship and health promotion toward quality built environments, designed to be functional, safe, adaptive, and responsive to the various needs of people. One necessary solution to the knowledge gap, however, is to infuse more of the right kind of information and ideas

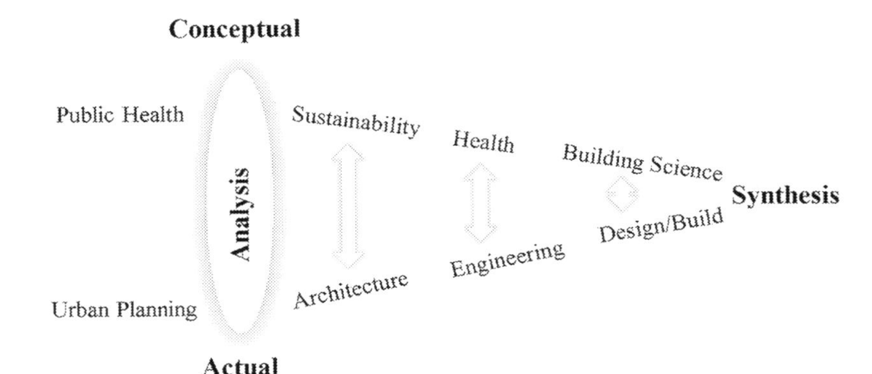

Fig. 9.5 Transdisciplinary combination of topics for comprehending urban knowledge.

into the classrooms. The goal is to develop responsible citizens, who are knowledgeable and adamant about improving their quality of life through the enhancement of the built environment (Ginny, 1990). It is necessary to emphasize that the best approach for including it into the curriculum at various levels is by blending-integrating with the existing curriculum content, rather than the creation of new courses or stand-alone units of study. All disciplines are now mutual resources for learning in this world of continual potentials. The content is learned through an experience that implies students undergo integration of experiences and this gives rise to an element of ambiguity, which educators need to be ready to face. Such education provides the students with opportunities to investigate a topic in detail with deeper knowledge and awareness. This invites teachers and learners to break the boundaries between traditional subjects and disciplines and move toward a transdisciplinary approach to exploring topics.

A transdisciplinary combination of topics allows comprehending whole systems in their complexity, as well as the interplay between natural and man-made determinants as is evident in the phenomena of IB as part of the smart city concept. This integration is the lens of phenomena through which all sets of conceptual, actual, analysis, and synthesis perspectives can be realized as shown in Fig. 9.5. It is a unique platform, in terms of assimilating disciplines such as engineering, computer science, building science, sustainability science, and public health. Narrowing down, it combines knowledge and blends skills and tools from design methodologies, simulation and software, social innovation, and circular economy. So, instead of studying each of the above subjects separately, it will be more beneficial to integrate them into a holistic system.

9.8 Intelligence for pandemic response

Megacities, which the UN defines as more than 10 million citizens, face increasing risks in environmental and population health, despite their economic prosperity as central hubs for growth and innovation. The UN predicts that there will be 43 megacities by 2030, with the majority of them in developing countries (Lai et al., 2020). Because cities in general and megacities, in particular, are major centers of global travel and commerce, including for migrants, the risk of disease transmission between cities is heightened by the transnational flow of people and products. These movements accelerate the speed at which pathogens can stretch across national boundaries. Regarding long-term sustainability, the effective response to public health emergencies requires strong, urban-intelligence leadership.

UI plays the same role in a public health crisis as it does in any crisis like the COVID-19 pandemic which has drawn significant global attention. The question to be addressed is how urban planners respond to such crises? What are the major intelligence factors that will influence the built environment? It seems that the existing design of community facilities does not provide sufficient protection for occupants and workers against contracting diseases, in particular, viruses. Also, public transport including airports, train stations, bus terminals, harbors, etc., is another matter of design concern, as these places are ideal breeding grounds for the transmission of diseases. Is there a need to rethink designing new facilities and spaces or the need is to redesigning the existing facilities to allow people to move around and preventing them from contracting diseases? It is important to note in this regard that UI requires all components of intelligence including AI to provide data gathering and intelligence analysis; HI by policymakers to determine whether and how to act on that intelligence; nature intelligence to provide spaces for overall well-being.

Due to the COVID-19, policymakers need to use their HI in managing the crises by setting up low-tech public health measures which may be seen in the context of privacy and personal space. This includes keeping people staying safe in places as well as maintaining a physical distance away from others in what is known in epidemiology as "quarantine," "social distancing," and "isolation." Considering the impact of social distancing on urban design, architects may design smaller venues and more open spaces to alleviate density. On the NI domain, enjoying the fresh air and continuous exercise is equally important to overall well-being. Because of this, architects can design creative spaces that can be called safe in the future. Within an urban environment, the imaginative integration of built and natural features can help to create a fast-recovery environment that is unique and interesting enough for people to lead healthy lives.

AI techniques can help policymakers and the health public community understand the pandemic and accelerate research on preparedness and response by rapidly analyzing large volumes of data. In this regard, several critical intelligence components for a pandemic response should be taken into consideration. Movement is an important component where near real-time human mobility data during a pandemic is valuable since regional population movement may unmask abnormal behavior during a critical event. This includes data about ground transportation, air flights, and mobile pings. During a pandemic, all these data provide more spatially specific indications for containment strategy and inform contact tracing. The second component is facilities, where city agencies collect and manage information on large data inventory of critical facilities and resources. The third component is people census and community surveys including quantifying local population characteristics and neighborhood health baseline measures (Dalziel et al., 2018). In the context of gathering data during a pandemic, the public will probably know that they are sacrificing part of their privacy by giving up personal information in exchange for the greater good.

Technology may play an important role by making sure the number of individuals inside a space like a shop or room is expected to ensure the minimal distance between people. With today's electronics, it is possible to automate this process easily. Robots and drones can be deployed in response and delivery tasks such as delivering materials, cleaning and disinfection, monitoring, and exposure tasks in hospitals.

9.9 Intelligence for postpandemic experience

The world today faces a unique moment in history: the opportunity to study and work remotely in the current electronic environment in which to create, disseminate, and use new practices. As schools and businesses begin their back-to-work transition, many are already rethinking their space design to better contain evolving opportunities and the merge of remote and workplaces within the new hybrid model. This may be the first time that people do not need to exist in their workplaces to finish their tasks. This ability to envision the hybrid model in different analogies is directly knitted to intelligence.

The pandemic accelerated not only technological but also societal changes. Adoption of trends like work-from-home, online retail, curbside pickup, entertainment-as-a-service, and telemedicine accelerated and created deep behavioral changes. The majority of these trends rely heavily on AI and ML to gather information, analyze, and solve problems (Bera, 2021). As a result of the paired human-centered solution of Industry 5.0 and Society 5.0, as well as the COVID-19 recession, digital transformation has soared forward, with a large-scale shift to remote working and e-commerce, driving a flow in work-from-home arrangements and a new market for remote work (WEF, 2020). Prepandemic, there was a common set of expectations of how people work, move around the built environment, how they walk down sidewalks, enter places, travel, and come and go from work (Goldstein and Kramer, 2020). Now, things have changed. The deep-seated behaviors people had before the pandemic will need to change, as will the interaction models they have had with the people and environments. This is a challenge that designers are particularly well-suited for. It is up to them to invent new spatial typologies and imagine the new products and services that support these environments, all to elevate the human experience.

From a design perspective, it is important to think about the users' experience as they live and work in places especially in public buildings, offices, hotels, and schools. Simplicity in design promotes the perception of cleanliness and the reset of a new aesthetic. Seamless surfaces, hard flooring, simple bedding materials, and limited furnishings can be designed elegantly, yet still, be easy to clean and maintain (Ito, 2020). Technology can be invaluable in the effort to instill trust in buildings. With immediate retrofits to existing buildings, measures can be taken that help filter and destroy bacteria and viruses from the indoor environments. Among other actions engineers recommend, promoting disinfection technologies which can help purify the air and contribute to a safe and healthy indoor environment. Another way to make the office environment healthier is to bring plant life indoors. This can be part of a broader effort toward adding elements of nature to the workplace, known as biophilia (Tranel, 2020).

The pandemic, an extreme wake-up call, has made a robust case for digital transformation and the debate has never been more pertinent. Today, homes are becoming permanent communication hubs and a life-long feature of the working community. Working from home, every day or most of the days provides wide options of living. This may also bring many of the environmental challenges associated with workplace delight and productivity, such as IAQ, thermal comfort, visual comfort, and noise pollution. For those who design and develop homes around the world, this has introduced a new challenge. The experience of the pandemic will, no doubt, have an ongoing impact on many aspects of life along with the changing roles of homes and buildings. Increased acceptance of digital connectivity could reinforce smaller cities and rural areas, where released workers, telecommuters, and freelancers, or technically "digital nomads," could become more prevalent. This lifestyle has been made possible through several technologies, including low-cost Internet access through WiFi, smartphones, and content management applications such as Microsoft Teams, Zoom, WebEx, WhatsApp, Instagram, Facebook, and more. These digital tools are creating multifunctional activities and will enforce a long-term adjustment to the work and living location strategy.

References

Alirol, E., Getaz, L., Stoll, B., Chappuis, F., Loutan, L., 2011. Urbanisation and infectious diseases in a globalised world. Lancet Infect. Dis. 11 (2), 131–141.

Allen, A., 2009. Sustainable Cities or Sustainable Urbanization? https://www.ucl.ac.uk/sustainable-cities/results/gcsc-reports/allen.pdf.

Asakura, T., Mallee, H., Tomokawa, S., Moji, K., Kobayashi, J., 2015. The ecosystem approach to health is a promising strategy in international development: lessons from Japan and Laos. Glob. Health 11, 3. https://doi.org/10.1186/s12992-015-0093-0.

Bera, S., 2021. Artificial Intelligence to Reign Post-Pandemic World: 5 Picks. https://finance.yahoo.com/news/artificial-intelligence-reign-post-pandemic-122312728.html?fr=sycsrp_catchall.

Bluyssen, P.M., 2009. The Indoor Environment Handbook: How to Make Buildings Healthy and Comfortable. Earthscan, Sterling.

Brown, S.C., Mason, C.A., Lombard, J.L., Martinez, F., Plater-Zyberk, E., Spokane, A., 2009. The relationship of built environment to perceived social support and psychological distress in hispanic elders: the role of eyes on the street. J. Gerontol.: Soc. Sci. 64b (2), 234–246.

Brumfield, A., Cubillos, C., 2020. Cities and the Public Health: Our New Challenge in Urban Planning. https://www.gensler.com/research-insight/blog/cities-and-public-health-our-new-challenge-in-urban-planning.

Bullen, P.A., 2007. Adaptive reuse and sustainability of commercial buildings. Facilities 25 (1/2), 20–31.

Cadena, A., 2020. How Biophilic Design Can Promote Healing? https://www.innovadesign.com.au/articles/how-biophilic-design-can-promote-healing/.

Campernel, S., Brummett, W., 2010. Creating Environments of Support: A Handbook for Dementia Responsive Design. http://www.brummettarchitects.com/ourresearch.htm.

Castelli, G., Cesta, A., Diez, M., Padula, M., Ravazzani, P., Rinaldi, G., Savazzi, S., Spagnuolo, M., Strambini, L., Tognola, G., Campana, E.F., 2019. Urban intelligence: a modular, fully integrated, and evolving model for cities digital twinning. In: IEEE 16th International Conference on Smart Cities: Improving Quality of Life Using ICT and IoT and AI (HONET-ICT), pp. 033–037, https://doi.org/10.1109/HONET.2019.8907962.

Colmers, J.M., Fox, D.M., 2003. The politics of emergency health powers and the isolation of public health. Am. J. Public Health 93, 397–399.

Constable, H., 2020. How Do you Build a City for Pandemic? https://www.bbc.com/future/article/20200424-how-do-you-build-a-city-for-a-pandemic.

Dalziel, B.D., Kissler, S., Gog, J.R., Viboud, C., Bjørnstad, O.N., Metcalf, C.J.E., et al., 2018. Urbanization and humidity shape the intensity of influenza epidemics in U.S. cities. Science 362 (6410), 75–79.

Fisher, T., 2010. Frederick Law Olmsted and the Campaign for Public Health. https://placesjournal.org/article/frederick-law-olmsted-and-the-campaign-for-public-health/?cn-reloaded=1&cn-reloaded=1.

Fricke, E., Schulz, A.P., 2006. Design for changeability (DfC): principles to enable changes in systems throughout their entire lifecycle. Syst. Eng. 8 (4), 342–359.

Fudge, C., Fawkes, S., 2017. Science meets imagination–cities and health in the twenty-first century. Cities Health 1 (2), 101–106. https://doi.org/10.1080/23748834.2018.1462610.

GCEC, 2016. Seizing the Global Opportunity: Partnerships for Better Growth and Better Climate. The 2015 New Climate Economy Report. The Global Commission on the Economy and Climate, Washington, DC. https://www.ledevoir.com/documents/pdf/nce2015.pdf.

GCPH, 2013. The Built Environment and Health: An Evidence Review. Glasgow Centre for Population Healh. https://www.gcph.co.uk/assets/0000/4174/BP_11_-_Built_environment_and_health_-_updated.pdf.

Gianfredi, V., Buffoli, M., Rebecchi, A., Croci, R., Oradini-Alacreu, A., Stirparo, G., Marino, M., Odone, A., Capolongo, S., Signorelli, C., 2021. Association between urban greenspace and health: a systematic review of literature. Int. J. Environ. Res. Public Health 18 (10), 5137.

Ginny, G., 1990. Teaching about the Built Environment. https://www.ericdigests.org/pre-9217/built.htm.

GlobalABC, 2021. Building and Climate Change Adaptation. https://globalabc.org/resources/publications/buildings-and-climate-change-adaptation-call-action.

Goldstein, J., Kramer, D., 2020. Fusing Architecture and Technology to Reenter a New World. https://www.gensler.com/research-insight/blog/fusing-architecture-and-technology.

Habash, R., 2017. Green Engineering: Innovation, Entrepreneurship, and Design. CRC Taylor and Francis, Boca Raton.

Householder, T., 2016. Architecture Concepts Can Boost Mental Health. https://www.desmoinesregister.com/story/opinion/abetteriowa/2016/04/22/architecture-concepts-can-boost-mental-health/83341812/.

Huisman, E.R.C.M., Morales, E., van Hoof, J., Kort, H.S.M., 2012. Healing environment: a review of the impact of physical environmental factors on users. Build. Environ. 58, 70–80.

Ito, T., 2020. How Will Hotels Rebound From COVID-19? https://www.gensler.com/research-insight/blog/how-will-hotels-rebound-from-covid-19.

Jones, A.M., Wilenius, M., 2020. In greensight: healthier futures for urban cores in transition. Cities Health 4 (2), 168–179. https://doi.org/10.1080/23748834.2020.1767902.

Kellert, S.R., 2008. Dimensions, elements, and attributes of biophilic design. In: Kellert, S.R., Heerwagen, J., Mador, M. (Eds.), Biophilic Design: The Theory, Science, and Practice of Bringing Buildings to Life. John Wiley and Sons, Hoboken.

Kent, J., Thompson, S., 2012. Health and the built environment: exploring foundations for a new interdisciplinary profession. J. Environ. Public Health 2012, 1–12. https://doi.org/10.1155/2012/958175 (Article ID 958175).

Kumar, S., Preetha, G.S., 2012. Health promotion: an effective tool for global health. Indian J. Community Med. 73 (1), 5–12.

Lai, Y., Yeung, W., Celi, L.A., 2020. Urban intelligence for pandemic response: viewpoint. JMIR Public Health Surveill. 6 (2), 1–6.

Mantas, J., 2018. Why Artificial Intelligence Is Learning Emotional Intelligence. https://www.weforum.org/agenda/2018/09/why-artificial-intelligence-is-learning-emotional-intelligence.

Melton, P., 2020. Building that Last: Design for Adaptability, Deconstruction and Reuse. http://content.aia.org/sites/default/files/2020-03/ADR-Guide-final_0.pdf.

Miller, E., Burton, L.O., 2020. Redesigning aged care with a biophilic lens: a call to action. Cities Health 2020, 1–30. https://doi.org/10.1080/23748834.2020.1772557.

Pan, Y., Tian, Y., Liu, X., Gu, D., Hua, G., 2016. Urban big data and the development of city intelligence. Engineering 2095-8099. 2 (2), 171–178. https://doi.org/10.1016/J.ENG.2016.02.003.

Perdue, W.C., Stone, L.A., Gostin, L.O., 2003. The built environment and its relationship to the public's health: the legal framework. Am. J. Public Health 93 (9), 1390–1394.

Pinter-Wollman, N., Jelić, A., Wells, N.M., 2018. The impact of the built environment on health behaviours and disease transmission in social systems. Philos. Trans. R. Soc. B 373 (1753), 1–18. https://doi.org/10.1098/rstb.2017.0245 (Article ID: 20170245).

Ramirez-Rubio, O., Daher, C., Fanjul, G., Gascon, M., Mueller, N., Pajin, L., Plansencia, A., Rojaz-Rueda, D., Thondo, M., Nieuwenhuijsen, M.J., 2019. Urban health: an example of a "health in all policies" approach in the context of SDGs implementation. Glob. Health 15, 87. https://doi.org/10.1186/s12992-019-0529-z.

Robinson, R., 2014. Smart City Design Patterns: The Urban Technologist. https://theurbantechnologist.com/design-patterns/.

Rosen, G., 1993. A History of Public Health. Johns Hopkins University Press, Baltimore.

RTF, 2020. Therapeutic Architecture: Role of Architecture in Healing Process. https://www.re-thinkingthefuture.com/fresh-perspectives/a597-therapeutic-architecture-role-of-architecture-in-healing-process/.

Spula, I., 2017. Designing for Health and Wellness: The Next Great Challenge. https://www.architectmagazine.com/practice/designing-for-health-and-wellness-the-next-great-challenge_o.

Sullivan, W., Chang, C.-Y., 2011. Mental health and the built environment. In: Dannenberg, A. L., Frumkin, H., Jackson, R.J. (Eds.), Making Healthy Places. Springer.

Tranel, B., 2020. The Future Workplace Will Embrace a Hybrid Reality. https://www.gensler.com/research-insight/blog/the-future-workplace-will-embrace-a-hybrid-reality.

WEF, 2020. The Future of Jobs Report. https://www.weforum.org/reports/the-future-of-jobs-report-2020//reports.weforum.org/future-of-jobs-2016/employment-trends/.

Wood, S., 2020. Architecture and Pandemics. https://www.kent.ac.uk/news/culture/25808/expert-comment-architecture-and-pandemics.

Youssef, O., 2021. Therapeutic Architecture Design Index. https://content.aia.org/sites/default/files/2016-04/DH-Therapeutic-architecture-design-index.pdf.

Index

Note: Page numbers followed by *f* indicate figures and *t* indicate tables.

C

D

E

F

G

H

M

N

O

P

Q

R

S

Printed in the United States
by Baker & Taylor Publisher Services